METHODS IN MOLECULAR BIOLOGY

Series Editor
John M. Walker
School of Life and Medical Sciences
University of Hertfordshire
Hatfield, Hertfordshire, AL10 9AB, UK

For further volumes:
http://www.springer.com/series/7651

Protein Lipidation

Methods and Protocols

Edited by

Maurine E. Linder

*Department of Molecular Medicine,
College of Veterinary Medicine, Cornell University, Ithaca, NY, USA*

Humana Press

Editor
Maurine E. Linder
Department of Molecular Medicine
College of Veterinary Medicine
Cornell University
Ithaca, NY, USA

ISSN 1064-3745 ISSN 1940-6029 (electronic)
Methods in Molecular Biology
ISBN 978-1-4939-9531-8 ISBN 978-1-4939-9532-5 (eBook)
https://doi.org/10.1007/978-1-4939-9532-5

This Humana imprint is published by the registered company Springer Science+Business Media, LLC part of Springer Nature.
The registered company address is: 233 Spring Street, New York, NY 10013, U.S.A.

Preface

Posttranslational modifications of proteins diversify the proteome and can be regulated or regulatory. In eukaryotic cells, protein lipidation encompasses a variety of lipid modifications that contribute to protein localization and function, with fatty acylation and isoprenylation among the most abundant. Fatty acids are attached to proteins through thioester linkage at cysteine (*S*-acylation, typically *S*-palmitoylation), amide linkage at lysine or an amino-terminal glycine or cysteine (*N*-acylation), and oxyester linkage to serine (*O*-acylation). A stable modification for some proteins, fatty acylation is reversible for others. Isoprenylation of proteins occurs at C-terminal cysteine motifs in which a C15-farnesyl or C20-geranylgeranyl isoprenoid is added through a stable thioether linkage. Proteins ending in a so-called CaaX box—cysteine, followed by two aliphatic amino acids, and a terminal amino acid that dictates whether the modification is farnesylation or geranylgeranylation—undergo subsequent posttranslational processing, proteolytic cleavage of the CaaX motif, and carboxymethylation of the isoprenylated cysteine.

This volume is designed to meet the needs of investigators working in the field of protein lipidation, as well as those who encounter protein lipidation in the course of their research on a particular cellular process or favorite protein. A major goal is to provide the resources to detect a lipid modification, either of an individual protein or globally using a proteomic strategy. Experimental analysis of protein lipidation has been challenging due to the hydrophobic nature of the modifications. The development of click chemistry-based probes and the use of acyl-exchange chemistries, combined with improvements in mass spectrometry, have simplified the detection of lipid modifications and greatly expanded the lipidated proteomes. This is particularly true for proteins fatty-acylated at cysteine residues, with current estimates that up to 10% of the proteins encoded by the human genome are substrates for *S*-fatty acylation. Part I of this volume describes applications of these chemically based strategies to identify substrates for protein lipidation. An enhanced acyl-switch method that facilitates protein recovery reports steady-state levels of protein *S*-acylation. Metabolic labeling of cells with alkyne- or azide-containing fatty acids or isoprenoids, followed by bioorthogonal Cu(I)-catalyzed Huisgen 1,3-dipolar cycloaddition reaction to conjugate the lipid-modified proteins to fluorophores or affinity tags (biotin), enables lipid-modified proteins to be captured and identified. Mass spectroscopy approaches are described, which not only facilitate cataloguing of fatty-acylated proteins but also identify the sites of attachment and dynamics using alkyne probes. An alternative approach is also described for direct detection of the fatty acid on peptides.

Both *S*-acylation and *N*-acylation of proteins are reversible modifications, and fatty acid removal is mediated by enzymes. Members of the metabolic serine hydrolase superfamily include thioesterases that remove palmitate from cysteine residues in proteins. In Part II, a method is described to screen members of the α/β hydrolase domain (ABHD)-containing proteins for acylprotein thioesterase activity using an acyl-exchange, gel-shift assay. The use of fluorescent probes to monitor depalmitoylating activity in vitro and in cells is also addressed in this section. Enzymatic removal of lysine-linked fatty acylation is mediated by members of the sirtuin family. A method to assay sirtuin defatty-acylase activity is presented, as well as a chemical proteomic strategy to globally profile sirtuin defatty-acylase substrates.

This section also describes metabolic labeling strategies to monitor protein biogenesis and palmitate turnover.

Part III of this volume addresses the DHHC *S*-acyltransferases, a sizable family of enzymes with 23 members encoded by the human genome. The size and diversity of the family makes identification of protein function for a specific family member challenging. A siRNA knockdown strategy is described to manipulate the expression of individual DHHC proteins, enabling the interrogation of their function in cells. Methods to purify and assay DHHC proteins are also presented, as is a bioinformatics approach to identify functionally and structurally relevant residues and motifs. In Part IV, the SwissPalm 2 database is presented, a valuable online resource that has catalogued the datasets from numerous palmitoyl-proteome studies, curated literature, and provided tools to compare datasets.

In Part V of this volume, the focus shifts from fatty acylation that occurs on the cytoplasmic face of membranes to that occurring in the lumen of the secretory pathway. Wnt and Hedgehog proteins and the peptide hormone ghrelin are modified with fatty acids prior to secretion from cells. Wnt proteins signal by binding to Frizzled receptors on neighboring cells. A method to assay the interaction of a lipidated Wnt protein with its receptor is described. Fatty acylation in the lumen of the secretory pathway is carried out by several members of the MBOAT (membrane-bound *O*-acyltransferase) family. Methods to assay the enzyme activity of GOAT (ghrelin *O*-acyltransferase), Hedgehog acyltransferase, and Porcupine fatty acyltransferase are presented.

Biophysical and structural characterization of membrane proteins requires expression and purification methods that deliver high yields and uniformly modified proteins, a topic addressed in Parts VI and VII of this volume. Production of fully processed Ras using an engineered insect cell expression system is described, as well as purification of a Rho GTPase labeled with a clickable fatty acid that is then used to monitor its protein–protein interactions. Using yeast as an expression system, methods to purify and assay the human and yeast orthologs of the zinc metalloprotease STE24 that is involved in postprenylation processing are presented. Finally, purification of the visual signal transduction complex for structural studies is described. A method for reconstitution of the complex into lipid nanodiscs is also presented, a method that is applicable to many of the integral membrane proteins that catalyze the addition of lipids to proteins.

Ithaca, NY, USA *Maurine E. Linder*

Contents

PART III zDHHC S-ACYLTRANSFERASES

PART IV ONLINE RESOURCE

PART V FATTY ACYLATION IN THE LUMEN OF THE SECRETORY PATHWAY

PART VI PRENYLATION AND POST-PRENYLATION PROCESSING

Contributors

LAURENCE ABRAMI • *Global Health Institute, School of Life Sciences, EPFL, Lausanne, Switzerland*

MINA AHMADI • *Department of Chemistry, University of Minnesota, Minneapolis, MN, USA*

JAMES JOHN ASCIOLLA • *Cell Biology Program, Memorial Sloan Kettering Cancer Center, New York, NY, USA; Biochemistry, Cell Biology and Molecular Biology Graduate Program, Weill-Cornell Graduate School of Medical Sciences, New York, NY, USA*

MATHIEU BLANC • *Global Health Institute, School of Life Sciences, EPFL, Lausanne, Switzerland*

JI CAO • *Department of Chemistry and Chemical Biology, Cornell University, Ithaca, NY, USA*

RICHARD A. CERIONE • *Departments of Chemistry and Chemical Biology and Molecular Medicine, Cornell University, Ithaca, NY, USA*

MELANIE CHEUNG SEE KIT • *Department of Chemistry, Chemical Biology Program, University of Michigan, Ann Arbor, MI, USA*

ELIZABETH R. CLEVERDON • *Department of Chemistry, Syracuse University, Syracuse, NY, USA*

FABRICE P. A. DAVID • *Global Health Institute, Gene Expression Core Facility, SV-IT, School of Life Sciences, EPFL, Lausanne, Switzerland*

ROBIN A. DENHARDT-ERIKSSON • *Global Health Institute, Laboratory of Computational Systems Biotechnology, EPFL, Lausanne, Switzerland*

ROBERT J. DESCHENES • *Department of Molecular Medicine, Morsani College of Medicine, University of South Florida, Tampa, FL, USA*

BRYAN C. DICKINSON • *Department of Chemistry, The University of Chicago, Chicago, IL, USA*

MARK D. DISTEFANO • *Department of Chemistry, University of Minnesota, Minneapolis, MN, USA*

MATTHEW DREW • *Protein Expression Laboratory, NCI RAS Initiative, Cancer Research Technology Program, Frederick National Laboratory for Cancer Research, Frederick, MD, USA*

JON W. ERICKSON • *Departments of Chemistry and Chemical Biology and Molecular Medicine, Cornell University, Ithaca, NY, USA*

DOMINIC ESPOSITO • *Protein Expression Laboratory, NCI RAS Initiative, Cancer Research Technology Program, Frederick National Laboratory for Cancer Research, Frederick, MD, USA*

PETER FRANK • *Protein Expression Laboratory, NCI RAS Initiative, Cancer Research Technology Program, Frederick National Laboratory for Cancer Research, Frederick, MD, USA*

MASAKI FUKATA • *Division of Membrane Physiology, Department of Molecular and Cellular Physiology, National Institute for Physiological Sciences, National Institutes of Natural Sciences, Okazaki, Japan; Department of Physiological Sciences, School of Life Science, SOKENDAI (The Graduate University for Advanced Studies), Okazaki, Japan*

YUKO FUKATA • *Division of Membrane Physiology, Department of Molecular and Cellular Physiology, National Institute for Physiological Sciences, National Institutes of Natural*

Sciences, Okazaki, Japan; Department of Physiological Sciences, School of Life Science, SOKENDAI (The Graduate University for Advanced Studies), Okazaki, Japan

YANG GAO • *Department of Chemistry and Chemical Biology, Cornell University, Ithaca, NY, USA*

WILLIAM GILLETTE • *Protein Expression Laboratory, NCI RAS Initiative, Cancer Research Technology Program, Frederick National Laboratory for Cancer Research, Frederick, MD, USA*

JENNIFER GREAVES • *Faculty of Health and Life Sciences, Centre for Sport, Exercise and Life Sciences, Coventry University, Coventry, UK*

CARISSA GROSE • *Protein Expression Laboratory, NCI RAS Initiative, Cancer Research Technology Program, Frederick National Laboratory for Cancer Research, Frederick, MD, USA*

HOWARD C. HANG • *Laboratory of Chemical Biology and Microbial Pathogenesis, The Rockefeller University, New York, NY, USA*

RAMI N. HANNOUSH • *Department of Early Discovery Biochemistry, Genentech, South San Francisco, CA, USA*

VASSILY HATZIMANIKATIS • *Laboratory of Computational Systems Biotechnology, EPFL, Lausanne, Switzerland*

PIERS A. HEMSLEY • *Division of Plant Sciences, School of Life Sciences, University of Dundee, Dundee, Scotland, UK; Cell and Molecular Sciences, The James Hutton Institute, Dundee, Scotland, UK*

JUN YOUNG HONG • *Department of Chemistry and Chemical Biology, Cornell University, Ithaca, NY, USA*

JAMES L. HOUGLAND • *Department of Chemistry, Syracuse University, Syracuse, NY, USA*

CHRISTINE A. HRYCYNA • *Department of Chemistry, Purdue University, West Lafayette, IN, USA*

ERH-TING HSU • *Department of Chemistry, Purdue University, West Lafayette, IN, USA*

CHARLOTTE H. HURST • *Division of Plant Sciences, School of Life Sciences, University of Dundee, Dundee, Scotland, UK; Cell and Molecular Sciences, The James Hutton Institute, Dundee, Scotland, UK*

YUHUAN JI • *Department of Biochemistry, Center for Biomedical Mass Spectrometry, Boston University School of Medicine, Boston, MA, USA*

TAKASHI KANADOME • *Division of Membrane Physiology, Department of Molecular and Cellular Physiology, National Institute for Physiological Sciences, National Institutes of Natural Sciences, Okazaki, Japan*

RAHUL S. KATHAYAT • *Department of Chemistry, The University of Chicago, Chicago, IL, USA*

CHENG LIN • *Department of Biochemistry, Center for Biomedical Mass Spectrometry, Boston University School of Medicine, Boston, MA, USA*

HENING LIN • *Department of Chemistry and Chemical Biology, Howard Hughes Medical Institute, Ithaca, NY, USA*

MAURINE E. LINDER • *Department of Molecular Medicine, College of Veterinary Medicine, Cornell University, Ithaca, NY, USA*

MARTIN IAN P. MALGAPO • *Department of Molecular Medicine, College of Veterinary Medicine, Cornell University, Ithaca, NY, USA*

BRENT R. MARTIN • *Department of Chemistry, Chemical Biology Program, University of Michigan, Ann Arbor, MI, USA*
HEATHER MCCLAFFERTY • *Centre for Discovery Brain Sciences, College of Medicine and Veterinary Medicine, University of Edinburgh, Edinburgh, UK*
DAVID A. MITCHELL • *Department of Molecular Medicine, Morsani College of Medicine, University of South Florida, Tampa, FL, USA*
AARON H. NILE • *Department of Early Discovery Biochemistry, Genentech, South San Francisco, CA, USA*
AKIYUKI NISHIMURA • *Department of Molecular Medicine, College of Veterinary Medicine, Cornell University, Ithaca, NY, USA*
LAURA C. PENDLETON • *Department of Molecular Medicine, Morsani College of Medicine, University of South Florida, Tampa, FL, USA*
SHELLEY PERKINS • *Protein Expression Laboratory, NCI RAS Initiative, Cancer Research Technology Program, Frederick National Laboratory for Cancer Research, Frederick, MD, USA*
RODRIGO QUIROGA • *Centro de Investigaciones en Físico Química de Córdoba (INFIQC), CONICET, Departamento de Química Teórica y Computacional, Facultad de Ciencias Químicas, Universidad Nacional de Córdoba, Córdoba, Argentina*
KALPANA RAJANALA • *Cell Biology Program, Memorial Sloan Kettering Cancer Center, New York, NY, USA*
SEKAR RAMACHANDRAN • *Departments of Chemistry and Chemical Biology and Molecular Medicine, Cornell University, Ithaca, NY, USA*
MARILYN D. RESH • *Cell Biology Program, Memorial Sloan Kettering Cancer Center, New York, NY, USA; Biochemistry, Cell Biology and Molecular Biology Graduate Program, Weill-Cornell Graduate School of Medical Sciences, New York, NY, USA*
MICHAEL J. SHIPSTON • *Centre for Discovery Brain Sciences, College of Medicine and Veterinary Medicine, University of Edinburgh, Edinburgh, UK*
MICHELLE A. SIEBURG • *Department of Chemistry, Syracuse University, Syracuse, NY, USA*
NICOLE A. SPIEGELMAN • *Department of Chemistry and Chemical Biology, Cornell University, Ithaca, NY, USA*
KIALL FRANCIS SUAZO • *Department of Chemistry, University of Minnesota, Minneapolis, MN, USA*
EMMANUELLE THINON • *Laboratory of Chemical Biology and Microbial Pathogenesis, The Rockefeller University, New York, NY, USA; Molecular Cell Biology of Autophagy, The Francis Crick Institute, London, UK*
NICHOLAS C. O. TOMKINSON • *WestCHEM, Department of Pure and Applied Chemistry, University of Strathclyde, Glasgow, UK*
DIONNE TURNBULL • *Division of Plant Sciences, School of Life Sciences, University of Dundee, Dundee, Scotland, UK*
JAVIER VALDEZ TAUBAS • *Centro de Investigaciones en Química Biológica de Córdoba (CIQUIBIC), CONICET, Departamento de Química Biológica Ranwel Caputto, Facultad de Ciencias Químicas, Universidad Nacional de Córdoba, Córdoba, Argentina*
F. GISOU VAN DER GOOT • *Global Health Institute, School of Life Sciences, EPFL, Lausanne, Switzerland*
JEFFREY S. VERVACKE • *Department of Chemistry, University of Minnesota, Minneapolis, MN, USA*

NORIHIKO YOKOI • *Division of Membrane Physiology, Department of Molecular and Cellular Physiology, National Institute for Physiological Sciences, National Institutes of Natural Sciences, Okazaki, Japan; Department of Physiological Sciences, School of Life Science, SOKENDAI (The Graduate University for Advanced Studies), Okazaki, Japan*
SHUAI ZHANG • *Department of Chemistry and Chemical Biology, Cornell University, Ithaca, NY, USA*

Part I

Detection of Lipid Modifications

Chapter 1

Determination of Protein *S*-Acylation State by Enhanced Acyl-Switch Methods

Charlotte H. Hurst, Dionne Turnbull, and Piers A. Hemsley

Abstract

S-Acylation is increasingly being recognized as an important dynamic posttranslational modification of cysteine residues in proteins. Various approaches have been described for assaying protein *S*-acylation with acyl-switch approaches being the most common and accessible. However, these approaches can be time-consuming with low reproducibility as a result of multiple protein precipitation/resuspension cleanup steps. Here we present a faster, cleaner, and more sensitive acyl-switch approach for detecting the *S*-acylation state of any protein, from any cell or tissue type, that can be detected by western blotting. In the case of acyl-RAC, the procedure is now performed without protein precipitation, greatly increasing speed and improving sample handling in the assay. This also allows for more samples to be processed simultaneously and opens the way for medium-throughput assays. Overall, maleimide scavenging improves the reliability of determination and quantification of protein *S*-acylation state by acyl-switch methods.

Key words *S*-Acylation, *S*-Palmitoylation, *S*-Acylated, Palmitoylated, *N*-Ethylmaleimide, Acyl-switch, Maleimide, 2,3-Dimethyl 1,3-butadiene, Biotin, Diels–Alder

1 Introduction

S-Acylation is an important posttranslational modification occurring on cysteine residues in many proteins. *S*-Acylation has a number of described roles, from controlling protein localization within the cell, anchoring soluble proteins to membranes, affecting protein stability and controlling the activation state of proteins. Current assays for assessing protein *S*-acylation commonly rely on two main approaches: orthogonal labeling with alkyne- or azide-functionalized fatty acids or hydroxylamine-mediated acyl-exchange assays. Acyl-exchange assays rely on *N*-ethylmaleimide to block free cysteine sulfhydryls and prevent nonspecific detection [1]. This is followed by chemical cleavage of *S*-acyl groups to reveal free cysteine sulfhydryls that can be either labeled with (cleavable) biotin and selectively enriched using biotin-binding beads (typically streptavidin derivatives, acyl-biotin exchange; ABE) [2] or directly

Maurine E. Linder (ed.), *Protein Lipidation: Methods and Protocols*, Methods in Molecular Biology, vol. 2009, https://doi.org/10.1007/978-1-4939-9532-5_1,

captured on sulfhydryl-reactive resin (acyl-resin assisted capture; acyl-RAC) [3]. Free sulfhydryls can alternatively be labeled with PEG in PEG-shift/Acyl-PEG Exchange assays [4]. Acyl-exchange assays have historically involved numerous precipitation-based cleanup steps that lead to sample loss, protein aggregation and smearing of protein bands on gels. We recently described a faster, cleaner and more sensitive acyl-exchange assay specifically to address the issues we encountered with certain large, difficult or aggregation-prone proteins [5], but the method is equally applicable to all *S*-acylated proteins. Our novel approach is to use Diels–Alder 4+2 cycloaddition chemistry followed by phase partitioning to scavenge NEM from aqueous solution [6] rather than protein precipitation. As a result, Acyl-RAC assays can now be performed without any precipitation steps being required, while ABE assays only require one precipitation step. In our hands this method greatly improves sensitivity and reproducibility while reducing hands-on time. Although not described here, the cleanup step to remove NEM should be equally applicable to PEG-shift/Acyl-PEG Exchange assays, or other assays in other fields requiring the removal of NEM from aqueous solution.

2 Materials

Prepare all solutions using ultrapure water (at least 18 MΩ cm at 25 °C) and analytical grade reagents. Prepare and store all reagents at room temperature (unless indicated otherwise). Follow all local waste disposal regulations. The protocol assumes the ability to detect and quantify the protein of interest by SDS-PAGE and immunoblotting.

2.1 General Solutions, Reagents and Equipment

1. 1 M *N*-ethylmaleimide (NEM) in ethanol (*see* **Note 1**).
2. Lysis buffer: 100 mM Tris–HCl pH 7.2, 150 mM NaCl, 5 mM EDTA, 2.5% SDS (*see* **Note 2**), 10 mM *N*-ethylmaleimide, protease inhibitors (*see* **Notes 3** and **4**).
3. 1 M Hydroxylamine in water (adjust pH to 7.2 with 10 mM NaOH) (*see* **Note 5**).
4. 1 M NaCl in water.
5. 2,3-Dimethyl, 1,3-butadiene.
6. 2× reducing SDS loading buffer: 100 mM Tris–HCl pH 6.8, 4% w/v SDS, 0.2% bromophenol blue, 20% v/v glycerol, 200 mM β-mercaptoethanol (store in 500 μL aliquots, add β-mercaptoethanol fresh).
7. BCA protein concentration determination kit (*see* **Note 6**).
8. Cooled benchtop microfuge (capable of ~16,000 × *g*).

9. Cooled benchtop centrifuge (capable of >4000 × *g* using 15 mL Falcon tubes).
10. Shaking heat block (*see* **Note 7**).
11. End-over-end 1.5 mL microfuge tube mixer (*see* **Note 7**).
12. 37 °C water bath.

2.2 Acyl-Biotin Exchange (ABE)

1. Resuspension buffer: 100 mM Tris–HCl pH 7.2, 150 mM NaCl, 5 mM EDTA, 2.5% SDS, 8 M urea (stored at −20 °C in single use aliquots).
2. Chloroform.
3. Methanol.
4. Ultrapure water.
5. 1 mM biotin-HPDP dissolved in DMSO (prepare fresh immediately before use).
6. High-capacity Neutravidin-agarose beads.
7. Phosphate-buffered saline (PBS) pH 7.2: 137 mM NaCl, 2.7 mM KCl, 10 mM Na_2HPO_4, 1.8 mM KH_2PO_4. Prepare PBS by dissolving 8 g NaCl, 0.2 g KCl, 1.13 g Na_2HPO_4, 0.2 g KH_2PO_4, in 800 mL of H_2O. Adjust the pH to 7.2 ± 0.1 with 6 N HCl. Add H_2O to a final volume of 1 L.
8. ABE wash buffer: 1× PBS pH 7.2, 1% SDS.
9. Roller mixer table.

2.3 Acyl-Resin Assisted Capture (Acyl-RAC)

1. Acyl-RAC wash buffer: 100 mM Tris–HCl pH 7.2, 150 mM NaCl, 5 mM EDTA, 2.5% SDS (*see* **Note 8**).
2. Thiopropyl-Sepharose 6B beads.

3 Methods

Initial sample processing (Subheading 3.1) and sample analysis (Subheading 3.4) is the same regardless of Acyl-RAC or ABE protocol. At Subheading 3.2 follow either protocol 3.2 (ABE) or 3.3 (Acyl-RAC). Perform all steps at room temperature unless noted otherwise.

3.1 Sample Lysis and Free Thiol Blocking

1. If required prior to sample solubilization, homogenize sufficient sample or tissue as appropriate for the species in question to provide ~1–2 mg total protein under conditions that minimize proteolytic degradation (*see* **Notes 2**, **9** and **10**). For example, 15 *Arabidopsis thaliana* seedlings (10 days post-germination) are ground to a fine powder under liquid nitrogen in a microfuge tube using a plastic micropestle and typically yields ~2 mg total protein.

2. Add 500 μL lysis buffer to sample and mix gently at room temperature for 10 min until the tissue is thawed and/or fully solubilized in lysis buffer (*see* **Note 2**).
3. Centrifuge for 1 min at 16,000 × *g* to remove insoluble material.
4. Transfer supernatant to a fresh, labeled 1.5 mL microfuge tube.
5. Remove an aliquot to determine protein concentration by BCA assay.
6. Mix the remaining samples gently using end-over-end mixing.
7. Dilute samples to the same protein concentration, typically 0.25–2 mg/mL (*see* **Note 10**) in lysis buffer to a final volume of 1.1 mL.
8. Incubate samples at room temperature for 1 h with gentle end-over-end mixing.
9. Add 12 μL 2,3-dimethyl 1,3-butadiene (~100 mM final concentration) (*see* **Note 11**).
10. Mix samples for 1 h in a shaking hot block at 1500 rpm, 25 °C (*see* **Note 12**).
11. Briefly centrifuge to collect liquid.
12. Add one-tenth volume chloroform.
13. Shake samples for 5 min in a shaking hot block at 1500 rpm, 25 °C (*see* **Note 13**).
14. Centrifuge the samples in a microfuge at 16,000 × *g* for 1 min to achieve phase separation (*see* **Note 14**).

3.2 Acyl-Biotin Exchange

1. Split the sample into two equal aliquots of 500 μL in fresh, labeled 15 mL conical centrifuge tubes.
2. Add 500 μL 1 M hydroxylamine pH 7.2 to one aliquot (Hyd+) and 500 μL 1 M NaCl to the other (Hyd−). Mix well without foaming (*see* **Note 15**).
3. Add 100 μL biotin-HPDP solution to both aliquots.
4. Mix the samples gently at room temperature for 1 h on a roller mixer table.
5. Chloroform–methanol-precipitate the proteins [7]: add 1 vol chloroform to each aliquot and vortex. Add 4 vols methanol and vortex. Add 3 vols water and vortex.
6. Centrifuge at >4000 × *g* for at 15 °C until phase separation is achieved (*see* **Note 16**).
7. Remove and discard the clear upper phase without disturbing the protein pellet at the interface.

8. Add 4 vols methanol and gently mix. Take care not to lose any protein through dispersion up the side of the tube (*see* **Notes 17** and **18**).
9. Centrifuge at 4000 × *g* for 30 min at 15 °C.
10. Remove and discard supernatant and air-dry the pellet (*see* **Note 19**).
11. Resuspend the pellets in 200 μL resuspension buffer (*see* **Note 20**).
12. Centrifuge tubes at 4000 × *g* to collect and clear the sample at the bottom of the 15 mL falcon tube.
13. Remove 10 μL from each hyd+ and hyd− aliquot to a new tube for a loading control (LC+ and LC− respectively). Add 10 μL 2× SDS loading buffer to each LC tube. Store samples at −20 °C until use.
14. Transfer 185 μL of each sample to a fresh labeled (EX+ and EX−) 1.5 mL microfuge tube and dilute with 1275 μL 1× PBS. Avoid transferring any insoluble pellet.
15. Add 40 μL of a 50% Neutravidin-agarose beads suspension to each EX+ and EX− and incubate at room temperature for 1 h with gentle mixing (*see* **Note 21**).
16. Collect the beads by centrifugation at ~1000 × *g*.
17. Wash the beads ×3 with 1 mL ABE wash buffer and then ×2 with 1 mL 1× PBS for 5 min each.

3.3 Acyl-RAC

1. Split the sample into two equal aliquots of 500 μL in fresh, labeled 1.5 mL microfuge tubes.
2. Add 500 μL 1 M hydroxylamine pH 7.2 to one aliquot (hyd+) and 500 μL 1 M NaCl to the other (hyd−). Mix well without foaming (*see* **Note 15**).
3. Remove 10 μL from each hyd+ and hyd− aliquot to a new tube for a loading control (LC+ and LC− respectively) and incubate at room temperature for 1 h. Add 10 μL 2× SDS loading buffer to each LC tube. Store samples at −20 °C until use.
4. Add 40 μL of a 50% Thiopropyl-Sepharose beads suspension to each EX+ and EX− and incubate at room temperature for 1 h with gentle mixing (*see* **Note 22**).
5. Collect the beads by centrifugation at ~1000 × *g*.
6. Wash the beads three times with 1 mL Acyl-RAC wash buffer.

3.4 Sample Elution and Analysis

1. Aspirate the beads to dryness after the final wash taking care not to remove any beads (*see* **Note 23**).
2. Elute the samples from the beads by adding 25 μL 2× reducing SDS loading buffer.

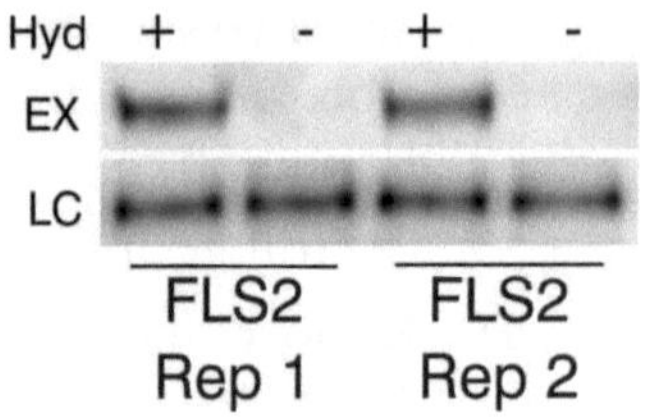

Fig. 1 *S*-Acylation state of *Arabidopsis thaliana* FLS2 as determined by western blotting of samples processed by acyl-biotin exchange. The same initial sample was split in two and processed in parallel as technical replicates (Rep 1 and Rep 2). Hydroxylamine-treated samples: Hyd+, NaCl treated samples: Hyd−, EX: eluate from neutravidin beads, LC: loading control from before binding to neutravidin beads

3. Heat samples at 37 °C for 30 min (*see* **Note 24**).
4. Analyze samples by SDS-PAGE and immunoblotting as appropriate for the protein of interest (*see* **Notes 25** and **26**).
5. For a protein to be considered *S*-acylated, signal should be detected in the EX+ lane but not the EX− lane. Equal signal should be detected in both LC+ and LC− lanes as shown in Fig. 1 (*see* **Note 27**).

4 Notes

1. *N*-Ethylmaleimide solutions should be prepared fresh in dry ethanol. NEM is unstable in aqueous solutions and hydrolysis is accelerated by elevated pH.
2. If cells/tissue used produces viscous DNA contamination from lysis of nuclei by SDS, substitute 0.5% Triton X-100 or IGEPAL CA-630 for SDS in lysis buffer. Using ice-cold 0.5% Triton X-100 lysis buffer, lyse the cells on ice for 1 min with gentle mixing, centrifuge at 5000 × *g* 4 °C for 1 min to pellet nuclei, and add the supernatant to one-third volume of 10% SDS containing 10 mM NEM.
3. *S*-Acylation is very sensitive to the presence of reducing agents in buffers. Do not include reducing agents, particularly thiol-based ones, in lysis buffer unless absolutely essential. If essential, use the absolute minimum or consider substituting for TCEP (Tris(2-carboxyethyl)phosphine hydrochloride) as this phosphine based reducing agent is reported to be compatible with *S*-acylation at up to 5 mM [8]. NEM will also react with TCEP [9] and thiol-based reducing agents, eliminating them from solution and reducing the effective NEM concentration. If using reducing agents, lysis buffer should not contain NEM; rather, NEM should be added after reducing agents have had their desired effect. Reducing agent concentration should also

be accounted for when adding NEM. Some protein quantification assays, such as BCA assays proposed for use here, are also sensitive to the use of reducing agents.

4. Use protease inhibitors appropriate to the species/system under study. In some cases the use of phosphatase inhibitors may also be appropriate.
5. Make hydroxylamine solutions immediately before use. Hydroxylamine solutions are unstable and degrade over time. As a result of degradation the pH of the solution increases, leading to undesirable side reactions.
6. Other protein concentration determination methods are available and suitable but must be compatible with the reagents used, in particular detergents.
7. Sample mixing requires both prolonged vortexing and gentle mixing. A range of lab-based equipment is suitable and specific types are not essential to the protocol.
8. This is the same buffer as lysis buffer but without NEM or protease inhibitors.
9. The variety of sample types to be analyzed makes detailed instructions at this stage impossible. As a general rule use species-appropriate protocols designed for maximum disruption of cells/tissue for denaturing protein extraction.
10. Typically, 250–1000 μg of total protein is required to perform an *S*-acylation assay depending on abundance of the protein of interest. We recommend starting with 10–20 times the mass of protein from which the protein of interest can be reliably and quantitatively detected by western blot.
11. 2,3-Dimethyl 1,3-butadiene is volatile, flammable and has an unpleasant odor and should only be used in a fume hood. Store 2,3-dimethyl 1,3-butadiene at 4 °C to minimize volatilization.
12. Samples should be vigorously mixed (vortexed) to ensure dispersal of 2,3-dimethyl 1,3-butadiene throughout the aqueous phase.
13. Samples should form a visible emulsion but not foam. Adjust shaking speed accordingly.
14. Do not disturb the thin white layer of detergent formed at the chloroform–aqueous interphase.
15. We occasionally observe a white precipitate forming in some samples when using high protein concentrations, cool buffers or if lab temperatures are low. Briefly warming the sample to 37 °C with gentle mixing will resolubilize the precipitate. Treat all samples identically even if they do not contain visible precipitate. No deleterious or unexpected effects or variations on downstream processing or results have been observed as a result of this precipitate.

16. We prefer to centrifuge for 20–30 min at 4000 × *g* using a swinging bucket rotor to ensure that the interphase remains undisturbed. Centrifugation using a fixed angle rotor allows for faster processing at higher speeds (protocol tested at 20,000 × *g* in a 45° fixed angle rotor) but care must be taken to avoid disturbing the interphase. Following centrifugation the upper phase should be completely clear; if it is still showing evidence of cloudiness repeat the centrifugation step.
17. Vigorous addition of methanol by pipette followed by gentle swirling is usually sufficient to achieve mixing. Take care not to disperse sample up the side of the tube.
18. For dilute samples incubation at −20 °C for 1 h to overnight can improve sample recovery. In our experience samples containing >100 μg/mL protein do not require this additional step for quantitative recovery.
19. Do not overdry the pellet as this will hinder resolubilization. The pellet is typically dry of methanol and chloroform within 10 min. Some water will remain behind and the pellet should appear damp but this will not interfere with downstream steps.
20. To aid resuspension, samples can be alternately heated to 37 °C for 10 min in a water bath and mixed on a roller mixer table for 10 min. Careful vortexing can also help but should be done with care to avoid sample loss. Improper resuspension can result in either sample loss or "smearing" of samples on SDS-PAGE gels.
21. Prepare High-capacity Neutravidin beads by washing 3× with >10 bead volumes PBS and resuspending as a ~50% suspension in PBS for use.
22. Prepare Thiopropyl-Sepharose beads by hydrating in Acyl-RAC wash buffer for 15 min followed by three washes with >10 bead volumes PBS and resuspending as a ~50% suspension in Acyl-RAC wash buffer for use.
23. Aspirate buffer from the tubes until the buffer meniscus is just above the beads. Using a 10 μL pipette tip mounted on a p20 pipette set to 20 μL draw up buffer. Expel the buffer with some force to resuspend the beads and place the pipette tip immediately against the base of the tube. Gently aspirate the remaining buffer, taking care to maintain the seal at the bottom of the tube, and wipe off any adherent beads from the pipette tip on the side of the tube.
24. For large, heavily glycosylated or aggregation-prone proteins, heating samples to 95 °C can result in smearing of samples on SDS-PAGE gels, particularly after the processing steps described here. Heating samples to 37 °C for 30 min or 65 °C for 15 min with occasional gentle mixing can help reduce

smearing. We routinely use 37 °C for 30 min for all proteins without issue and have found elution to be more consistent without affecting protein mobility on SDS-PAGE gels.

25. We typically load all EX samples on one gel and all LC on another in order EX+, EX− or LC+, LC− for each sample. *See* Fig. 1 for example running order.

26. For fully quantitative analysis load the entirety of the sample, including beads, using a 200 μL pipette tip. Loading the sample and beads will require a 1.5 mm thick SDS-PAGE gel to accommodate the ~40 μL sample volume. Fill in empty wells with 2× SDS loading buffer to maintain consistent lane widths across the gel. Use appropriate quantification equipment and software designed for quantification of western blots rather than scanned film exposures.

27. If no signal is found in EX+ lanes this indicates the protein of interest is not *S*-acylated. If a signal appears in both EX+ and EX− lanes this indicates improper blocking of free sulfhydryls. No signal in LC lanes indicates either sample loss or the protein of interest is not expressed highly enough for detection by western blot. In either of the two latter cases, the experiment must be repeated for a conclusion to be drawn.

References

1. Drisdel RC, Green WN (2004) Labeling and quantifying sites of protein palmitoylation. Biotechniques 36(2):276–285
2. Hemsley PA, Taylor L, Grierson CS (2008) Assaying protein palmitoylation in plants. Plant Methods 4(1):2
3. Forrester MT et al (2011) Site-specific analysis of protein *S*-acylation by resin-assisted capture. J Lipid Res 52(2):393–398
4. Yokoi N et al (2016) Identification of PSD-95 Depalmitoylating Enzymes. J Neurosci 36 (24):6431–6444
5. Hurst CH et al (2017) Maleimide scavenging enhances determination of protein *S*-palmitoylation state in acyl-exchange methods. Biotechniques 62(2):69–75
6. Rideout DC, Breslow R (1980) Hydrophobic acceleration of Diels-Alder reactions. J Am Chem Soc 102(26):7816–7817
7. Wessel D, Flugge UI (1984) A method for the quantitative recovery of protein in dilute solution in the presence of detergents and lipids. Anal Biochem 138(1):141–143
8. Ji Y et al (2013) Direct detection of *S*-palmitoylation by mass spectrometry. Anal Chem 85 (24):11952–11959
9. Kantner T, Watts AG (2016) Characterization of reactions between water-soluble trialkylphosphines and thiol alkylating reagents: implications for protein-conjugation reactions. Bioconjug Chem 27(10):2400–2406

Chapter 2

Detection of Heterogeneous Protein *S*-Acylation in Cells

Jennifer Greaves and Nicholas C. O. Tomkinson

Abstract

The use of synthetically synthesized azide and alkyne fatty acid analogs coupled with bioorthogonal Cu(I)-catalyzed Huisgen 1,3-dipolar cycloaddition reaction-based detection methods to study protein *S*-acylation reactions has replaced the traditional method of using in vivo metabolic radiolabeling with tritiated palmitic acid and has greatly facilitated our understanding of this essential cellular process. Here, we describe the chemical synthesis of myristic (C:14), palmitic (C16:0), and stearic (C18:0) acid-azide probes and detail how they may be utilized as chemical reporters for the analysis of *S*-acylation of exogenously expressed proteins in cells.

Key words Click chemistry, Fatty acid azide, Fatty acylation, Palmitoylation, *S*-Acylation

1 Introduction

Protein *S*-acylation (palmitoylation) describes the reversible post-translational attachment of fatty acids on to cysteine amino acids. Although palmitic acid (C16:0) is the most frequent fatty acid to modify proteins in this way, other shorter and longer chain saturated and unsaturated fatty acids may be added on to *S*-acylated proteins [1–6]. Historically, the only method to detect protein *S*-acylation was by in vivo metabolic radiolabeling using tritiated palmitate followed by fluorographic detection [7–9]. Despite being the primary method used for decades, this approach has several limitations, including a low signal intensity that requires long exposure times (up to 3 months) in order for *S*-acylation to be detected. Over the past decade, however, there has been a significant advance over this traditional methodology through the use of synthetically synthesized azide and alkyne fatty acid analogs as alternatives to tritiated palmitic acid [10–17]. These fatty acid analogs mimic endogenous fatty acids and can be utilized as reporters of cellular processes involving fatty acids, such as protein *S*-acylation [6]. Coupled with bioorthogonal Cu(I)-catalyzed Huisgen 1,3-dipolar cycloaddition ("Click") reaction-based detection

Maurine E. Linder (ed.), *Protein Lipidation: Methods and Protocols*, Methods in Molecular Biology, vol. 2009,
https://doi.org/10.1007/978-1-4939-9532-5_2,

methods, the use of azide and alkyne fatty acid analogs is a highly sensitive, quantitative and practical method for studying protein *S*-acylation reactions that can be adapted for high-throughput analysis [18]. Since fatty acyl chain diversity can also be incorporated during the synthesis of these chemical probes, these reporters facilitate our understanding of how acyl chain heterogeneity contributes to protein *S*-acylation [6].

This method describes the detection of S-acylation of EGFP-tagged synaptosomal-associated protein of 25 kDa (SNAP25), expressed in HEK293T cells along with hemagglutinin (HA)-tagged zDHHC3 as a catalyst, using chemically synthesized myristic (C14:0), palmitic (C16:0), and stearic (C18:0) acid azide probes. Detection of *S*-acylation of other substrate proteins may not necessarily require the coexpression of a zDHHC enzyme [19]. HEK293 cells are a well-established mammalian cell line used for heterologous protein expression due to their efficient transfection rate and subsequent high level expression of proteins from plasmid DNA [20]. In this method, HEK293T cells are transiently transfected using Lipofectamine 2000 transfection reagent (Invitrogen) with a plasmid encoding the *S*-acylated protein of interest. Following high-level expression, cells are labeled with the synthetically synthesized fatty acid-azide probes, which can become metabolically incorporated into *S*-acylated and other lipidated cellular proteins. The azido-fatty acid modification on cellular proteins is then conjugated to an alkyne infrared fluorophore by performing a Cu (I)-catalyzed "Click" reaction. Isolated cellular proteins are then resolved electrophoretically in vertical discontinuous sodium dodecyl sulfate–polyacrylamide gels according to the Laemmli method [21] and expression of exogenous proteins is confirmed by western blotting.

2 Materials

2.1 Reduction of Carboxylic Acids

1. Hotplate stirrer.
2. Cryogenic bath for cooling.
3. Spatula.
4. 500 mL B24 round bottom flask.
5. B24 suba seal.
6. Schlenk line.
7. 250 mL measuring cylinder.
8. Sintered filter funnel with side-arm.
9. B24 water condenser.
10. Tubing for water condenser.
11. Silicone oil bath.

12. Magnetic stirrer bar.
13. Pressure tubing for filtration.
14. 1 L separating funnel with stopper.
15. 2× 500 mL conical flask.
16. Rotary evaporator.

2.2 Bromination of Diols

1. Hotplate stirrer.
2. Spatula.
3. 2 neck 250 mL B24 round bottom flask.
4. B24 glass stopper.
5. B24 100 mL pressure-equalizing dropping funnel.
6. Silicone oil bath for heating.
7. 250 mL measuring cylinder.
8. Sintered filter funnel with side-arm.
9. B24 water condenser.
10. Tubing for water condenser.
11. Magnetic stirrer bar.
12. 500 mL separating funnel with stopper.
13. Pressure tubing for filtration.
14. 3× 250 mL conical flask.
15. 80 × 3 cm glass column with ball joint for chromatography.
16. Rotary evaporator.

2.3 Oxidation of Primary Alcohols

1. Hotplate stirrer.
2. Spatula.
3. 100 mL B24 round bottom flask.
4. B24 glass stopper.
5. 100 mL measuring cylinder.
6. Sintered filter funnel with side-arm.
7. Magnetic stirrer bar.
8. Pressure tubing for filtration.
9. 250 mL separating funnel with stopper.
10. 3× 250 mL conical flask.
11. 80 × 3 cm column with ball joint for flash chromatography.
12. 100 mL conical flask.
13. Rotary evaporator.

2.4 Azide Addition

1. Hotplate stirrer.
2. Schlenk line.

3. Spatula.
4. 50 mL B24 round bottom flask..
5. B24 suba seal.
6. 50 mL measuring cylinder.
7. Sintered filter funnel with side-arm.
8. B24 water condenser.
9. Tubing for water condenser.
10. Silicone oil bath.
11. Magnetic stirrer bar.
12. Pressure tubing for filtration.
13. 100 mL separating funnel with stopper.
14. 3× 100 mL conical flask.
15. Rotary evaporator.

2.5 Cell Culture Reagents and Equipment

1. Human Embryonic Kidney 293T (HEK293T) culture medium: DMEM, 2 mM L-glutamine, 10% fetal bovine serum; store at 4 °C.
2. Trypsin–EDTA (0.05%).
3. Lipofectamine 2000 (Invitrogen) transfection reagent; store at 4 °C.
4. Fatty-acid free cell culture medium (FAF culture medium): DMEM, 2 mM L-glutamine, 1 mg/mL of fatty-acid free bovine serum albumin.
5. Fatty-acid azides: Saturated fatty acid azides are stored at −20 °C as 100 mM stocks made up in DMSO (*see* **Note 1**).
6. Fatty-acid azide labeling medium: FAF culture medium, 100 μM fatty-acid azide.
7. Poly-D-lysine-coated 24-well cell culture plates.
8. T75 flasks.
9. Sterile 1.5 mL tubes.
10. Laminar flow tissue culture hood.
11. Neubauer chamber.
12. Light microscope.

2.6 Click Chemistry

1. Phosphate-buffered Saline (PBS; 10×): 1.37 M NaCl, 27 mM KCl, 100 mM Na_2HPO_4, 18 mM KH_2PO_4, pH 7.4 (*see* **Note 2**).
2. Lysis Buffer: 50 mM Tris, pH 8.0, 0.5% SDS. Store at 4 °C. Protease inhibitors should be added immediately before use.
3. LI-COR IRDye® 800CW Alkyne Infrared Dye (LI-COR Biotechnology UK Ltd., Catalogue number 929-60002):

Resuspend in dimethyl sulfoxide (DMSO) at a concentration of 4 mM and store stock solutions in aliquots at −20 °C for short-term or −80 °C for long-term storage (*see* **Note 3**).

4. Copper sulfate: 40 mM solution.
5. Tris[(1-benzyl-1*H*-1,2,3-triazol-4-yl)methyl]amine (TBTA): Resuspend in DMSO at a concentration of 100 mM and store stock solutions in small aliquots at −20 °C.
6. Ascorbic acid: 4 mM solution (*see* **Note 4**).
7. Acetone: 100% and 70% stock solutions stored at −20 °C (*see* **Note 5**).
8. End-over-end rotator.

2.7 SDS–Polyacrylamide Gel Electrophoresis

1. 4× SDS Lysis buffer: 200 mM Tris–HCl, pH 6.8, 40% glycerol, 8% sodium dodecyl sulfate (SDS) (*see* **Note 6**), 0.4% Bromophenol Blue, supplemented with 25 mM DTT (*see* **Note 7**).
2. 1 M Dithiothreitol (DTT) stock: in dH_2O and store in 1 mL aliquots at −20 °C.
3. 2× Resolving gel buffer: 750 mM Tris–HCl, pH 8.9, 4 mM ethylenediaminetetraacetic acid (EDTA), 0.2% SDS.
4. 2× Stacking gel buffer: 250 mM Tris–HCl, pH 6.8, 4 mM EDTA, 0.2% SDS.
5. 30% acrylamide–bisacrylamide solution, stored at 4 °C.
6. *N*,*N*,*N*′,*N*′-Tetramethyl-ethane-1,2-diamine (TEMED), stored at 4 °C.
7. 10% ammonium persulfate stock in H_2O and stored at 4 °C (*see* **Note 8**).
8. 10× Tris–glycine electrophoresis buffer: 250 mM Tris–HCl, 2.5 M glycine, 1% SDS (*see* **Note 9**).
9. Prestained molecular weight markers.
10. Bio-Rad Mini-PROTEAN Tetra Cell system.
11. Bio-Rad PowerPac Universal Power Supply.
12. Heat block.

2.8 Western Blotting

1. Nitrocellulose membrane, 0.45 μM pore size.
2. 3 mm Whatman blotting paper.
3. Bio-Rad Trans-Blot Cell.
4. Bio-Rad PowerPac Universal Power Supply.
5. 10× Transfer buffer: 480 mM Tris base, 390 mM glycine, 13 mM SDS (*see* **Note 10**).
6. PBS-T: PBS, 0.02% Tween 20 (*see* **Note 11**).

2.9 Detection of S-Acylated Proteins

1. Western blotting membrane incubation box.
2. Orbital rotator.
3. PBS-T: PBS, 0.02% Tween 20 (*see* **Note 11**).
4. Blocking Buffer: 5% nonfat milk powder in PBS-T (*see* **Note 12**).
5. Primary antibody: for example, Living Colors® Av Monoclonal anti GFP Antibody (JL-8), Clontech.
6. Secondary antibody diluent: 1% nonfat skimmed milk powder in PBS-T (*see* **Note 12**).
7. LI-COR IRDye® 680RD secondary antibody raised against the appropriate species of primary antibody (*see* **Note 13**). For use with the anti-GFP JL8 antibody, use LI-COR IRDye® 680RD goat anti-mouse IgG (H + L).
8. LI-COR Odyssey Imaging System (*see* **Note 14**).

3 Methods

Unless otherwise stated, all commercially available reagents are used as supplied without any further purification. All procedures should be carried out at room temperature, unless stated otherwise. Prepare all solutions using ultrapure Milli-Q water and store at room temperature, unless stated otherwise.

3.1 Fatty Acid Azide Synthesis

Use dry THF directly from a PureSolv MD 5 Solvent Purification System by Innovative Technology Inc., and handle under inert atmosphere. Carry out flash chromatography using Merck Kieselgel 60 H silica. Carry out analytical thin layer chromatography using aluminum-backed plates coated with Merck Kieselgel 60 GF254 and visualize using *p*-anisaldehyde.

Record nuclear magnetic resonance (NMR) spectra on a 400 MHz Ultrashield Magnet, Prodigy liquid nitrogen cryoprobe, AVIII console, and a Z420 HP workstation running TopSpin 3.X running at 400 MHz (^{1}H NMR) and 101 MHz (^{13}C NMR); a 500 MHz Ascend magnet, BBO multinuclear Smart probe, AVIIIHD500 console, and Z420 HP workstation running TopSpin 3.X running at 500 MHz (^{1}H NMR) and 126 MHz (^{13}C NMR); or a 600 MHz Ultrashield magnet, BBO multinuclear probe, AVII+ console, and a Z420 HP workstation running TopSpin 3.X running at 600 MHz (^{1}H NMR) or 151 MHz (^{13}C NMR). Chemical shifts are reported in parts per million (ppm) in the scale relative to $CDCl_3$, 7.26 ppm for ^{1}H NMR and 77.16 for ^{13}C NMR; DMSO-d_6, 2.50 ppm for ^{1}H NMR and 39.52 for ^{13}C NMR. Coupling constants are measured in Hertz (Hz).

Record low resolution mass spectra (LRMS) on an Agilent 6130 single quadrupole with APCI/ESI dual source, on a ThermoQuest Finnigan LCQ DUO electrospray, or on an Agilent

7890A GC system, equipped with a 30 m DB5MS column connected to a 5975C inert XL CI MSD with Triple-Axis Detector. Perform MALDI on an Axima-CFR from Kratos-Shimadzu. Record infrared spectra on an Agilent 5500a FTIR equipped with ATR (attenuated total reflectance) and report in cm^{-1}. In vacuo refers to evaporation under reduced pressure using a rotary evaporator connected to a diaphragm pump, followed by the removal of trace volatiles using a high vacuum (oil) pump. Determine melting points with a Gallenkamp SG92 melting point apparatus and are uncorrected.

3.2 Preparation of Fatty Acid Azides

Fatty acid azides followed the synthesis outlined in Fig. 1.

1. Preparation of 1,14-tetradecanediol **4** (*see* **Note 15**). Add solid $LiAlH_4$ (1.47 g, 38.7 mmol) to a solution of 1,14-tetradecanedioic acid (5.00 g, 19.35 mmol) in THF (194 mL) at 0 °C. Allow the reaction to warm to room

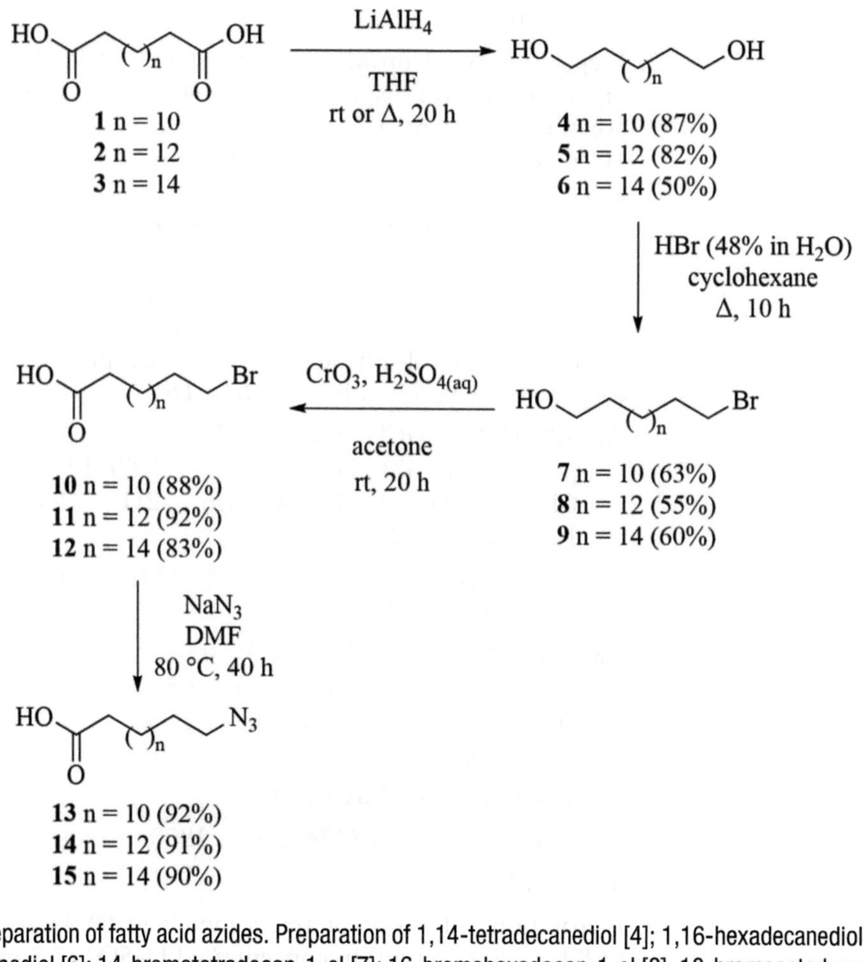

Fig. 1 Preparation of fatty acid azides. Preparation of 1,14-tetradecanediol [4]; 1,16-hexadecanediol [5]; 1,18-octadecanediol [6]; 14-bromotetradecan-1-ol [7]; 16-bromohexadecan-1-ol [8]; 18-bromooctadecan-1-ol [9]; 14-bromotetradecanoic acid [10]; 16-bromohexadecanoic acid [11]; 18-bromooctadecanoic acid [12]; 14-azidotetradecanoic acid [13]; 16-azidohexadecanoic acid [14]; 18-azidooctadecanoic acid [15]

temperature and stir for 20 h. Upon completion, add wet Na_2SO_4 portionwise until the grey suspension turns white. Stir the suspension at room temperature until the white solid is free-flowing, and add solid $MgSO_4$ (5 g). Filter the reaction and wash the filter cake with 5× 50 mL Et_2O. Evaporate the solvent in vacuo to give the product (3.88 g, 87%) as a white solid.

δ_H (600 MHz, DMSO-d_6) 4.29 (t, 2H, J 5.0 Hz, 2× OH), 3.35–3.38 (m, 4H, 2× CH_2OH), 1.36–1.42 (m, 4H, 2× CH_2CH_2OH), 1.22–1.29 (m, 20H, 10× CH_2).

δ_C (151 MHz, DMSO-d_6) 60.7, 32.5, 29.1, 29.0, 29.0, 28.9, 25.5.

LR-MS (MALDI-TOF) 253.2 ($[M + Na]^+$).

HR-MS calculated for $C_{14}H_{31}O_2$ ($[M + H]^+$) 231.2318, found 231.2318.

υ_{max} (thin film, cm^{-1}) 3410, 3351, 2921, 2891, 2850.

M_p 88–90 °C.

2. Preparation of 14-bromotetradecan-1-ol **7**. Add HBr (48% in H_2O, 41 mL) to a suspension of diol **4** (3.57 g, 15.5 mmol) in cyclohexane (41 mL). Heat the biphasic mixture to reflux for 10 h and then cool to room temperature. Separate the layers and extract the aqueous phase with CH_2Cl_2 (4× 30 mL). Wash the combined organics with $NaHCO_3$ (4× 20 mL of a saturated aqueous solution), brine (20 mL), dry over $MgSO_4$, filter and concentrate in vacuo. Purify the crude residue by column chromatography, eluting with 95:5 petrol–EtOAc then 70:30 petrol–EtOAc, to afford the product (2.85 g, 63%) as a pale yellow solid (*see* **Note 16**).

   δ_H (400 MHz, $CDCl_3$) 3.64 (t, 2H, J 6.6 Hz, CH_2OH), 3.40 (t, 2H, J 6.9 Hz, CH_2Br), 1.80–1.90 (m, 2H, CH_2CH_2Br), 1.53–1.60 (m, 2H, CH_2CH_2OH), 1.25–1.46 (m, 20H, 10× CH_2).

   δ_C (101 MHz, $CDCl_3$) 63.2, 34.2, 33.0, 33.0, 29.7, 29.7, 29.6, 28.3, 28.9.

   LR-MS (EI^+) 294.9 ($[M(^{81}Br)]^+$, 0.5%), 292.9 ($[M(^{79}Br)]^+$, 1%), 276.9 ($[M(^{81}Br)\text{-}H_2O]^+$, 3.5%), 292.9 ($[M(^{79}Br)\text{-}H_2O]^+$, 3%), 213.9 ($[M - Br]^+$, 2.5%).

   HR-MS calculatedfor $C_{14}H_{33}ON^{79}Br^+$($[M + NH_4]^+$) 310.1740, found 310.1744.

   υ_{max} (thin film, cm^{-1}) 3274, 2919, 2850.

   M_p 46–48 °C.

3. Preparation of 14-bromotetradecanoic acid **10**. Dissolve CrO_3 (3.89 g, 38.92 mmol) in concentrated H_2SO_4 (7.2 mL). Add cold H_2O (16.2 mL) slowly and stir the solution for 10 min. Add the resulting solution dropwise to a solution of alcohol

7 (2.85 g, 9.73 mmol) in acetone (243 mL). Stir the reaction mixture for 20 h before adding H_2O (100 mL) and CH_2Cl_2 (40 mL). Separate the layers and extract the aqueous phase with CH_2Cl_2 (4× 30 mL). Wash the combined organics with brine (30 mL), dry over $MgSO_4$, filter and concentrate in vacuo. Purify the crude residue by column chromatography, eluting with 90:10 petrol–EtOAc (+0.1% AcOH), to afford the product (2.63 g, 88%) as a white solid.

δ_H (400 MHz, $CDCl_3$) 3.41 (t, 2H, J 6.9 Hz, CH_2Br), 2.35 (t, 2H, J 7.5 Hz, CH_2CO_2H), 1.81–1.89 (m, 2H, CH_2CH_2Br), 1.59–1.68 (m, 2H, $CH_2CH_2CO_2H$), 1.24–1.45 (m, 18H, 9× CH_2).

δ_C (101 MHz, $CDCl_3$) 179.0, 34.2, 34.0, 33.0, 29.7, 29.7, 29.6, 29.4, 29.2, 28.9, 28.3, 24.8.

LR-MS (EI^+) 309.0 ($[M(^{81}Br)]^+$, 16%), 307.0 ($[M(^{79}Br)]^+$, 20%), 291.0 ($[M(^{81}Br)\text{-}H_2O]^+$, 13%), 289.0 ($[M(^{79}Br)\text{-}H_2O]^+$, 17%), 227.1 ($[M - Br]^+$, 24%).

HR-MS calculated for $C_{14}H_{26}O^{79}Br$ ($[M - H]^-$) 305.1122, found 305.1123.

HR-MS calculated for $C_{14}H_{26}O^{79}Br$ ($[M - H]^-$) 305.1122, found 305.1123.

M_p 63–66 °C.

4. Preparation of 14-azidotetradecanoic acid **13** (*see* **Note 15**). Add NaN_3 (636 mg, 9.78 mmol) to a solution of bromide **10** (500 mg, 1.63 mmol) in DMF (6.5 mL). Stir the reaction at 80 °C for 40 h and then cool to room temperature. Add a 1:1 mixture of EtOAc and H_2O (20 mL), separate the layers, and extract the aqueous phase with EtOAc (3× 10 mL). Wash the combined organics with brine (10 mL), dry over $MgSO_4$, filter and concentrate in vacuo. Purify the crude residue by column chromatography, eluting with 90:10 petrol–EtOAc (+0.1% AcOH), to afford the product (403 mg, 92%) as a white solid (*see* **Note 17**).

δ_H (400 MHz, $CDCl_3$) 3.25 (t, 2H, J 7.0 Hz, CH_2N_3), 2.34 (t, 2H, J 7.5 Hz, CH_2CO_2H), 1.55–1.68 (m, 4H, $CH_2CH_2N_3$, $CH_2CH_2CO_2H$), 1.23–1.38 (m, 18H, 9× CH_2).

δ_C (101 MHz, $CDCl_3$) 180.1, 51.6, 34.2, 29.7, 29.6, 29.6, 29.5, 29.4, 29.3, 29.2, 29.0, 26.9, 24.8.

LR-MS (ES^-) 268.1 ($[M - H]^-$, 100%), 240.1 ($[M - N_2 - H]^-$, 34%).

HR-MS calculated for $C_{14}H_{26}N_3O_2$ ($[M - H]^-$) 268.2031, found 268.2028.

υ_{max} (thin film, cm^{-1}) 3016, 2915, 2848, 2101, 1701.

M_p 38–40 °C.

5. Preparation of 1,16-hexadecanediol **5** (*see* **Note 15**). Add solid $LiAlH_4$ (1.33 g, 34.9 mmol) to a solution of 1,16-hexadecanedioic acid (5.00 g, 17.46 mmol) in THF (175 mL) at 0 °C. Allow the reaction to warm to room temperature and stir for 20 h. Upon completion, add wet Na_2SO_4 portionwise until the grey suspension turns white. Stir the suspension until the white solid is free-flowing and add solid $MgSO_4$ (5 g). Filter the reaction and wash the filter cake with Et_2O (5× 50 mL). Evaporate the solvent in vacuo to afford the product (3.68 g, 82%) as a white solid.

   δ_H (600 MHz, DMSO-d_6) 4.29 (t, 2H, J 4.9 Hz, 2× OH), 3.34–3.39 (m, 4H, 2× CH_2OH), 1.36–1.42 (m, 4H, 2× CH_2CH_2OH), 1.22–1.29 (m, 24H, 12× CH_2).

   δ_C (151 MHz, DMSO-d_6) 60.7, 32.5, 29.1, 29.0, 29.0, 29.0, 28.9, 25.5.

   δ_C (151 MHz, DMSO-d_6) 60.7, 32.5, 29.1, 29.0, 29.0, 29.0, 28.9, 25.5.

   HR-MS calculated for $C_{16}H_{35}O_2$ ($[M + H]^+$) 259.2632, found 259.2632.

   υ_{max} (thin film, cm^{-1}) 3414, 3353, 2919, 2891, 2848.

   M_p 91–94 °C.

6. Preparation of 16-bromohexadecan-1-ol **8**. Add HBr (48% in H_2O, 36 mL) to a suspension of diol **5** (3.48 g, 13.5 mmol) in cyclohexane (36 mL). Heat the biphasic mixture to reflux for 10 h before cooling to room temperature. Separate the layers and extract the aqueous phase with CH_2Cl_2 (4× 30 mL). Wash the combined organics with $NaHCO_3$ (4× 20 mL of a saturated aqueous solution), brine (20 mL), dry over $MgSO_4$, filter and concentrate in vacuo. Purify the crude residue by column chromatography, eluting with 95:5 petrol–EtOAc then 70:30 petrol–EtOAc, to afford the product (2.40 g, 55%) as a pale yellow solid (*see* **Note 16**).

   δ_H (400 MHz, $CDCl_3$) 3.64 (t, 2H, J 6.6 Hz, CH_2OH), 3.41 (t, 2H, J 6.9 Hz, CH_2Br), 1.80–1.91 (m, 2H, CH_2CH_2Br), 1.52–1.62 (m, 2H, CH_2CH_2OH), 1.25–1.46 (m, 24H, 12× CH_2).

   δ_C (101 MHz, $CDCl_3$) 63.3, 34.2, 33.0, 33.0, 29.8, 29.8, 29.7, 29.6, 28.9, 28.3, 25.9.

   LR-MS (EI^+) 320.9 ($[M(^{81}Br)]^+$, 0.5%), 318.9 ($[M(^{79}Br)]^+$, 0.5%), 304.9 ($[M(^{81}Br)\text{-}H_2O]^+$, 1%), 302.9 ($[M(^{79}Br)\text{-}H_2O]^+$, 1%), 241.8 ($[M - Br]^+$, 1.5%).

   HR-MS calculated for $C_{16}H_{37}{}^{79}BrON$ ($[M + NH_4]^+$) 338.2053, found 338.2056.

   υ_{max} (thin film, cm^{-1}) 3274, 2917, 2850.

   M_p 54–56 °C.

7. Preparation of 16-bromohexadecanoic acid **11**. Dissolve CrO_3 (3.00 g, 30.0 mmol) in concentrated H_2SO_4 (5.5 mL). Add cold H_2O (12.5 mL) slowly and stir the solution at room temperature for 10 min. Add the resulting solution dropwise to a solution of alcohol **8** (2.40 g, 7.50 mmol) in acetone (188 mL). Stir the reaction mixture for 20 h before adding H_2O (100 mL) and CH_2Cl_2 (40 mL). Separate the layers and extract the aqueous phase with CH_2Cl_2 (4× 30 mL). Wash the combined organics with brine (30 mL), dry over $MgSO_4$, filter and concentrate in vacuo. Purify the crude residue by column chromatography, eluting with 90:10 petrol–EtOAc (+0.1% AcOH), to afford the product (2.30 g, 92%) as a white solid.

   δ_H (400 MHz, $CDCl_3$) 3.41 (t, 2H, J 6.9 Hz, CH_2Br), 2.35 (t, 2H, J 7.5 Hz, CH_2CO_2H), 1.80–1.90 (m, 2H, CH_2CH_2Br), 1.59–1.68 (m, 2H, $CH_2CH_2CO_2H$), 1.25–1.48 (m, 22H, 11× CH_2).

   δ_C (101 MHz, $CDCl_3$) 178.6, 34.2, 33.9, 33.0, 29.8, 29.7, 29.6, 29.4, 29.2, 28.9, 28.3, 24.8.

   LR-MS (EI^+) 337.0 ($[M(^{81}Br)]^+$, 5%), 335.0 ($[M(^{79}Br)]^+$, 7%), 317.0 ($[M(^{81}Br)\text{-}H_2O]^+$, 3%), 315.0 ($[M(^{79}Br)\text{-}H_2O]^+$, 5%), 257.1 ($[M - Br]^+$, 8.5%), 237.1 ($[M - Br - H_2O]^+$, 22%).

   HR-MS calculated for $C_{16}H_{30}{}^{79}BrO$ ($[M\text{–}H]^-$) 333.1435, found 333.1430.

   υ_{max} (thin film, cm^{-1}) 3034, 2917, 2850, 1696.

   M_p 72–74 °C.

8. Preparation of 16-azidohexadecanoic acid **14** (*see* **Note 15**). Add NaN_3 (582 mg, 8.96 mmol) to a solution of bromide **11** (500 mg, 1.49 mmol) in DMF (6 mL). Stir the reaction at 80 °C for 40 h before cooling to room temperature. Add a 1:1 mixture of EtOAc and H_2O (20 mL), separate the layers and extract the aqueous phase with EtOAc (3× 10 mL). Wash the combined organics with brine (10 mL), dry over $MgSO_4$, filter and concentrate in vacuo. Purify the crude residue by column chromatography, eluting with 90:10 petrol–EtOAc (+0.1% AcOH), to afford the product (402 mg, 91%) as a white solid (*see* **Note 17**).

   δ_H (400 MHz, $CDCl_3$) 3.25 (t, 2H, J 7.0 Hz, CH_2N_3), 2.34 (t, 2H, J 7.5 Hz, CH_2CO_2H), 1.55–1.67 (m, 4H, $CH_2CH_2N_3$, $CH_2CH_2CO_2H$), 1.24–1.40 (m, 22H, 11× CH_2).

   δ_C (101 MHz, $CDCl_3$) 180.0, 51.7, 34.2, 29.8, 29.7, 29.7, 29.6, 29.6, 29.4, 29.3, 29.3, 29.2, 29.2, 29.0, 26.9, 24.8.

   LR-MS (ES^-) 268.2 ($[M - N_2 - H]^-$, 26%), 296.2 ($[M - H]^-$, 100%).

HR-MS calculated for $C_{16}H_{32}NO_2$ ([M − N_2 + H]$^+$) 270.2440, found 270.2433.

υ_{max} (thin film, cm^{-1}) 3014, 2915, 2848, 2103, 1701.

M_p 48–50 °C.

9. Preparation of 1,18-octadecanediol **6** (*see* **Note 15**). Add solid $LiAlH_4$ (1.44 g, 37.96 mmol) to a vigorously stirred solution of dimethyl octadecanedioate (5.00 g, 14.9 mmol). Heat the reaction to reflux and stir at reflux for 20 h. Upon completion, cool the reaction to room temperature and add wet Na_2SO_4 portionwise until the grey suspension turns white. Stir the suspension until the white solid is free-flowing, and solid $MgSO_4$ (5 g). Filter the reaction and wash the filter cake with Et_2O (5× 50 mL). Evaporate the solvent in vacuo to afford the product (4.15 g, 99%) as a white solid.

   δ_H (600 MHz, DMSO-d_6) 4.30 (t, 2H, J 5.1 Hz, 2× OH), 3.34–3.39 (m, 4H, 2× CH_2OH), 1.35–1.43 (m, 4H, 2× CH_2CH_2OH), 1.23 (app. br. s., 28H, 14× CH_2).

   δ_C (151 MHz, DMSO-d_6) 60.6, 32.3, 28.8, 28.8, 28.7, 25.3.

   LR-MS (MALDI-TOF) 293.3 ([M + Li]$^+$), 309.3 ([M + Na]$^+$).

   HR-MS calculated for $C_{18}H_{39}O_2$ ([M + H]$^+$) 287.2945, found 287.2946.

   υ_{max} (thin film, cm^{-1}) 3416, 3353, 2919, 2891.

   M_p 103–106 °C.

10. Preparation of 18-bromooctadecan-1-ol **9**. Add HBr (48% in H_2O, 21 mL) to a suspension of diol **6** (2.26 g, 7.9 mmol) in cyclohexane (21 mL). Heat the biphasic mixture to reflux for 10 h before cooling to room temperature. Separate the layers and extract the aqueous phase with CH_2Cl_2 (4× 20 mL). Wash the combined organics with $NaHCO_3$ (4× 20 mL of a saturated aqueous solution), brine (20 mL), dry over $MgSO_4$, filter and concentrate in vacuo. Purify the crude residue by column chromatography, eluting with 95:5 petrol–EtOAc then 70:30 petrol–EtOAc, to afford the product (1.65 g, 60%) as a pale yellow solid (*see* **Note 16**).

   δ_H (400 MHz, $CDCl_3$) 3.64 (t, 2H, J 6.6 Hz, CH_2OH), 3.40 (t, 2H, J 6.9 Hz, CH_2Br), 1.81–1.89 (m, 2H, CH_2CH_2Br), 1.53–1.60 (m, 2H, CH_2CH_2OH), 1.25–1.46 (m, 28H, 14× CH_2).

   δ_C (101 MHz, $CDCl_3$) 63.3, 34.2, 33.0, 33.0, 29.8, 29.8, 29.7, 29.6, 28.9, 28.3, 25.9.

   LR-MS (ES$^+$) 332.5 (M(^{81}Br)-H_2O, 0.5%), 330.5 (M(^{79}Br)-H_2O, 0.5%).

   HR-MS calculated for $C_{18}H_{41}{}^{79}BrON$ ([M + NH_4]$^+$) 366.2366, found 336.2368.

υ_{max} (thin film, cm^{-1}) 3274, 2917, 2850.

M_p 59–61 °C.

11. Preparation of 18-bromooctadecanoic acid **12**. Dissolve CrO_3 (1.42 g, 14.2 mmol) in concentrated H_2SO_4 (3.5 mL). Add cold H_2O (7.9 mL) slowly and stir the solution for 10 min. Add the resulting solution dropwise to a solution of alcohol **9** (1.65 g, 4.73 mmol) in acetone (120 mL). Stir the reaction for 20 h before adding H_2O (60 mL) and CH_2Cl_2 (40 mL). Separate the layers and extract the aqueous phase with CH_2Cl_2 (4× 20 mL). Wash the combined organics with brine (20 mL), dry over $MgSO_4$, filter and concentrate in vacuo. Purify the crude residue by column chromatography, eluting with 90:10 petrol–EtOAc (+0.1% AcOH), to afford the product (1.43 g, 83%) as a white solid (*see* **Note 18**).

    δ_H (400 MHz, $CDCl_3$) 3.41 (t, 2H, J 6.9 Hz, CH_2Br), 2.35 (t, 2H, J 7.5 Hz, CH_2CO_2H), 1.81–1.90 (m, 2H, CH_2CH_2Br), 1.59–1.68 (m, 2H, $CH_2CH_2CO_2H$), 1.25–1.46 (m, 26H, 13× CH_2).

    δ_C (101 MHz, $CDCl_3$) 179.6, 34.2, 34.1, 33.0, 29.8, 29.7, 29.7, 29.6, 29.4, 29.2, 28.9, 28.3, 24.8.

    LR-MS (EI^+) 365.0 ($[M(^{81}Br)]^+$, 9%), 363.0 ($[M(^{79}Br)]^+$, 12%), 347.0 ($[M(^{81}Br)\text{-}H_2O]^+$, 5%), 345.0 ($[M(^{79}Br)\text{-}H_2O]^+$, 7%), 284.0 ($[M - Br]^+$, 12%), 265.1 ($[M - Br - H_2O]^+$, 29%).

    HR-MS calculated for $C_{18}H_{34}{}^{79}BrO_2$ ($[M - H]^-$) 361.1748, found 361.1741.

    υ_{max} (thin film, cm^{-1}) 3034, 2915, 2850, 1696.

    M_p 77–80 °C.

12. Preparation of 18-azidooctadecanoic acid **15** (*see* **Note 15**). Add NaN_3 (537 mg, 8.3 mmol) to a solution of bromide **12** (500 mg, 1.38 mmol) in DMF (5.5 mL). Stir the reaction at 80 °C for 40 h before cooling to room temperature. Add a 1:1 mixture of EtOAc and H_2O (30 mL), separate the layers, and extract the aqueous phase with EtOAc (3× 10 mL). Wash the combined organics with brine (10 mL), dry over $MgSO_4$, filter and concentrate in vacuo. Purify the crude residue by column chromatography, eluting with 90:10 petrol–EtOAc (+0.1% AcOH), to afford the product (403 mg, 90%) as a white solid (*see* **Note 16**).

    δ_H (400 MHz, $CDCl_3$) 3.25 (t, 2H, J 7.0 Hz, CH_2N_3), 2.34 (t, 2H, J 7.5 Hz, CH_2CO_2H), 1.55–1.67 (m, 4H, $CH_2CH_2N_3$, $CH_2CH_2CO_2H$), 1.25–1.39 (m, 26H, 13× CH_2).

δ_C (101 MHz, $CDCl_3$) 180.1, 51.6, 34.2, 29.8, 29.7, 29.7, 29.6, 29.6, 29.4, 29.4, 29.3, 29.2, 29.0, 26.7, 24.8.

LR-MS (ES^-) 296.3 ($[M - N_2 - H]^-$, 24%), 324.1 ($[M - H]^-$, 100%).

HR-MS calculated for $C_{18}H_{34}N_3O_2$ ($[M - H]^-$) 324.2657, found 324.2649.

υ_{max} (thin film, cm^{-1}) 3040, 2915, 2850, 2098, 1696.

M_p 56–58 °C.

3.3 HEK293T Cell Culture and Seeding

1. Warm trypsin–EDTA and HEK293T Culture Medium to 37 °C in a prewarmed water bath.
2. In a laminar flow tissue culture cabinet, remove the medium from HEK293T cells cultured in a 75 cm^2 flask and discard.
3. Wash cells briefly with 3 mL of prewarmed trypsin–EDTA and immediately discard.
4. Add 3 mL of prewarmed trypsin–EDTA to the cell monolayer and incubate at 37 °C for 5 min.
5. Detach cell monolayer by gently tapping the side of the flask and wash the remaining cells off the flask surface by pipetting up and down 5–10 times. Care should be taken to ensure all cells are harvested.
6. Transfer the contents of the flask to a sterile falcon tube and add 7 mL of prewarmed HEK293T Culture Medium. Pipet up and down five times to mix.
7. To reseed cells into a new 75 cm^2 flask, dilute cells 1:10 by mixing 1 mL of the cells with 11 mL of HEK293T Culture Medium.
8. Determine cell density using a Neubauer Chamber. Resuspend cells in HEK293T Culture Medium at a density of approximately 10^6 cells/mL and seed 500 μL/well into poly-D-lysine-coated 24-well cell culture plates.
9. Incubate cells overnight at 37 °C in a humidified atmosphere containing 95% air and 5% CO_2.

3.4 HEK293T Cell Transfection

1. Transfect HEK293T approximately 24 h after seeding.
2. For each well of HEK293T cells to be transfected, set up two tubes each containing 50 μL of serum-free DMEM. In Tube 1 add up to 3 μg plasmid DNA (e.g., 0.8 μg EGFP-SNAP25B + 1.6 μg HA-ZDHHC3). In Tube 2 add two volumes of Lipofectamine 2000 reagent per μg of plasmid DNA used (e.g., 4.8 μL Lipofectamine 2000). Incubate for 5 min at room temperature (*see* **Note 19**).
3. Mix the contents of the two tubes and incubate the DMEM–DNA–Lipofectamine 2000 mixture for 20 min at room temperature.

4. Add the transfection complex to one well of the 24-well plate of HEK293T cells.
5. Incubate cells overnight at 37 °C.

3.5 Metabolic Labeling of Transfected Cells with Azido-Fatty Acid

1. Serum-starve transfected HEK293T cells by removing and discarding the culture medium and replacing it with 500 μL/well of FAF culture medium (*see* **Note 20**). Incubate cells at 37 °C in a humidified atmosphere containing 95% air and 5% CO_2 for 30 min.
2. Add 500 μL/well of fatty-acid azide labeling medium and incubate cells for 4 h cells at 37 °C in a humidified atmosphere containing 95% air and 5% CO_2 (*see* **Note 21**).

3.6 Cell Lysis

1. Wash cells twice on ice by gently aspirating off the medium and replacing it with 1 mL ice-cold PBS.
2. Remove the PBS and add 100 μL of Lysis Buffer (containing protease inhibitors) to each well. Incubate cells on ice for 10 min with gentle agitation (*see* **Note 22**).

3.7 Click Chemistry and Protein Precipitation

1. Remove aliquots of 4 mM Alkyne Infrared Dye and 100 mM TBTA from the freezer and bring up to room temperature.
2. Make up 40 mM ascorbic acid solution by dissolving 0.07 g of ascorbic acid in 10 mL of H_2O and mixing with a magnetic stirrer (*see* **Note 23**).
3. In a 1.5 mL tube, mix 0.12 μL 4 mM Alkyne IR800 dye, 10 μL 40 mM $CuSO_4$, 0.4 μL 100 mM TBTA, and 69.48 μL dH_2O (*see* **Note 24**).
4. Add the cell lysate to the click reaction cocktail and vortex to mix.
5. Add 20 μL of 40 mM ascorbic acid to the lysate (*see* **Note 25**).
6. Vortex and incubate for 1 h at room temperature with end-over-end rotation.
7. Precipitate proteins using ice-cold acetone. Add 3 volumes (600 μL) of ice-cold acetone to the tubes and vortex. Incubate at −20 °C for a minimum of 20 min (*see* **Note 26**).
8. Centrifuge at 20,000 × *g* for 5 min at 4 °C to pellet proteins.
9. Remove and discard supernatant. Add 1 mL ice-cold 70% acetone and vortex tube to wash pellet.
10. Centrifuge at 20,000 × *g* for 5 min at 4 °C to pellet proteins.
11. Remove and discard supernatant and allow pellet to air-dry for 5 min.
12. Resuspend pellet in 100 μL of 1 SDS Lysis Buffer containing 25 mM DTT (*see* **Notes 26** and **27**).

3.8 SDS–Polyacrylamide Gel Electrophoresis and In-Gel Analysis

1. Clean and dry two 1 mm spacer plates and two short plates by washing with 70% ethanol and rinsing in H_2O. Allow to dry.
2. Assemble the gel cassette by placing the casting frame upright with the pressure cams in the open position and facing forward on a flat surface. With the labeling on the spacer plate up, place a short plate on top and slide the two glass plates into the casting frame with the short plate facing the front (*see* **Note 28**). Secure the glass cassette sandwich in the casting frame by engaging the pressure cams. Place the casting frame onto the casting gasket, pressure cams facing outward, and lock onto the casting stand by engaging the spring-loaded lever of the casting stand onto the spacer plate (*see* **Note 29**).
3. Prepare the resolving gel monomer. For two 12% gels, mix 5 mL of Resolving gel buffer, 4 mL of 30% acrylamide–bisacrylamide solution and 1 mL H_2O. Add 200 μL of 10% ammonium persulfate and 8 μL of TEMED and mix. Carefully pipet 4.5 mL of the resolving gel monomer solution between the glass plates and immediately overlay with 500 μL of propan-1-ol. Allow the gel to polymerize for 45 min. Pour off the propan-1-ol and rinse the gel surface completely with distilled water. Dry the top of the resolving gel with filter paper.
4. Prepare the stacking gel monomer. Mix 4 mL of Stacking gel buffer, 1.3 mL of 30% acrylamide–bisacrylamide solution and 2.7 mL H_2O. Add 200 μL of 10% ammonium persulfate and 10 μL of TEMED and mix. Carefully pour on top of the resolving gel monomer until the top of the short plate is reached and insert a 15-well comb in the gel cassette, aligning the ridge with the top of the short plate. Allow the gel to polymerize for 45 min (*see* **Note 30**).
5. Place the gel cassettes into the clamping frame with the short plates pointed inward and clamp into place. Place the clamping frame in the tank and fill the inner and outer chamber with 1 L of 1× Tris–glycine electrophoresis buffer.
6. Heat the samples in a heat block for 5 min at 100 °C to denature proteins. Centrifuge the samples briefly to bring down the condensate.
7. Load 5–20 μL/well of each sample into the wells with a Hamilton syringe or a pipette using gel loading tips. Apply 5 μL of prestained protein standards to one well.
8. Place the lid on the Mini-PROTEAN Tetra tank ensuring to align the color-coded banana plugs and jacks. Insert the electrical leads into the suitable power supply and electrophorese at 180 V constant until the dye front reaches the bottom of the gel.

9. Turn off the power supply, disconnect the electrical leads, and remove the tank lid. Pour off and discard the running buffer.
10. Remove the gel cassettes from the assembly and separate the glass plates to remove the gels from the gel cassette. Place the gels in an incubation box containing H_2O.
11. To visualize *S*-acylation of your expression protein via in-gel analysis, place the gel on the scanning surface of a LI-COR Odyssey Infrared Imagining system taking care not to trap bubbles underneath.
12. Scan the gel in the 800 nm channel using the Image Studio Software Acquire ribbon (*see* **Note 31**).

3.9 Electrophoretic Transfer of Proteins to Nitrocellulose Membranes

1. For each gel to be transferred, cut one nitrocellulose membrane and two Whatman filter papers to the dimensions of the gel (*see* **Notes 32** and **33**).
2. Soak the gel, nitrocellulose membrane and filter paper in 1× transfer buffer.
3. Prepare the gel sandwich: Open the gel holder cassette from the Bio-Rad Trans-Blot Electrophoretic Transfer Cell with the grey side down. Place a prewetted fiber pad on the grey side of the cassette. Place the wet filter paper on the fiber pad and place the gel on top. Place the nitrocellulose membrane on top of the gel and the final piece of filter paper on top of the membrane (*see* **Note 34**).
4. Add a fiber pad on top and close and lock the gel holder cassette with the white latch.
5. Place the gel holder cassette in the tank, ensuring the grey side is toward the plate cathode, and completely fill the tank with transfer buffer.
6. Place the lid on the Transfer Cell and insert the electrical leads into the suitable power supply and run at 120 mA overnight.
7. Upon completion, disassemble the blotting sandwich and place the membrane in an incubation box containing PBS-T (*see* **Note 35**).

3.10 Western Blotting

1. Wash membranes once in PBS-T.
2. Place the membranes in an incubation box and block the membranes in Blocking buffer for 45 min with gentle agitation.
3. Prepare the primary antibody by diluting in PBST using the manufacturer's recommended dilution for Western blot applications (*see* **Note 36**). Use enough antibody solution to completely cover the membrane. Incubate the membrane in diluted primary antibody for between 1 h at room temperature and overnight at 4 °C, with gentle agitation (*see* **Note 37**).

4. Wash the membranes by pouring off the primary antibody solution and rinsing the membranes 5× in PBS-T for 5 min each, with vigorous agitation (*see* **Note 38**).
5. Prepare the secondary antibody by diluting the IRDye secondary antibody 1:10,000 in secondary antibody diluent. Use enough antibody solution to completely cover the membrane. Incubate the membrane in diluted secondary antibody for 45 min at room temperature with gentle agitation.
6. Wash the membranes by pouring off and discarding the secondary antibody solution. Rinse the membranes five times in PBS-T for 5 min each, with vigorous agitation.
7. Rinse membranes in PBS to remove residual Tween 20 (*see* **Note 39**).
8. Image the blots using a LI-COR Odyssey Infrared Imagining system by placing the membranes on the scanning surface and taking care to ensure no bubbles are trapped underneath. Scan the membranes simultaneously in the 700 nm and 800 nm channels using the Image Studio Software Acquire ribbon.

4 Notes

1. Longer chain saturated fatty acid azide stocks (C18:0 and longer) may need warming until the solution becomes clear before use.
2. Add 100 mL of 10× PBS to 900 mL of water for use.
3. IRDyes should be protected from light and repeated freeze–thaw cycles should be avoided.
4. Prepare immediately before use.
5. Acetone is flammable and should be stored in a spark-free freezer.
6. SDS precipitates at low temperatures; if this occurs, warm buffer in a water bath before use.
7. Add 250 μL of 4× SDS Lysis buffer to 725 μL of water and add 25 μL of 1 M DTT.
8. The ammonium persulfate solution can be stored at 4 °C for 7 days.
9. Add 100 mL of 10× Tris–glycine electrophoresis buffer to 900 mL of water.
10. Add 250 mL of 10× Transfer buffer to 1750 mL of water and add 500 mL of methanol.
11. Add 2 mL of 10% Tween 20 to 1 L of PBS.
12. Blocking Buffer should be made up fresh each time.

13. Store the IRDye secondary antibody in darkness and minimize exposure to light.
14. For dual imaging the system requires two infrared channels with 685 and 785 nm lasers.
15. Perform reaction under a N_2 atmosphere.
16. For the monobromination, reaction time and vigorous stirring were crucial for success to prevent bis-bromination.
17. For azide addition the reaction should be monitored carefully by TLC to ensure the reaction has reached completion. Addition of additional azide was found to be problematic with this process.
18. For the reduction of the bis-carboxylic acids the reaction generally proceeded efficiently at room temperature; however, this was not reliable with 1,18-octadecanedioic acid, therefore for reproducibility we used 1,18-dimethyl octadecanedioate as the substrate, performing the reduction at reflux in THF.
19. To scale up the transfection, multiply by the number of wells to be transfected.
20. For incubation times of up to 8 h, it is not necessary to use aseptic techniques.
21. Depending on the expression level and extent of *S*-acylation of the protein of interest, it may be necessary to increase the incubation time beyond 4 h. In this case, aseptic techniques should be used especially in the case of overnight incubations. Prepare the DMEM containing L-Glutamine supplemented with 1% fatty-acid free BSA and 100 μM fatty acid azide and in a laminar flow hood use a syringe to pass the medium through a 0.22 μm sterile filter.
22. Cell lysates may be frozen by directly transferring the plates to −20 °C, allowing the procedure to be continued at a later date.
23. The ascorbic acid solution should be made up fresh each time. Ascorbic acid may take several minutes to dissolve.
24. The total volume per lysate should equal 80 μL. If performing multiple click reactions, scale the reaction volume up accordingly and aliquot 80 μL of the click reaction mix per experimental replicate into separate 1.5 mL tubes.
25. Final concentrations are 2.5 μM azide dye, 2 mM $CuSO_4$, 0.2 mM TBTA, and 4 mM ascorbic acid. Total volume including lysate is 200 μL.
26. Samples may be stored at −20 °C, allowing the procedure to be continued at a later date.
27. Vigorous vortexing may be required to ensure the pellet is fully resuspended in the SDS Lysis Buffer.

28. Ensure that the plates are flush on a level surface to minimize the risk of leaking.
29. Ensure the grey casting stand gaskets are clean and dry before use.
30. Gels in a cassette sandwich may be made ahead and stored at 4 °C for a few days. Ensure that the gels remain hydrated by wrapping the glass plates first in tissue paper soaked in water and then in cling film.
31. If the signal is too strong, the saturated pixels will appear as cyan in the image. In this case, rescan using a lower intensity setting.
32. Cut membranes to 6 × 9 cm.
33. Always wear gloves when handling membranes to prevent contamination.
34. Remove any air bubbles which may have formed by using a glass tube to roll out any air bubbles.
35. The membranes can be stored in PBS-T for up to 48 h in the dark at 4 °C.
36. For the anti-GFP JL8 antibody, dilute 1:3000.
37. Optimal incubation times will vary depending on the primary antibody.
38. The diluted primary antibody solution may be stored at −20 °C and reused.
39. If the blot is prepared in advance, air-dry the blot and store in the dark until ready to image.

References

1. Muszbek L, Laposata M (1993) Covalent modification of proteins by arachidonate and eicosapentaenoate in platelets. J Biol Chem 268:18243–18248
2. Hallak H, Muszbek L, Laposata M, Belmonte E, Brass LF, Manning DR (1994) Covalent binding of arachidonate to G protein alpha subunits of human platelets. J Biol Chem 269(7):4713–4716
3. Veit M, Reverey H, Schmidt MF (1996) Cytoplasmic tail length influences fatty acid selection for acylation of viral glycoproteins. Biochem J 318(Pt 1):163–172
4. Muszbek L, Haramura G, Cluette-Brown JE, Van Cott EM, Laposata M (1999) The pool of fatty acids covalently bound to platelet proteins by thioester linkages can be altered by exogenously supplied fatty acids. Lipids 34: S331–S337
5. Wilson JP, Raghavan AS, Yang YY, Charron G, Hang HC (2011) Proteomic analysis of fatty-acylated proteins in mammalian cells with chemical reporters reveals S-acylation of histone H3 variants. Mol Cell Proteomics 10: M110.001198
6. Greaves J, Munro KR, Davidson SC, Wojno J, Riviere M, Smith TK, Tomkinson NCO, Chamberlain LH (2017) Molecular basis of fatty acid selectivity in the zDHHC family of S-acyltransferases revealed by click chemistry. Proc Natl Acad Sci U S A 114(8): E1365–E1374
7. Schmidt MF, Bracha M, Schlesinger MJ (1979) Evidence for covalent attachment of fatty acids to Sindbis virus glycoproteins. Proc Natl Acad Sci U S A 76:1687–1691
8. Schmidt MFG, Schlesinger MJ (1979) Fatty acid binding to vesicular stomatitis virus glycoprotein: a new type of post-translational

modification of the viral glycoprotein. Cell 117:813–819

9. Schlesinger M, Magee A, Schmidt M (1980) Fatty acid acylation of proteins in cultured cells. J Biol Chem 255:10021–10024
10. Gao X, Hannoush RN (2017) A Decade of click chemistry in protein palmitoylation: impact on discovery and new biology. Cell Chem Biol 25(3):236–246
11. Hang HC, Geutjes E, Grotenbreg G, Pollington AM, Bijlmakers MJ, Ploegh HL (2007) Chemical probes for the rapid detection of fatty-acylated proteins in mammalian cells. J Am Chem Soc 129(10):2744–2745
12. Hang HC, Linder ME (2011) Exploring protein lipidation with chemical biology. Chem Rev 111(10):6341–6358. FASEB J. 2008 Mar;22(3):721-32
13. Hannoush RN, Arenas-Ramirez N (2009) Imaging the lipidome: Omega-alkynyl fatty acids for detection and cellular visualization of lipid-modified proteins. ACS Chem Biol 4 (7):581–587
14. Hannoush RN, Sun J (2010) The chemical toolbox for monitoring protein fatty acylation and prenylation. Nat Chem Biol 6(7):498–506
15. Kostiuk MA, Corvi MM, Keller BO, Plummer G, Prescher JA, Hangauer MJ, Bertozzi CR, Rajaiah G, Falck JR, Berthiaume LG (2008) Identification of palmitoylated mitochondrial proteins using a bio-orthogonal azido-palmitate analogue. FASEB J 22 (3):721–732
16. Yap MC, Kostiuk MA, Martin DD, Perinpanayagam MA, Hak PG, Siddam A, Majjigapu JR, Rajaiah G, Keller BO, Prescher JA, Wu P, Bertozzi CR, Falck JR, Berthiaume LG (2010) Rapid and selective detection of fatty acylated proteins using omega-alkynyl-fatty acids and click chemistry. J Lipid Res 51(6):1566–1580
17. Charron G, Zhang MM, Yount JS, Wilson J, Raghavan AS, Shamir E, Hang HC (2009) Robust fluorescent detection of protein fatty-acylation with chemical reporters. J Am Chem Soc 131(13):4967–4975
18. Ganesan L, Shieh P, Bertozzi CR, Levental I (2017) Click-Chemistry Based High Throughput Screening Platform for Modulators of Ras Palmitoylation. Sci Rep 7:41147
19. Werno W, Chamberlain LH (2015) S-acylation of the insulin-responsive aminopeptidase (IRAP): quantitative analysis and identification of modified cysteines. Sci Rep 5:12413
20. Graham FL, Smiley J, Russell WC, Nairn R (1977) Characteristics of a human cell line transformed by DNA from human adenovirus type 5. J Gen Virol 36(1):59–74
21. Laemmli UK (1970) Cleavage of structural proteins during the assembly of the head of bacteriophage T4. Nature 227 (5259):680–685

Chapter 3

Optimization of Metabolic Labeling with Alkyne-Containing Isoprenoid Probes

Mina Ahmadi, Kiall Francis Suazo, and Mark D. Distefano

Abstract

Protein prenylation, found in eukaryotes, is a posttranslational modification in which one or two isoprenoid groups are added to the C terminus of selected proteins using either a farnesyl group or a geranylgeranyl group. Prenylation facilitates protein localization mainly to the plasma membrane where the prenylated proteins, including small GTPases, mediate signal transduction pathways. Changes in the level of prenylated proteins may serve a critical function in a variety of diseases. Metabolic labeling using modified isoprenoid probes followed by enrichment and proteomic analysis allows the identities and levels of prenylated proteins to be investigated. In this protocol, we illustrate how the conditions for metabolic labeling are optimized to maximize probe incorporation in HeLa cells through a combination of in-gel fluorescence and densitometric analysis.

Key words Protein prenylation, Farnesylation, Geranylgeranylation, Metabolic labeling, Isoprenoid analog, In-gel fluorescence, Lovastatin

1 Introduction

Protein prenylation, found in eukaryotes [1] is a posttranslational modification in which one or two isoprenoid groups are added to the C terminus of selected proteins using either a farnesyl group (C15) or a geranylgeranyl group (C20) [2]. This irreversible covalent modification is catalyzed by three prenyltransferase enzymes including farnesyl transferase (FTase) and geranylgeranyl transferase type I (GGTase-I) and geranylgeranyl transferase type II (GGTase-II) [3–5]. FTase and GGTase-I catalyze the attachment of a single farnesyl group or geranylgeranyl group to a cysteine residue located within a CaaX motif, respectively in which "C" is cysteine, "a" is an aliphatic amino acid and the "X" residue is critical for determining which isoprenoid group is attached and hence which enzyme is involved [2]. In contrast, GGTase-II catalyzes the addition of two geranylgeranyl groups to two cysteine residues in sequences including CXC and CC. In addition to the known

Maurine E. Linder (ed.), *Protein Lipidation: Methods and Protocols*, Methods in Molecular Biology, vol. 2009,
https://doi.org/10.1007/978-1-4939-9532-5_3,

canonical CaaX motif that can be recognized by FTase, recent studies have showed that a longer $C(x)_3X$ motif can also recognized by both yeast and mammalian FTases [6]. Protein prenylation by FTase and GGTase-I typically is followed by the removal of the aaX residues promoted by Ras-converting enzyme (Rce1) or Ste24 [7, 8]. In the final step, the mature protein is generated by the action of carboxylmethyltransferase (Ste14), which catalyzes the transfer of a methyl group from S-adenosyl methionine to the carboxylate of the C-terminal prenyl cysteine residue to yield a C-terminal methyl ester [9]. Prenylation facilitates protein localization mainly to the plasma membrane where the prenylated proteins including small GTPases mediate signal transduction pathways [10]. Changes in the level of prenylated proteins may be critical in a variety of diseases including Parkinson's disease [11], Alzheimer's disease [12], neurodegeneration [13], viral infections [14], and some types of cancers [15].

To improve understanding of the process of protein prenylation, as well as identify protein substrates that are lipidated through this process, synthetic small molecule analogs of the isoprenoids are being used. Photoactivatable [16, 17] and fluorescent [18] isoprenoid analogs have been used to examine protein–protein interactions of the prenylated proteins and to develop convenient assays for prenyltransferase activity, respectively. In addition, there has been significant interest in identifying prenylated proteins via metabolic labeling using modified isoprenoid probes followed by enrichment and proteomic analysis. In this approach, the isoprenoid probe is functionalized with a bioorthogonal functional group (azide or alkyne) that allows conjugation of a fluorophore or an affinity handle for pulldown and enrichment of labeled proteins for subsequent identification [19]. Several strategies using this approach have emerged and been used to identify the prenylome in mammalian [20] and parasitic systems [21], as well as prenylated proteins with antiviral activity [14]. We have previously demonstrated that isoprenoid analogs (Fig. 1) C15AlkOP and

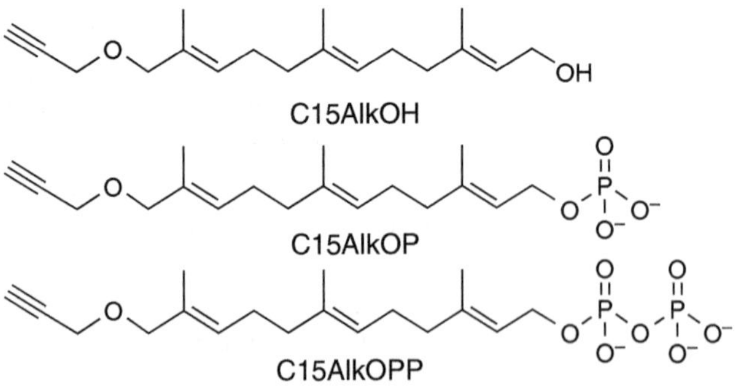

Fig. 1 Isoprenoid phosphate analogs functionalized with alkyne moieties for metabolic labeling of prenylated proteins

C15AlkOPP served as good substrates for metabolic labeling in COS-7 cells whereas C15AlkOH was incorporated with much lower efficiency [22]. In that study, we showed how labeling efficiency can be optimized by varying the probe concentration. While some incorporation can be observed using 1 μM alkyne-functionalized-probe, 10 μM probe gives significantly higher labeling without significantly raising the level of background labeling which occurs at higher probe concentrations [23]. Importantly, probe incorporation can also be enhanced by inhibition of the synthesis of the endogenous isoprenoid substrates farnesyl and geranylgeranyl diphosphate. Hence, a common strategy to accomplish this is to employ statins that inhibit HMG-CoA reductase upstream in the mevalonate biosynthetic pathway [24]. In this protocol, we illustrate how the concentration of lovastatin can be optimized to maximize probe incorporation (but avoid toxicity) in HeLa cells through a combination of in-gel fluorescence and densitometric analysis.

2 Materials

2.1 Cell Culture

1. Dulbecco's Modified Eagle's Medium (DMEM).
2. Fetal bovine serum (FBS).
3. Penicillin–streptomycin (Pen/strep): 10,000 U penicillin, 10 mg/mL streptomycin (10× solution).
4. DMEM culture medium: DMEM, 10% FBS, 1× Pen/strep.
5. Phosphate-buffered saline (PBS): 8.1 mM Na_2HPO_4, 1.5 mM KH_2PO_4, pH 7.4, 137 mM NaCl, 2.7 mM KCl.
6. 2.5% trypsin media (10×), no EDTA (Invitrogen, product number 15090-046).
7. Versene: 1× PBS + 0.6 mM Na_2EDTA.
8. 100 mm culture dish.

2.2 Metabolic Labeling

1. Isoprenoid probe analogs: C15AlkOH, C15AlkOP, and C15AlkOPP [22, 25].
2. Lovastatin (Cayman Chemical).

2.3 Cell Harvest, Lysis, Protein Assay

1. 1× PBS.
2. Cell scraper.
3. Lysis buffer: 1× PBS, 1% sodium dodecyl sulfate (SDS), 2.4 μM phenylmethylsulfonyl fluoride (PMSF), 85 kU/mL benzonase nuclease, 1.5% (v/v) protease inhibitor cocktail.
4. BCA protein assay kit (Thermo Scientific).
5. Absorbance plate reader.

2.4 Click Reaction on Labeled Lysates

1. 1 mM TAMRA-PEG3-azide in DMSO (Broadpharm).
2. 50 mM Tris(2-carboxyethyl)phosphine (TCEP) in DMSO.
3. 10 mM Tris[(1-benzyl-1*H*-1,2,3-triazol-4-yl)methyl]amine (TBTA) in DMSO.
4. 50 mM Copper (II) sulfate ($CuSO_4$) in water.
5. ProteoExtract® protein precipitation kit (Calbiochem).

2.5 Gel Electrophoresis and In-Gel Fluorescence

1. Laemmli buffer (1×): 125 mM dithiothreitol (DTT), 2% SDS, 10% glycerol, 0.02% bromophenol blue, 50 mM Tris–HCl, pH 6.8.
2. 12% SDS-PAGE gel (3 × 4 in.).
3. Fluorescence scanner (Typhoon FLA 9500).
4. Coomassie Blue staining solution: 0.2% (w/v) Coomassie Brilliant Blue R250, 10% (v/v) acetic acid, 45% (v/v) methanol, 45% water.
5. Gel destaining solution: 10% (v/v) acetic acid, 30% (v/v) methanol, 60% (v/v) water.

3 Method

3.1 Cell Culture and Metabolic Labeling

1. Maintain HeLa cells in DMEM culture medium and incubate at 37 °C under 5% CO_2.
2. When cells near confluence, passage cells (1:10 split). Wash cells with PBS and detach from the dish using 1 mL of 0.25% trypsin in Versene. Incubate for 5 min at 37 °C under 5% CO_2, collect the cells and dilute with DMEM culture medium.
3. For metabolic labeling, seed 9×10^5 cells in 100-mm culture dishes in 10 mL of DMEM culture medium.
4. After 24 h (40–60% confluency), replace the medium in each plate with 5 mL of fresh culture medium containing the desired concentration of lovastatin (0, 0.1, 1.0, 10, or 25 μM, *see* **Note 1**).
5. After 6 h of lovastatin pretreatment, replace the medium with 5 mL of fresh medium in the presence or absence of lovastatin. Add the isoprenoid probes according to the desired final concentrations. In this protocol, a final concentration of 10 μM was used for all three probes. Incubate cells at 37 °C under 5% CO_2 for 24 h.

3.2 Cell Harvest, Lysis, Protein Assay

1. Harvest cells 24 h after the addition of the isoprenoid probes. Place the cell culture dishes on ice. Remove the medium by suction and wash adherent cells with 5 mL of cold 1× PBS twice. Gently scrape the cells into 1 mL ice-cold PBS using a

cell scraper. Collect the cells in 1.5 Eppendorf tubes using a 1-mL pipette and pellet by centrifugation for 5 min at 180 × *g*. Remove the supernatant by suction. Store the cell pellets at −80 °C.

2. Add Lysis buffer (240–300 μL, *see* **Note 2**) to each cell pellet. Lyse the cells by sonication for 6–8 times for 2 s at 10-min intervals.
3. Determine the protein concentration in the cell lysates using BCA protein assay kit and following the manufacturer's protocol.

3.3 Click Reaction with TAMRA-PEG3-N_3

Perform reactions in 1.5 mL Eppendorf tubes and keep covered to avoid exposure to light.

1. Dilute aliquots of protein lysates (100 μg) with 1× PBS containing 1% SDS to final volume of 92.5 μL. Add 2.5 μL of 1 mM TAMRA-PEG3-N_3 (final concentration, 25 μM), 2 μL of 50 mM TCEP (final concentration, 1 mM), and 1 μL of 10 mM TBTA (final concentration, 100 μM) to each reaction tube. Vortex each sample briefly prior to adding 2 μL of 50 mM $CuSO_4$ (final concentration, 1 mM). Incubate samples at room temperature in the dark (covered with aluminum foil) on a rotor mixer for 1 h.
2. After the reaction, spin down sample tubes to capture the reaction mixtures at the bottom of the tube. Precipitate protein using a ProteoExtract precipitation kit following the manufacturer's protocol. Recover proteins as white pellets at the bottom of the tube and allow to dry (*see* **Note 3**).

3.4 Gel Electrophoresis and Densitometry Analysis

1. Dissolve protein pellets in 45 μL of 1× Laemmli loading buffer and boil at 95 °C for 5–10 min. Apply each sample (15 μL) to a 12% SDS-PAGE gel and electrophorese at 120 V.
2. Analyze gels for TAMRA fluorescence (542/568 nm excitation/emission).
3. Stain gels using Coomassie Blue staining solution for 30 min, followed by incubation with destaining solution for 2–3 h. Scan gels using white light transillumination to image total proteins in the gel.
4. Quantify isoprenoid analog labeling of proteins using ImageJ, 32-bit type images with brightness/contrast adjusted accordingly; in some cases, such as when C15AlkOH is used, it is necessary to increase the brightness/contrast to allow fainter bands to be visualized. For each lane, the area of interest is selected using a rectangular selection tool. In Fig. 2, a rectangle including the 25 kDa region (where most small GTPases appear) was used. To normalize the

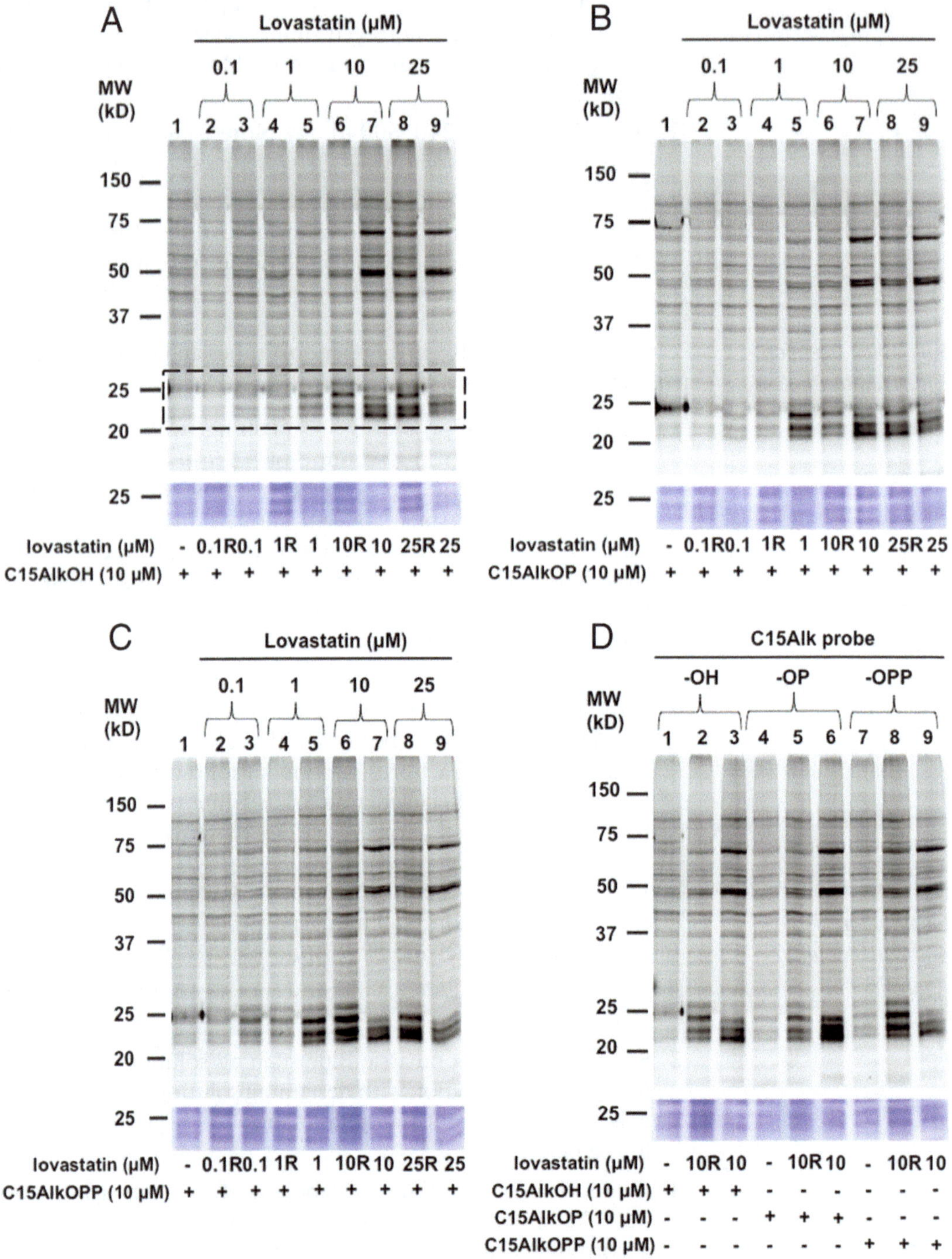

Fig. 2 In-gel fluorescence of HeLa cells treated with varying lovastatin concentrations and metabolically labeled with isoprenoid phosphate analogs C15AlkOH (**a**), C15AlkOP (**b**), C15AlkOPP (**c**) and a comparison of probe labeling at 10 μM with lovastatin at 10 μM (**d**). In (**a–c**), the lovastatin is either removed (e.g., 10R) or present in the incubation medium with the isoprenoid probe. Quantification using densitometry is performed on the bands in the 25 kDa region (indicated by the dashed rectangle) where many different prenylated small GTPases migrate. The intensities of these bands are normalized to the total protein in the same region of the Coomassie Blue-stained gel (lower panels)

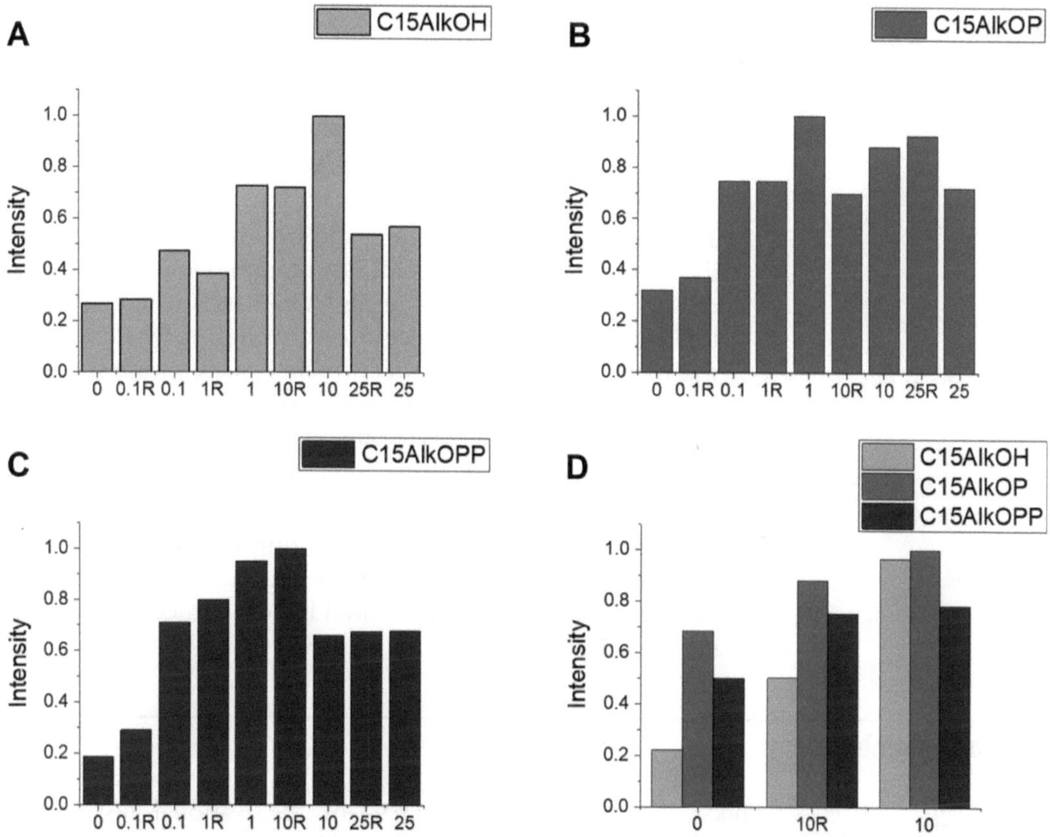

Fig. 3 Densitometric analysis of the fluorescence intensities of the labeled proteins in the presence of varying lovastatin concentrations from HeLa cells metabolically labeled with isoprenoid analogs C15AlkOH (**a**), C15AlkOP (**b**), C15AlkOPP (**c**), and the comparison of probe labeling at 10 μM with lovastatin at 10 μM (**d**). Fluorescence intensities are evaluated using the 25 kDa region where many prenylated proteins reside. The optimal concentration of lovastatin for treatment of HeLa cells is 10 μM

fluorescence intensity to the total protein in each lane, the intensity from the fluorescence image is divided by the intensity of the same region of the Coomassie Blue-stained image (lower panels). For each gel, the normalized data is transformed to numerical values of 0–1, by dividing all data by the maximum value in each set.

Data from different gels have variable intensities due to brightness and contrast settings to best visualize the labeled protein bands. The graphical representations of the data are plotted using the graphing program Origin® 2016 version 93E (Fig. 3).

4 Notes

1. Lovastatin is toxic at some concentrations to sensitive cell lines (e.g., COS-7). Hence it is necessary to optimize the concentration used. No significant toxicity was observed at 25 μM lovastatin on HeLa cells.
2. The volume of the lysis buffer can be adjusted depending on the amount of cells collected. For example, lower amounts of cells were recovered when using COS-7 cells where lovastatin toxicity was observed. Hence, a lower volume of the lysis buffer was used (200 μL). The target protein concentration should be between 1.5 and 2 mg/mL to be consistent with the protocol suggested for the click reaction.
3. Protein pellets should not be allowed to overdry, as this results in difficulty in redissolving them in 1× Laemmli loading buffer. To avoid this difficulty, after decanting the wash buffer, any remaining wash buffer is aspirated using suction through a glass pipette. The protein pellet should then immediately be dissolved in loading buffer.

Acknowledgments

This work was supported in part by the National Institutes of Health (RF1AG056976 and GM084152) and by the National Science Foundation grant (CHE-1308655).

References

1. Marakasova ES, Akhmatova NK, Amaya M, Eisenhaber B, Eisenhaber F, van Hoek ML et al (2013) Prenylation: from bacteria to eukaryotes. Mol Biol 47:622–633
2. Zhang FL, Casey PJ (1996) Protein prenylation: molecular mechanisms and functional consequences. Annu Rev Biochem 65:241–269.
3. Casey PJ, Solski PA, Der CJ, Buss JE (1989) p21ras is modified by a farnesyl isoprenoid. Proc Natl Acad Sci U S A 86:8323–8327
4. Casey PJ, Thissen JA, Moomaw JF (1991) Enzymatic modification of proteins with a geranylgeranyl isoprenoid. Proc Natl Acad Sci U S A 88:8631–8635
5. Seabra MC, Goldstein JL, Südhof TC, Brown MS (1992) Rab geranylgeranyl transferase. A multisubunit enzyme that prenylates GTP-binding proteins terminating in Cys-X-Cys or Cys-Cys. J Biol Chem 267:14497–14503
6. Blanden MJ, Suazo KF, Hildebrandt ER, Hardgrove DS, Patel M, Saunders WP et al (2017) Efficient farnesylation of an extended C-terminal C(x)3X sequence motif expands the scope of the prenylated proteome. J Biol Chem 293:2770–2785
7. Boyartchuk VL, Ashby MN, Rine J (1997) Modulation of ras and a-factor function by carboxyl-terminal proteolysis. Science 80 (275):1796–1800
8. Fujimura-Kamada K, Nouvet FJ, Michaelis S (1997) A novel membrane-associated metalloprotease, Ste24p, is required for the first step of NH2-terminal processing of the yeast a-factor precursor. J Cell Biol 136:271–285
9. Huyer G, Kistler A, Nouvet FJ, George CM, Boyle ML, Michaelis S (2006) *Saccharomyces cerevisiae* a-factor mutants reveal residues critical for processing, activity, and export. Eukaryot Cell 5:1560–1570

10. Schafer WR, Rine J (1992) Protein prenylation: genes, enzymes, targets, and functions. Annu Rev Genet 26:209–237
11. Liu Z, Meray RK, Grammatopoulos TN, Fredenburg RA, Cookson MR, Liu Y et al (2009) Membrane-associated farnesylated UCH-L1 promotes α-synuclein neurotoxicity and is a therapeutic target for Parkinson's disease. Proc Natl Acad Sci U S A 106:4635–4640
12. Jeong A, Suazo KF, Wood WG, Distefano MD, Li L (2018) Isoprenoids and protein prenylation: implications in the pathogenesis and therapeutic intervention of Alzheimer's disease. Crit Rev Biochem Mol Biol 53:279–310
13. Li H, Kuwajima T, Oakley D, Nikulina E, Hou J, Yang WS et al (2016) Protein Prenylation constitutes an endogenous brake on axonal growth. Cell Rep 16:545–558
14. Charron G, Li MMH, MacDonald MR, Hang HC (2013) Prenylome profiling reveals S-farnesylation is crucial for membrane targeting and antiviral activity of ZAP long-isoform. Proc Natl Acad Sci 110:11085–11090
15. Resh MD (2012) Targeting protein lipidation in disease. Trends Mol Med 18:206–214
16. Kale TA, Raab C, Yu N, Dean DC, Distefano MD (2001) A photoactivatable prenylated cysteine designed to study isoprenoid recognition. J Am Chem Soc 123:4373–4381
17. Vervacke JS, Funk AL, Wang Y-C, Strom M, Hrycyna CA, Distefano MD (2014) Diazirine-containing photoactivatable isoprenoid: synthesis and application in studies with isoprenylcysteine carboxyl methyltransferase. J Org Chem 79:1971–1978
18. Dozier JK, Khatwani SL, Wollack JW, Wang Y-C, Schmidt-Dannert C, Distefano MD (2014) Engineering protein farnesyltransferase for enzymatic protein labeling applications. Bioconjug Chem 25:1203–1212
19. Tate EW, Kalesh KA, Lanyon-hogg T, Storck EM, Thinon E (2015) Global profiling of protein lipidation using chemical proteomic technologies. Curr Opin Chem Biol 24:48–57
20. DeGraw AJ, Palsuledesai C, Ochocki JD, Dozier JK, Lenevich S, Rashidian M et al (2010) Evaluation of alkyne-modified isoprenoids as chemical reporters of protein prenylation. Chem Biol Drug Des 76:460–471
21. Suazo KF, Schaber C, Palsuledesai CC, Odom John AR, Distefano MD (2016) Global proteomic analysis of prenylated proteins in *Plasmodium falciparum* using an alkyne-modified isoprenoid analogue. Sci Rep 6:38615
22. Suazo KF, Hurben AK, Liu K, Xu F, Thao P, Sudheer C et al (2018) Metabolic labeling of prenylated proteins using alkyne-modified isoprenoid analogues. Curr Protoc Chem Biol 10 (3):e46
23. Palsuledesai CC, Ochocki JD, Kuhns MM, Wang YC, Warmka JK, Chernick DS et al (2016) Metabolic labeling with an alkyne-modified isoprenoid analog facilitates imaging and quantification of the prenylome in cells. ACS Chem Biol 11:2820–2828
24. Goldstein JL, Brown MS (1990) Regulation of the mevalonate pathway. Nature 343:425–430
25. Hosokawa A, Wollack JW, Zhang Z, Chen L, Barany G, Distefano MD (2007) Evaluation of an alkyne-containing analogue of farnesyl diphosphate as a dual substrate for protein-prenyltransferases. Int J Pept Res Ther 13:345–354

Chapter 4

Chemical Proteomic Analysis of *S*-Fatty Acylated Proteins and Their Modification Sites

Emmanuelle Thinon and Howard C. Hang

Abstract

Protein *S*-fatty-acylation, the covalent addition of a long-chain fatty acid, predominantly palmitate (*S*-palmitoylation), to cysteine, is a highly dynamic and regulated process that controls protein function and localization of membrane-associated proteins in eukaryotes. The analysis of *S*-fatty acylated peptides by mass spectrometry remains challenging due to the hydrophobic and potentially labile thioester linkage of the *S*-fatty acylated peptides.

Here we describe an optimized protocol for the global analysis of *S*-palmitoylated proteins based on the combination of an alkyne-tagged chemical reporter of palmitoylation, alk-16 with hydroxylamine-selective hydrolysis of thioester bonds. This protocol decreased the number of false positive proteins and was applied to identify *S*-fatty acylation sites, providing modification sites for 44 proteins out of the 106 *S*-fatty acylated proteins identified.

Key words *S*-palmitoylation, *S*-fatty acylation, Chemical proteomics, Protein identification, Site identification

1 Introduction

Protein *S*-fatty acylation, is the reversible attachment of a long-chain fatty acid, primarily palmitate (C16:0, *S*-palmitoylation), to cysteine via a thioester bond [1]. This modification promotes protein association to membranes and is essential to regulate the trafficking and function of membrane-associated proteins in eukaryotes. Since the mass spectrometry analysis of *S*-fatty acylated peptides remains challenging, due to the hydrophobic and potentially labile thioester linkage of the *S*-fatty acylated peptides, alternative chemical proteomics methods have been developed [2, 3]. One method involves metabolic labeling with alkyne-tagged palmitic acid, alk-16, and subsequent bioorthogonal tagging of proteins with biotin, for enrichment and proteomics analysis [4]. This method was applied for the global identification of *S*-fatty acylated proteins in a wide range of cell lines and organisms [5]. However,

Maurine E. Linder (ed.), *Protein Lipidation: Methods and Protocols*, Methods in Molecular Biology, vol. 2009,
https://doi.org/10.1007/978-1-4939-9532-5_4,

alk-16 does not only label *S*-fatty acylated proteins in cells but can also be incorporated into other protein modifications such as *O*-palmitoylated, *N*-myristoylated, or GPI-anchored proteins [6, 7]. To distinguish between *S*- and *N*,*O*-acylation, a hydroxylamine (NH_2OH) treatment, which selectively hydrolyzes thioesters bonds of alk-16 labeled proteins, can be performed (Fig. 1) [8, 9]. Quantitative comparison of alk-16 labeled samples plus/minus NH_2OH treatment, as described in the protocol below, provides a list of *S*-fatty acylated proteins with decreased number of false-positive results. The combination of alk-16 labeling with hydroxylamine-selective hydrolysis of thioester bonds can be applied to identify *S*-fatty acylation sites (Fig. 1). Several methods

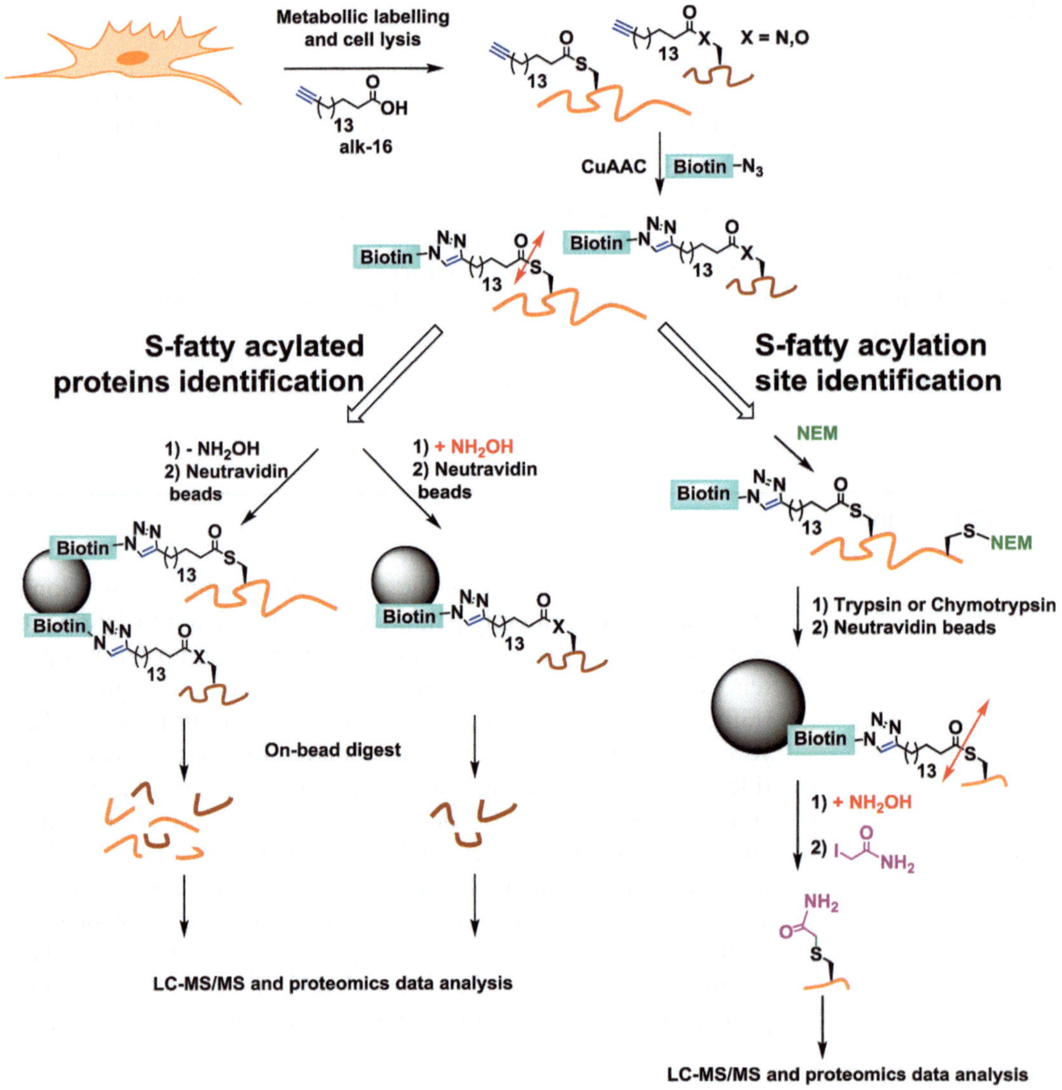

Fig. 1 Workflow for chemical proteomics of *S*-fatty acylated proteins

will be described, based on chymotrypsin/trypsin digestion before or after enrichment. Combining these methods, we have identified *S*-fatty acylation sites for 44 of the 106 *S*-fatty acylated proteins identified in HeLa cells [10]. *S*-Fatty acylation sites are typically located adjacent to a transmembrane domain in integral membrane proteins or in a soluble domain of proteins. Chymotrypsin digestion will facilitate the identification of sites in integral membrane proteins as the amino acids surrounding the *S*-fatty acylation site are likely to be hydrophobic such as tryptophan, phenylalanine, and tyrosine residues which are chymotrypsin cleavage sites. Trypsin digestion will facilitate the identification of *S*-fatty acylation sites found in soluble domains of proteins since these domains are likely to comprise several arginine/lysine specific to trypsin digest. For chymotrypsin digestion, digestion after enrichment gave the highest number of identified sites, probably due to the low solubility of the modified peptides. For trypsin digestion, digestion before or after enrichment gave similar results.

2 Materials

Prepare all solutions using ultrapure water (prepared by purifying deionized water, to attain a sensitivity of 18 MΩ cm at 25 °C) or LC-MS grade water. Use analytical grade reagents for cell lysis and click chemistry. Following the last protein precipitation, take extra care not to contaminate the samples (PEG, detergent, keratin, etc.): work on a clean bench, use mass spectrometry grade reagents, use filtered tips and Eppendorf® LoBind microcentrifuge tubes.

2.1 alk-16 Labeled Cell Lysates

1. HeLa cells (*see* **Note 1**).
2. Dulbecco's Modified Eagle's Medium (DMEM).
3. 10% charcoal/dextran-treated fetal bovine serum (FBS).
4. 50 mM alk-16 stock solution in DMSO: dissolve 10 mg of alk-16 in 713 μL DMSO. Prepare 50 μL aliquots and store at −20 °C (*see* **Note 2**).

2.2 Cell Lysis and Copper(I)-Catalyzed Alkyne-Azide Cycloaddition (CuAAC)

1. Lysis buffer: 4% sodium dodecyl sulfate (SDS), 150 mM NaCl, 50 mM HEPES pH 7.4, EDTA-free protease inhibitor, benzonase. Prepare fresh.
2. 0% SDS HEPES buffer: 150 mM NaCl, 50 mM HEPES pH 7.4.
3. 10 mM azide-biotin in DMSO. Store at −20 °C.
4. 50 mM $CuSO_4$ in water. Store aliquots at −20 °C (*see* **Note 3**).
5. 50 mM Tris(2-carboxyethyl)phosphine hydrochloride (TCEP) in water. The pH of the solution does not need to be adjusted. Store aliquots at −20 °C (*see* **Note 3**).

6. 2 mM Tris[(1-benzyl-1*H*-1,2,3-triazol-4-yl)methyl]amine (TBTA) in DMSO:*t*-butanol 1:4 (*see* **Note 4**).
7. CuAAC reactant solution: 5 μL 10 mM azide-biotin, 10 μL 50 mM $CuSO_4$, 10 μL 50 mM TCEP, 25 μL 2 mM TBTA (*see* **Note 5**).
8. 0.5 M EDTA in water, pH8.

2.3 Global Analysis of S-Fatty Acylated Proteins by Chemical Proteomics

1. Ice-cold chloroform ($CHCl_3$), methanol (MeOH), and water.
2. 4% SDS HEPES buffer: 4% SDS, 150 mM NaCl, 50 mM HEPES pH 7.4, 1 mM EDTA.
3. 1.33 M NH_2OH solution: 1.33 M NH_2OH pH 7.4, 1 mM EDTA in water. Prepare fresh before use (*see* **Note 6**).
4. 0% SDS HEPES buffer+ EDTA: 150 mM NaCl, 50 mM HEPES pH 7.4, 1 mM EDTA.
5. 0.4% SDS HEPES buffer: 0.4% SDS, 150 mM NaCl, 50 mM HEPES pH 7.4.
6. PBS (phosphate-buffered saline) 137 mM NaCl, 2.7 mM KCl, 10 mM Na_2HPO_4, 1.8 mM KH_2PO_4 pH 7.4.
7. 1% SDS in PBS.
8. 4 M Urea in PBS. Prepare fresh before use.
9. AMBIC: 50 mM ammonium bicarbonate in water. The pH of the solution does not need to be adjusted. Prepare fresh before use.
10. 100 mM TCEP in AMBIC pH 8. Adjust the pH to 8. Aliquot and store at −20 °C.
11. 10 mM TCEP in AMBIC pH 8: Dilute 100 mM TCEP in AMBIC tenfold. Prepare fresh before use.
12. 50 mM iodoacetamide in AMBIC. Prepare fresh before use and protect from light.
13. Trypsin solution: typically 0.05 μg Trypsin/250 μg of lysate in 50 μL AMBIC. Prepare fresh before use.
14. 1% formic acid (FA), 15% acetonitrile in H_2O. Prepare fresh before use.
15. 1% FA in H_2O. Prepare fresh before use.

2.4 Identification (ID) of S-Fatty Acylation Sites

1. 200 mM TCEP pH 7.4 in H_2O. Prepare fresh before use.
2. 1 M *N*-ethylmaleimide (NEM) in ethanol. Prepare fresh before use.
3. 100 mM Tris–HCl, 10 mM $CaCl_2$, pH 8.
4. 1% Rapigest (*see* **Note 7**) in 100 mM Tris–HCl, 10 mM $CaCl_2$ pH 8. Dissolve 10 mg of Rapigest in 100 μL 100 mM Tris–HCl, 10 mM $CaCl_2$ pH 8.

5. Trypsin (20 μg) or chymotrypsin (25 μg), dissolved in 25 μL resuspension buffer recommended by the supplier. Prepare fresh before use.
6. 0.2% SDS in 100 mM Tris–HCl pH 8.
7. 0.5% SDS in PBS.
8. 4 M urea in PBS.
9. 10× TEA buffer: 1.5 M NaCl, 500 mM triethanolamine pH 7.4.
10. TEA buffer: 150 mM NaCl, 50 mM triethanolamine pH 7.4.
11. 2 M NH_2OH pH 7.4 in H_2O supplemented with 5 mM EDTA.
12. NH_2OH solution: 25 μL 1% Rapigest (*see* **Note 7**), 2.5 μL 0.5 M EDTA, 1.25 μL 10× TEA buffer pH 7.4, 21.25 μL 2 M NH_2OH pH 7.4.
13. 250 mM iodoacetamide in water. Prepare fresh and protect from light.
14. Zeba spin desalting columns, 7 kDa molecular weight cutoff.

2.5 ID of S-Fatty Acylation Sites: Trypsin Digest After NeutrAvidin Enrichment

1. 0.2% SDS in PBS.
2. 1 μg trypsin dissolved in 25 μL resuspension buffer recommended by the supplier. Prepare fresh before use.
3. Other reagents are similar to Subheading 2.4.

2.6 ID of S-Fatty Acylation Sites: Chymotrypsin Digest After NeutrAvidin Enrichment

1. TEA buffer pH 8: 150 mM NaCl, 50 mM triethanolamine pH 8.
2. 2 M urea in PBS, 1 mM $CaCl_2$.
3. 2 μg chymotrypsin in 200 μL 2 M urea–PBS, 1 mM $CaCl_2$. Prepare fresh before use.
4. 6 M urea.
5. Other reagents are similar to Subheading 2.4.

3 Methods

Carry out all procedures at room temperature unless otherwise specified. Subheadings 3.1 and 3.2 are common to the global identification of *S*-fatty acylated proteins and the identification of *S*-fatty acylation sites.

3.1 alk-16 Labeled Lysates

1. Seed HeLa cells 24 h before treatment in a 10-cm dish. Cells should be 80–90% confluent at the time of lysis (4–24 h following alk-16 treatment) (*see* **Note 8**).

2. Pre-warm FBS and DMEM to 37 °C.
3. Dissolve 10 μL of alk-16 (50 mM alk-16 stock solution in DMSO) in 1 mL FBS and add 9 mL of DMEM (final concentration 50 μM alk-16) (*see* **Note 9**).
4. Replace cell culture medium with medium containing alk-16 (**step 3**).
5. 4–24 h following treatment, trypsinize cells, transfer to a 1.5 mL Eppendorf, centrifuge at 300 × *g*, remove the supernatant and wash the cell pellet with 2× 1 mL ice-cold PBS.
6. Snap freeze the cell pellet and store at −80 °C for up to 12 months.

3.2 Cell Lysis and CuAAC

1. Thaw cells on ice. Add 100 μL lysis buffer and pipette up and down to resuspend the cell pellet. Measure protein concentration (*see* **Note 10**).
2. Dilute 500 μg of lysate with 0% SDS HEPES buffer to obtain 450 μL (*see* **Notes 11** and **12**).
3. Add 50 μL of freshly prepared CuAAC reactant solution and stir for 1 h.
4. Add 10 μL of 0.5 M EDTA to quench the CuAAC reaction and briefly vortex the sample (*see* **Note 13**). The quenched sample is used for Subheadings 3.3 or 3.4 or 3.5.

3.3 Global Analysis of S-Fatty Acylated Proteins by Chemical Proteomics

1. Precipitate the protein sample by $CHCl_3$–MeOH precipitation (*see* **Note 14**).
2. Resuspend the pellet in 125 μL 4% SDS HEPES buffer. Split the sample in two ("$-NH_2OH$" sample and "$+NH_2OH$" sample). Add 187.5 μL of 1.33 M NH_2OH to the "$+NH_2OH$" sample and add 187.5 μL of 0% SDS HEPES + EDTA to the "$-NH_2OH$" sample. Vortex for 1 h at room temperature.
3. Precipitate the protein samples by $CHCl_3$–MeOH precipitation (*see* **Note 14**).
4. Resuspend the pellet in 62.5 μL 4% SDS HEPES buffer. Add 187.5 μL of 0.67 M NH_2OH to the "$+NH_2OH$" sample and add 187.5 μL of 0% SDS HEPES + EDTA to the "$-NH_2OH$" sample. Boil the samples for 10 min.
5. Precipitate the protein samples by $CHCl_3$–MeOH precipitation (*see* **Note 14**).
6. Resuspend the pellet in 50 μL 4% SDS HEPES buffer. Vortex for 10 min at room temperature. Once the pellet is resuspended, add 450 μL 0% SDS HEPES pH 7.4 buffer, vortex briefly and spin down at 10,000 × *g* for 5 min.

7. Prepare 2× 15 μL of low-binding NeutrAvidin beads by washing three times with 0.4% SDS–HEPES buffer. Add supernatant from each sample from **step 6** to 15 μL NeutrAvidin beads. Incubate at room temperature with end-over-end rotation for 90 min.
8. Wash the beads with 1 mL 1% SDS in PBS three times, then wash twice with 1 mL 4 M urea in PBS, and finally, wash with 1 mL AMBIC five times (*see* **Note 15**).
9. Add 50 μL of 10 mM TCEP in AMBIC, vortex for 30 min at room temperature. Centrifuge the samples for 1 min at 4000 × *g* and discard the supernatant.
10. Wash the beads with 1 mL AMBIC.
11. Add 50 μL of 10 mM iodoacetamide in AMBIC. Briefly vortex, centrifuge the samples for 5 s at 4000 × *g*, and leave in the dark for 30 min. Centrifuge the samples for 1 min at 4000 × *g* and discard the supernatant. Wash the beads with 1 mL AMBIC.
12. Add 50 μL trypsin solution. Incubate at 37 °C overnight.
13. Centrifuge samples at 4000 × *g* for 1 min. Transfer the supernatant to a clean Eppendorf. Wash the beads with 50 μL 1% FA, 15% acetonitrile in H_2O and 50 μL 1% FA in H_2O, and combine with the first wash (*see* **Note 16**). Concentrate samples using a speed-vac and store at −80 °C prior to analysis.
14. Resuspend the peptide pellets in 5% acetonitrile, 1% FA in H_2O for LC-MS analysis.
15. Analyze the samples on a nano-LC-MS/MS (*see* **Note 17**).
16. Search the data using a proteomics software such as Maxquant (*see* **Note 18**).

3.4 ID of S-Fatty Acylation Sites: Protease Digestion Before NeutrAvidin Enrichment

1. Following quenching of the CuAAC reaction with EDTA (Subheading 3.2 **step 4**), add 200 mM TCEP to a final concentration of 10 mM. Vortex for 10 min. Add 1 M NEM to a final concentration of 20 mM. Vortex for 1 h.
2. Remove the excess of NEM by three $CHCl_3$–MeOH precipitations (*see* **Note 14**).
3. Resuspend the samples (typically 2 mg) in 100 μL 1% Rapigest in 100 mM Tris, 10 mM $CaCl_2$. Add 900 μL 100 mM Tris, 10 mM $CaCl_2$ to dilute the sample to 0.1% Rapigest.
4. Add trypsin (20 μg) or chymotrypsin (25 μg) to the sample and incubate overnight at 37 °C for trypsin or room temperature for chymotrypsin.
5. Boil for 10 min to inactivate the enzyme. Add 20 μL of 10% SDS to a final concentration 0.2% SDS in preparation for NeutrAvidin enrichment.

6. Prepare high capacity NeutrAvidin beads (typically 100 μL beads/2 mg lysate) by washing three times with 0.2% SDS in 100 mM Tris pH 8. Add digested samples to the beads and incubate with end-over-end rotation for 2 h.
7. Discard the supernatant and wash the beads sequentially with 0.5% SDS in PBS three times, 4 M urea in PBS three times, PBS five times, and TEA buffer pH 7.4 twice.
8. Hydrolyze thioester-linked alk-16 labeled peptides with three times 50 μL NH_2OH solution for 30 min at 37 °C. After each elution, transfer the supernatant to an Eppendorf tube, add 5 μL of 250 mM iodoacetamide, vortex briefly and keep in the dark. Combine eluates and vortex for an additional 15 min in the dark.
9. Add 2 μL FA and remove excess of salt by C18-stage tipping (*see* **Note 19**).
10. Resuspend peptides in 90 μL AMBIC, 10 μL acetonitrile. Add 11 μL 250 mM iodoacetamide in H_2O and vortex for 90 min in the dark.
11. Precipitate Rapigest by adding 2 μL TFA and 5 μL FA and incubating the samples at 37 °C for 30 min. Spin down for 10 min at 8000 × *g* and process the supernatant by C18 stage-tipping (*see* **Note 19**). Keep the pellet at −80 °C prior to LC-MS/MS analysis.

3.5 ID of S-Fatty Acylation Sites: Trypsin Digestion After NeutrAvidin Enrichment

1. Following quenching of the CuAAC reaction with EDTA, add 200 mM TCEP to a final concentration of 10 mM. Vortex for 10 min. Add 1 M NEM to a final concentration of 20 mM. Vortex for 1 h.
2. Remove excess of NEM by three $CHCl_3$–MeOH precipitations (*see* **Note 14**).
3. Resuspend samples (typically 2 mg) in 2% SDS in PBS and dilute to 0.2% SDS (1 mg/mL final protein concentration). Pre-wash High capacity NeutrAvidin beads (typically 100 μL beads/2 mg lysate) with 0.2% SDS in PBS and add digested samples. Incubate with end-over-end rotation for 2 h. Wash beads three times with 1 mL 0.5% SDS/PBS, three times with 1 mL 4 M Urea/PBS, five times with 1 mL PBS, and twice with TEA buffer pH 7.4.
4. Hydrolyze thioester-linked alk-16 labeled peptides with 3× 50 μL NH_2OH solution for 30 min at 37 °C. After each elution, transfer the supernatant to an Eppendorf tube, add 5 μL of 250 mM iodoacetamide, vortex briefly and keep in the dark. Combine eluates and vortex for an additional 15 min in the dark.

5. Desalt the sample using a 7 kDa MW cutoff Zeba column (Thermo Fisher), according to the supplier's protocol. Elute the sample in TEA buffer pH 8.
6. Add 250 mM iodoacetamide to a final concentration of 10 mM. Incubate in the dark for 1 h. Desalt the sample using a 7 kDa MW cutoff Zeba column and elute in AMBIC.
7. Dilute with AMBIC up to 1 mL and add 1 μg trypsin. Incubate at 37 °C overnight.
8. Concentrate samples in a speed-vac until ~150 μL remains.
9. Precipitate Rapigest by adding 2 μL TFA and 5 μL FA. Incubate the samples at 37 °C for 30 min. Spin down for 10 min at 8000 × *g*. Process the supernatant by C18 stage-tipping. Keep the pellet at −80 °C prior to LC-MS/MS analysis.

3.6 ID of S-Fatty Acylation Sites: Chymotrypsin Digestion After NeutrAvidin Enrichment

1. Following quenching of the CuAAC reaction with EDTA (Subheading 3.2 **step 3**), add 200 mM TCEP to a final concentration of 10 mM. Vortex for 10 min. Add 1 M NEM to a final concentration of 20 mM. Vortex for 1 h.
2. Remove excess of NEM by three $CHCl_3$–MeOH precipitations (*see* **Note 14**).
3. Resuspend samples (typically 2 mg) in 2% SDS in PBS and dilute to 0.2% SDS (1 mg/mL final protein concentration). Pre-wash High capacity NeutrAvidin beads (typically 100 μL beads/2 mg lysate) with 0.2% SDS in PBS and add digested samples. Incubate with end-over-end rotation for 2 h. Wash beads three times with 1 mL 0.5% SDS in PBS, three times with 1 mL 4 M Urea in PBS, five times with 1 mL PBS, and twice with 100 μL 2 M Urea in PBS.
4. On-bead digest proteins by incubating beads at room temperature overnight with 2 μg chymotrypsin in 200 μL 2 M urea in PBS, 1 mM $CaCl_2$. Wash the beads four times with PBS and twice with TEA buffer pH 8. Hydrolyze thioester-linked alk-16 labeled peptides with 3× 50 μL NH_2OH solution for 30 min at 37 °C. After each elution, transfer the supernatant to an Eppendorf tube, add 5 μL of 250 mM iodoacetamide, vortex briefly and keep in the dark. Combine eluates and vortex for an additional 15 min in the dark.
5. Add 2 μL FA and remove excess salt by C18-stage tipping (*see* **Note 19**).
6. Resuspend peptides in 90 μL AMBIC, 10 μL acetonitrile. 11 μL iodoacetamide (250 mM stock in H_2O) and vortex for 90 min in the dark.
7. Add 20 μL of 6 M urea and vortex briefly. Precipitate Rapigest by adding 2 μL TFA and 5 μL FA and incubating the samples at 37 °C for 30 min. Spin down for 10 min at 8000 × *g* and

process the supernatant by C18 stage-tipping. Keep the pellet at −80 °C prior to LC-MS/MS analysis.

8. Resuspend the samples in 5 μL 1:1:1 FA–75% ACN–1% TFA and then dilute by adding 25 μL 1% TFA.
9. Analyze the samples on a nanoLC-MS/MS (*see* **Note 20**).
10. Search the data using a proteomics software such as Maxquant (*see* **Note 21**).

4 Notes

1. HeLa cells are used here but alk-16 labeling can be applied to any eukaryotic cells. The labeling time and growth media will need to be changed accordingly.
2. Alk-16 (also known as 17-ODYA) is commercially available.
3. Do not re-freeze aliquots. Use one aliquot for each experiment.
4. TBTA from several suppliers can result in high background. To test the background, compare labeling with azido-fluorophore (such as azido-rhodamine) of cell lysates obtained for cells treated with DMSO as a control or alk-16. By in-gel fluorescence, a fluorescent signal should be observed only for the alk-16 sample [11].
5. Prepare just before use. Briefly mix after adding each reagent in the same order as written.
6. Prepare fresh before use.
7. Rapigest is a detergent compatible with proteomics analysis because it can be precipitated prior to LC-MS/MS analysis.
8. To increase alk-16 labeling, notably for site-identification, perform alk-16 labeling for 24 h. alk-16 can be metabolized in cells and incorporated into other residues. Longer labeling might increase alk-16 incorporation into other amino acid residues or functional groups. However, this should not be an issue as hydroxylamine (NH_2OH) will distinguish between alk-16 incorporated into cysteine via a thioester vs. other linkages.
9. Alk-16 can be difficult to dissolve and should be adsorbed on BSA to increase its solubility. Add 60 μL of 0.01 N NaOH solution to 12 μL alk-16 (50 mM in DMSO). Stir at 70 °C for 5 min and add 150 μL of warm (37 °C) 5% BSA in PBS. Stir at 37 °C for 5 min and add 11.78 mL 10% FBS in DMEM. Filter-sterilize the solution and use fresh.
10. We use the BCA assay to measure protein concentration; however, any other assays would be suitable. Follow supplier's protocol to measure protein concentration.

11. Five-hundred micrograms of total protein is usually sufficient for the global analysis of *S*-fatty acylated proteins. For site identification, we recommend starting with 2 mg of lysate, which might require more than one 10-cm dish of 80–90% confluent cells. The amount and volumes of reagents for the CuAAC reaction should be scaled up accordingly.
12. We recommend performing the CuAAC reaction in 0.2–0.5% SDS. A high concentration of SDS (>0.5% SDS) inhibits the CuAAC reaction, while a low concentration of SDS might result in protein precipitation. We also recommend performing the CuAAC with a protein concentration of 1–2 mg/mL.
13. EDTA will quench the CuAAC reaction. Addition of EDTA is essential to prevent protein degradation during the NH_2OH treatment.
14. Perform the $CHCl_3$–MeOH precipitation by adding 4 volumes ice-cold MeOH, 1 volume ice-cold $CHCl_3$, 3 volumes ice-cold H_2O. Briefly vortex the sample after adding each reagent. Centrifuge the sample at 10,000 × *g* for 5 min at 4 °C, and remove the top and bottom layers. The protein precipitate should be at the interface of the two layers. Add 4 volumes MeOH and centrifuge at 10,000 × *g* for 1 min at 4 °C. Carefully remove the supernatant, and air-dry the protein pellet for 5–10 min.
15. Carefully remove the supernatant. Do not completely remove the supernatant to prevent loss of beads. Leave approximately 25–50 μL solution on top of the beads.
16. An optional C18 clean-up step can be added to remove any excess of salts [12].
17. We analyze our samples using the following instrument and conditions; nonetheless other instruments are suitable. Tryptic or chemotryptic peptides were desalted on a trap column prior to separation on a 12 cm/75 μm reversed phase C18 column (Nikkyo Technos Co., Ltd. Japan). A 120 min method increasing from 10% B to 45% B in 77 min (A: 0.1% Formic Acid, B: Acetonitrile/0.1% Formic Acid) were delivered at 200 nL/min. The liquid chromatography setup was coupled to an Orbitrap XL (Thermo, San Jose, CA, USA) mass spectrometer operated in top-8-CID-mode. Mass spectra were collected in a 300–1800 *m/z* mass range using 60,000 resolution.
18. We processed our data with MaxQuant version 1.5.3.8 [13], using the following parameters; nonetheless, other proteomics software might be suitable. The peptides were identified from the MS/MS spectra searched against the database using the Andromeda search engine [14]. Cysteine carbamidomethylation and cysteine modified by NEM were used as variable modifications. For the identification, the false discovery rate

was set to 0.01 for peptides, proteins and sites, the minimum peptide length allowed was seven amino acids, and the minimum number of unique peptides allowed was set to one. Other parameters were used as pre-set in the software. "Unique and razor peptides" mode was selected; this calculates ratios from unique and razor peptides. Label-free quantification experiments in MaxQuant were performed using the built-in label free quantification algorithm.

19. The published protocol [12] was modified to recover hydrophobic peptides. 6 layers of C18 (3 M Empore) were activated with 100 μL ACN 1% FA and equilibrated with 100 μL H_2O 1% FA. The sample was added, and the tip was centrifuged. Peptides were eluted with 100 μL 70% ACN 5% FA 25% H_2O, and 100 μL 90% isopropanol 5% FA 5% H_2O. Eluates were combined, and peptides were dried in a speed-vac.
20. Peptides were analyzed by LC–MS/MS(Ultimate 3000 nano-HPLC system coupled to a Q-Exactive Plus mass spectrometer, Thermo Scientific). Peptides were separated on a C18 column (12 cm/75 μm, 3 μm beads, Nikkyo Technologies) at 200 nL/min with a gradient increasing from 1% buffer B/99% buffer A to 95% buffer B/10% buffer A in 120 min (buffer A, 0.1% formic acid; buffer B, 0.1% formic acid in acetonitrile; gradient : 1% B for 10 min, 1–6% B in 4 min, 6–50% B in 77 min, 50–95% in 1 min, remaining at 95% B for 17 min, 95–1% in 1 min, and 1% for 9 min) and were analyzed in a data-dependent acquisition manner. MS spectra were recorded at 17,500 resolution with m/z 100 as lowest mass. Normalized collision energy was set at 27, with AGC target and maximum injection time being 2×10^5, and 60 ms, respectively.
21. The data were processed with MaxQuant version 1.5.3.8 [13], and the peptides were identified from the MS/MS spectra searched against the database using the Andromeda search engine [14]. Acetyl on protein N-terminus, oxidation of Met, Cysteine carbamidomethylation, and Cysteine modified by NEM were used as variable modifications. For the identification, the false discovery rate was set to 0.01 for peptides, proteins and sites, the minimum peptide length allowed was five amino acids, and the minimum number of unique peptides allowed was set to one. Up to 10 miscleavages were allowed for chymotrypsin. Other parameters were used as preset in the software.

Acknowledgments

This work was supported by a Marie Skłodowska-Curie Individual Fellowship ("QuantPalm_immunity") to E.T. H.C.H. acknowledges grant support from NIH-NIGMS R01GM087544.

References

1. Chamberlain LH, Shipston MJ (2015) The physiology of protein *S*-acylation. Physiol Rev 95:341–376. https://doi.org/10.1152/physrev.00032.2014
2. Peng T, Thinon E, Hang HC (2016) Proteomic analysis of fatty-acylated proteins. Curr Opin Chem Biol 30:77–86. https://doi.org/10.1016/j.cbpa.2015.11.008
3. Lanyon-Hogg T, Faronato M, Serwa RA, Tate EW (2017) Dynamic protein acylation: new substrates, mechanisms, and drug targets. Trends Biochem Sci 42:566–581. https://doi.org/10.1016/j.tibs.2017.04.004
4. Thinon E, Hang HC (2015) Chemical reporters for exploring protein acylation. Biochem Soc Trans 43:253–261. https://doi.org/10.1042/BST20150004
5. Blanc M, David F, Abrami L et al (2015) SwissPalm: protein palmitoylation database. F1000Res 4:261. https://doi.org/10.12688/f1000research.6464.1
6. Gao X, Hannoush RN (2014) Single-cell imaging of Wnt palmitoylation by the acyltransferase porcupine. Nat Chem Biol 10:61–68. https://doi.org/10.1038/nchembio.1392
7. Wright MH, Clough B, Rackham MD et al (2014) Validation of N-myristoyltransferase as an antimalarial drug target using an integrated chemical biology approach. Nat Chem 6:112–121. https://doi.org/10.1038/nchem.1830
8. Martin BR, Cravatt BF (2009) Large-scale profiling of protein palmitoylation in mammalian cells. Nat Methods 6:135–138. https://doi.org/10.1038/nmeth.1293
9. Foe IT, Child MA, Majmudar JD et al (2015) Global analysis of Palmitoylated proteins in toxoplasma gondii. Cell Host Microbe 18:501–511. https://doi.org/10.1016/j.chom.2015.09.006
10. Thinon E, Fernandez JP, Molina H, Hang HC (2018) Selective enrichment and direct analysis of protein S-Palmitoylation sites. J Proteome Res 17:1907–1922. https://doi.org/10.1021/acs.jproteome.8b00002
11. Wilson JP, Raghavan AS, Yang Y-Y et al (2011) Proteomic analysis of fatty-acylated proteins in mammalian cells with chemical reporters reveals S-acylation of histone H3 variants. Mol Cell Proteomics 10:M110.001198. https://doi.org/10.1074/mcp.M110.001198
12. Rappsilber J, Ishihama Y, Mann M (2003) Stop and go extraction tips for matrix-assisted laser desorption/ionization, Nanoelectrospray, and LC/MS sample pretreatment in proteomics. Anal Chem 75:663–670. https://doi.org/10.1021/ac026117i
13. Cox J, Mann M (2008) MaxQuant enables high peptide identification rates, individualized p.P.B.-range mass accuracies and proteome-wide protein quantification. Nat Biotechnol 26:1367–1372. https://doi.org/10.1038/nbt.1511
14. Cox J, Neuhauser N, Michalski A et al (2011) Andromeda: a peptide search engine integrated into the MaxQuant environment. J Proteome Res 10:1794–1805. https://doi.org/10.1021/pr101065j

Chapter 5

Direct Analysis of Protein *S*-Acylation by Mass Spectrometry

Yuhuan Ji and Cheng Lin

Abstract

Dynamic and reversible protein *S*-acylation, most commonly occurring as *S*-palmitoylation, plays an important role in protein/membrane association and the regulation of intracellular signaling via cycles of palmitoylation and depalmitoylation. Direct analysis of protein *S*-acylation by mass spectrometry (MS) offers several benefits over indirect detection methods in that it can definitively determine the location and nature of the acyl modification, and is not prone to false discoveries. However, characterization of acyl proteins is challenging because of the tendency of acyl loss during sample preparation and tandem MS analysis. In this chapter, we present a sample preparation protocol that preserves labile acyl modifications and an LC-MS/MS workflow for detection of *S*-acylation with high confidence and sensitivity.

Key words *S*-palmitoylation, Reversed phase liquid chromatography–mass spectrometry (RPLC-MS), Tandem mass spectrometry (MS/MS)

1 Introduction

Protein fatty acylation refers to the covalent attachment of a fatty acid to an amino acid residue via either an amide linkage (*N*-acylation), a thioester linkage (*S*-acylation), or an oxyester linkage (*O*-acylation). Acylation most commonly occurs as *N*-myristoylation or *S*-palmitoylation [1–4]. *N*-Myristoylation modifies an amino group, usually on the N-terminal glycine, with a 14-carbon saturated acyl chain (myristate), whereas *S*-palmitoylation targets the sulfhydryl group of a cysteine with a 16-carbon saturated acyl chain (palmitate). The amide bond in *N*-myristoylation is stable and there are no known enzymes that remove myristate from the N-terminal myristoyl-glycine. By contrast, the thioester linkage in *S*-palmitoylation is reversible under physiological conditions, and there exist a family of palmitoyl acyl transferases that can attach a palmitoyl group to a cysteine residue, as well as several acylprotein thioesterases that catalyze the detachment of palmitate from *S*-palmitoylated proteins. Consequently, *S*-palmitoylation, like

Maurine E. Linder (ed.), *Protein Lipidation: Methods and Protocols*, Methods in Molecular Biology, vol. 2009,
https://doi.org/10.1007/978-1-4939-9532-5_5,

phosphorylation, may vary in different intracellular environments and is subject to change upon regulation. The dynamic and reversible nature of *S*-palmitoylation also presents a challenge for its characterization, necessitating the development of analytical methods that are capable of preserving the thioester bond, localizing the modification sites, and measuring the level of protein palmitoylation under various cellular conditions.

Mass spectrometry (MS) is a powerful tool for structural characterization of biomolecules owing to its high sensitivity, specificity, and speed of analysis. Complex biological samples can be analyzed by MS in conjunction with online liquid chromatographic (LC) separation [5]. Tandem mass spectrometry (MS/MS)-based protein sequencing is the major driving force behind the recent rapid advances in the field of proteomics [6]. Posttranslational modifications (PTMs) can be localized by identifying the characteristic mass shift associated with the PTM, or lack thereof, in a series of fragments [7, 8]. Several early studies investigated the potential of MS/MS in characterizing acylpeptides. Whereas collision-induced dissociation (CID) can generate diagnostic ions for N-terminally myristoylated peptides [9, 10], CID of *O*-octanoyl ghrelin produced only limited sequence coverage and resulted in loss of the octanoyl group in several fragment ions. In contrast, electron capture dissociation (ECD) could produce extensive peptide backbone cleavages and preserve the octanoyl modification, allowing detection and localization of the acylation site [11].

Characterization of *S*-palmitoylation is considerably more challenging, as this labile modification can fall off during sample preparation or upon ion activation [12]. MS characterization of palmitoylated proteins may be achieved by utilizing the acyl-biotinyl exchange (ABE) chemistry that replaces the palmitoyl group with a stable biotin label [13–15]. An alternative approach utilizes *m*etabolic *l*abeling with a palmitic acid analog that contains an alkynyl or azido group, which allows for introduction of a biotin tag via *c*lick *c*hemistry (MLCC) [16–20]. In both ABE and MLCC, biotinylated proteins can be purified and enriched with streptavidin agarose beads and digested for LC-MS/MS analysis. Although large-scale profiling of protein palmitoylation can be achieved with either ABE or MLCC, each has its own limitations. The ABE approach does not differentiate *S*-palmitoylation from other forms of *S*-acylation [21, 22] or thioester modifications with non-acyl groups. False positives and negatives could arise due to incomplete blockage of free cysteines and inadequate HA-induced hydrolysis, respectively. MLCC is generally limited to cell culture studies, as it is difficult and costly to scale up metabolic labeling to large organisms. There is also concern over the toxicity of the palmitate analogs and their metabolic degradation products leading to other forms of modification, such as *N*-myristoylation [17].

Both ABE and MLCC are indirect detection methods that do not reveal the identity of the fatty acid chain attached to a specific *S*-acylation site. It is also necessary to perform comparative and quantitative proteomics measurement to differentiate *S*-acylated proteins from copurified, contaminant proteins [18]. Analysis of *S*-acylated peptides in their native state can be advantageous as it provides definitive evidence of the location and nature of the acyl modification. Several challenges exist for direct analysis of protein *S*-acylation by LC-MS. These include the tendency of acyl loss during sample preparation and tandem MS analysis, the need to purify and enrich acylated peptides, and potential complications caused by *S*- to *N*-acyl transfer [23]. We have previously investigated the stability of *S*-palmitoylated peptide standards under different incubation and fragmentation conditions, and identified conditions under which palmitoylation is largely preserved. We showed that palmitoyl loss can be minimized with careful choices of the reducing agent, digestion buffer, temperature, and detergent during the sample preparation process, and developed an LC-MS/MS workflow for successful analysis of palmitoyl peptides [12]. We have recently applied this workflow to analyze biological samples containing in vivo palmitoylated proteins, and showed that our method has sensitivity comparable to the conventional radioactive metabolic labeling approach with a significantly reduced analysis time, and the ability to localize the modification sites without the need of laborious mutation studies.

In this chapter, the regulator of G-protein signaling 4 (RGS4) is chosen as the model system to illustrate our workflow for direct analysis of protein palmitoylation from protein expression to LC-MS/MS analysis. Palmitoylated RGS4 is expressed in insect *Spodoptera frugiperda* 9 (Sf9) cells, incubated with exogenous palmitic acid to increase the stoichiometry of palmitoylation in cells, and purified for analysis. A palmitoylated RGS4 standard (Palm-RGS4-Std) is prepared by incubating purified RGS4 with palmitoyl-CoA under conditions that promote autopalmitoylation of RGS4 at cysteine residues previously mapped as palmitoylation sites [24]. The same protocol should be applicable toward the analysis of proteins with other forms of *S*-acylation.

2 Materials

Use HPLC grade water for all experiments. Prepare and store all buffers at room temperature, unless stated otherwise.

2.1 Expression and Palmitoylation of RGS4 in Sf9 Cells

1. Sf9 cells (Thermo Fisher Scientific, Waltham, MA), stored in liquid nitrogen.
2. SF-900™ II SFM Media (Thermo Fisher Scientific, Waltham, MA), stored at 4 °C.

3. His-RGS4 baculovirus, RGS4 tagged with hexahistidine at the N-terminus [24].
4. Palmitic acid stock solution: 2 mM palmitic acid, 6.7% fatty acid free bovine serum albumin (BSA) in SF-900™ II SFM Media, stored at −20 °C (*see* **Note 1**).
5. Shaking incubator (New Brunswick Scientific).

2.2 Enrichment of His-RGS4

1. Phosphate buffered saline (PBS) buffer: 0.01 M PBS, 0.138 M NaCl, 0.0027 M KCl, dissolved in deionized water, pH 7.4.
2. Lysis buffer: 300 mM NaCl, 25 mM sucrose, 20 mM imidazole, 0.5% CHAPS in PBS.
3. cOmplete™, EDTA-free Protease Inhibitor Cocktail (Roche).
4. Ni-NTA magnetic agarose beads, stored at 4 °C (Qiagen).
5. Elution buffer: 250 mM imidazole, 300 mM NaCl, 25 mM sucrose, 0.5% CHAPS in PBS.
6. Micro BCA protein assay kit (Pierce).
7. Microcentrifuge.
8. Branson Sonifier 250 (Branson Ultrasonic Corp).
9. 12-Tube magnet (Qiagen).

2.3 In Vitro Palmitoylation to Generate Palm-RGS4-Std

1. Palmitoyl-CoA labeling solution: 1 mM palmitoyl-CoA in 50 mM NaHepes (pH 7.8), 0.1% RapiGest (Waters, Milford, MA).

2.4 Buffer Exchange and Trypsin Digestion

1. Amicon Ultra-0.5 centrifugal filter MWCO 10 K (Millipore).
2. 50 mM Tris–HCl (pH 7.4).
3. 50 mM Tris–HCl (pH 7.4), 0.05% RapiGest.
4. 0.1 μg/μL Trypsin Gold (Promega) in 50 mM Tris–HCl (pH 7.4).
5. 10% trifluoroacetic acid (TFA).
6. Centrifuge (Eppendorf, Hamburg, Germany).
7. SpeedVac (Thermo Fisher Scientific, Waltham, MA).

2.5 Preparation of POROS R1 50 Microcolumns and Enrichment of Hydrophobic Peptides

1. HPLC syringe needle (Hamilton, Reno, NV).
2. 2.5 cm GFC glass fiber discs (Whatman, Clifton, NJ).
3. GELoader tip (Eppendorf, Hamburg, Germany).
4. POROS R1/50 (Applied Biosystems, Framingham, MA).
5. Acetonitrile (ACN).
6. 20% ACN, 0.1% TFA.
7. 40% ACN, 0.1% TFA.

8. 60% ACN, 0.1% TFA.
9. 80% ACN, 0.1% TFA.

2.6 Reversed Phase (RP) LC-MS

1. nanoACQUITY UPLC system (Waters).
2. Analytical nanoLC column: nanoACQUITY BEH300 C4 column (1.7 μm, 150 μm ID × 100 mm) (Waters).
3. Solvent A: 5% ACN, 0.1% formic acid (FA).
4. Solvent B: 95% ACN, 0.1% FA.
5. Strong needle wash: 80% ACN, 10% isopropanol (IPA), 0.1% FA.
6. Weak needle wash: 40% ACN, 0.1% FA.
7. Pump seal wash: 10% methanol.
8. Triversa Nanomate system (Advion BioSystems, Inc).
9. LTQ-Orbitrap XL instrument (Thermo Fisher Scientific).
10. Proteome Discoverer software (Thermo Fisher Scientific).

3 Methods

Perform all procedures at room temperature unless specified otherwise.

3.1 Expression and Palmitoylation of RGS4 Proteins in Sf9 Cells

1. Maintain Sf9 cells with SF-900™ II SFM Media in a sterile Erlenmeyer flask in a shaking incubator at 125 rpm, at 28 °C.
2. When cell growth reaches the mid-logarithmic phase (about 2.5×10^6 cells/mL) with a cell viability >90%, infect the cells with His-RGS4 virus (*see* **Note 2**) at multiplicity of infection (MOI) of 5 (*see* **Note 3**).
3. 48 h after infection, harvest the cells by centrifugation at 1000 × *g* for 30 min. Cell pellets can be either stored at −80 °C for in vitro palmitoylation (Subheading 3.2) or subjected to in vivo metabolic labeling as described in the next three steps.
4. For metabolic labeling with palmitic acid, prepare the labeling medium by diluting 2 mM palmitic acid stock solution to 30 μM with Sf-900™ II SFM media. For example, add 75 μL of 2 mM palmitic acid to 5 mL of Sf-900™ II SFM media, mix well. Scale up or down proportionally if needed.
5. Resuspend the cell pellet in **step 3** in palmitic labeling medium at approximately 2.5×10^6 cells/mL. Incubate for 1 h in a shaking incubator at 125 rpm, at 28 °C.
6. Harvest the cells by centrifuging at 1000 × *g* for 30 min. Store the cell pellets at −80 °C if not used immediately.

3.2 Enrichment of His-RGS4

1. Wash the cell pellet with PBS buffer twice.
2. Resuspend the cell pellet in 10 volumes of lysis buffer supplemented with cOmplete™, EDTA-free Protease Inhibitor Cocktail.
3. Sonicate the sample using Branson Sonifier 250 equipped with a microtip. Set the duty cycle to 20%, and the output control to 2. Keep cell lysate on ice during sonication.
4. Keep cell lysate on ice for an additional 30 min.
5. Centrifuge the lysate at 21,000 × *g* for 20 min at 4 °C to pellet the cellular debris. Transfer the resultant supernatant, which contains the His-RGS4, into a fresh tube.
6. Incubate 500 μL of the supernatant per 50 μL of the slurry of Ni-NTA Magnetic Agarose Beads on an end-over-end shaker for 2 h at 4 °C.
7. Collect the Ni-NTA Magnetic Agarose Beads using a QIAGEN 12-Tube Magnet. Remove the supernatant.
8. Wash the beads three times with 1 mL of lysis buffer.
9. Add 100 μL of elution buffer for every 50 μL Ni-NTA Magnetic Agarose Beads; mix the suspension on an end-over-end shaker for 30 min at 4 °C.
10. Collect the Ni-NTA Magnetic Agarose Beads using a QIAGEN 12-Tube Magnet, and transfer the elution to a fresh tube.
11. Perform BCA protein assay to determine the protein concentration.
12. Store the elution at −80 °C if not used immediately.

3.3 In Vitro Palmitoylation to Generate Palm-RGS4-Std

1. Express His-RGS4 following Subheading 3.1, **steps 1–3**. Do not incubate with palmitic acid.
2. Purify His-RG4 following Subheading 3.2, **steps 1–7**.
3. Incubate His-RGS4 beads with 20 μL of palmitoyl-CoA labeling solution.; mix the suspension on an end-over-end shaker at 30 °C for 3 h (*see* **Note 4**).
4. Complete the enrichment of His-RGS4 following Subheading 3.2, **steps 8–12**.

3.4 Buffer Exchange and Trypsin Digestion

1. Wash the Amicon Ultra-0.5 centrifugal filter (MWCO 10 K) with 400 μL of 50 mM Tris–HCl (pH 7.4) by centrifuging the device at 14,000 × *g* for 20 min.
2. Add 20 μg of the His-RGS4 eluate from either Subheading 3.3 (RGS4-palm-std) or Subheading 3.2 (His-RGS4 palmitoylated in cells) to the filter and adjust the final volume to 200 μL with 50 mM Tris–HCl (pH 7.4).
3. Centrifuge the device at 14,000 × *g* for 10 min.

4. Add 100 μL of 50 mM Tris–HCl (pH 7.4, *see* **Note 5**), 0.05% RapiGest (*see* **Notes 6–8)** to the filter, and pipet all sample into a fresh 1.5 mL Eppendorf tube. The total sample volume is usually 120 μL.
5. Add 10 μL of 0.1 μg/μL Trypsin Gold in 50 mM Tris–HCl (pH 7.4), and incubate the sample at 37 °C for 3 h (*see* **Notes 9** and **10**).
6. Stop the digestion by adding 7 μL of 10% TFA, to a final TFA concentration of about 0.5%. Keep it at room temperature for 10 min (*see* **Note 11**).
7. Dry down the digest using a SpeedVac.

3.5 Preparation of POROS R1 50 Microcolumns and Enrichment of Hydrophobic Peptides

1. Use a HPLC syringe needle to stamp out a disk from a glass microfiber filter.
2. Place the disk into the constricted end of a GELoader tip by pushing it out from the HPLC syringe needle with an LC capillary (*see* **Note 12**).
3. Weigh 50 mg POROS R1 50 resin into a 1.5 mL plastic tube, add 1 mL ACN to suspend the resin. The final concentration of this slurry is 5%.
4. Transfer 20 μL of the slurry prepared in **step 3** into an empty microcolumn prepared in **step 2**. Pack the resin by pushing air through the column using a plastic syringe. Repeat this step until the final bed volume is about 5 μL.
5. Equilibrate the microcolumn with 40 μL of 20% ACN, 0.1% TFA by air pressure, repeat three times.
6. Resuspend the digest (Subheading 3.3, **step 7**) with 80 μL 20% ACN, 0.1% TFA, vortex thoroughly and spin down briefly.
7. Load the sample slowly (about 1 drop/s) onto the microcolumn by air pressure, repeat three times.
8. Wash with 40 μL 20% ACN, 0.1% TFA by air pressure, repeat three times.
9. To a fresh tube, elute with 40 μL 40% ACN, 0.1% TFA by air pressure, repeat two times.
10. To a fresh tube, elute with 40 μL 60% ACN, 0.1% TFA by air pressure, repeat two times.
11. To a fresh tube, elute with 40 μL 80% ACN, 0.1% TFA by air pressure, repeat two times.

3.6 RPLC-ESI-MS/MS Analysis

The RPLC-MS/MS analysis is performed on a C4-nanoACQUITY UPLC system (*see* **Note 13**) connected to an LTQ-Orbitrap XL mass spectrometer through a Triversa Nanomate system.

1. Dry down the 60% ACN, 0.1% TFA elution of the His-RGS4 digest from Subheading 3.4, **step 10**, using a SpeedVac. Resuspend the sample with 20 μL 40% ACN, 0.1% FA.

2. Load 2 μL sample onto a nanoACQUITY UPLC system with 40% B. The flow rate is kept at 1.5 μL/min for the first 2 min, ramped down to 0.5 μL/min during the next minute, and kept constant at 0.5 μL/min for the rest of the LC run. The gradient is held at 40% B for 4 min, and increased to 100% B over 30 min. It is then held at 100% B for 5 min, followed by a ramp down to 40% B over 2 min, and maintained at 40% B for 29 min for column reequilibration (*see* **Note 14**).
3. The eluted peptides are subjected to RPLC-MS/MS analysis on an LTQ-Orbitrap XL mass spectrometer. The MS event cycle consists of one MS scan ($r = 60{,}000$ at m/z 400) and three data-dependent MS/MS scans ($r = 7500$), where the three most abundant precursor ions with charge state ≥ 2 were selected with an isolation window of ± 3 m/z for CID tandem MS analysis with the normalized collision energy set at 35% (*see* **Notes 15** and **16**). The MS data were processed using the Proteome Discoverer software.

4 Notes

1. 2 mM Palmitic acid stock solution is prepared by gradually adding 400 mM palmitic acid in DMSO into 6.7% BSA, Sf-900™ II SFM media, heating at 57 °C until well dissolved. The resulting solution was subjected to sterile filtering and stored at −20 °C for later use.
2. The His-RGS4 virus was generously provided by Prof. Elliott M. Ross at the University of Texas Southwestern Medical Center [24].
3. If infected successfully, cells will stop growing after infection, and have a lower density compared to uninfected cells. Infected cells will also have a larger size and enlarged nuclei.
4. Palmitoyl-CoA labeling solution contains 1 mM palmitoyl-CoA, a concentration sufficiently high to produce universal palmitoylation at all cysteine sites on RGS4 under suggested conditions, confirmed by LC-MS/MS analysis.
5. It is important to use Tris–HCl buffer (pH 7.4) instead of the commonly used ammonium bicarbonate (ABC) buffer during digestion. ABC buffer cannot maintain the pH at 7.4 at 37 °C over the course of digestion, because one of its degradation products, CO_2, has a lower solubility in aqueous solution than the other, ammonia. Preferential loss of CO_2 leads to a gradual increase in pH. Thioesters are susceptible to hydrolysis at pH >8, and it has been shown that palmitoyl loss can occur in ABC buffer in as short as 1 h at 37 °C [12].
6. The addition of detergent to the digestion buffer not only improves the solubility of hydrophobic proteins/peptides and

digestion efficiency, but also shields *S*-acylation from water and other nucleophiles, thus preventing palmitoyl loss due to thioester hydrolysis or cleavage by nucleophilic attacks.

7. The presence of detergent greatly reduces the extent of intermolecular palmitoyl transfer from the cysteine thiol to either the peptide N-terminus or the lysine side chain during sample preparation [23]. The *S*- to *N*-palmitoyl transfer can lead to erroneous reporting of *N*-palmitoylation.
8. For analysis of *S*-acylation, RapiGest is recommended among commercial MS-compatible detergents. Another commonly used acid-labile surfactant, ProteaseMAX, generates degradation products that can lead to artificial, in vitro modifications on cysteines mimicking *S*-palmitoylation and hydroxyfarnesylation [25]. The effect of other detergents has not been tested.
9. Since there are no known disulfide bonds in RGS4, reduction and alkylation are not performed here in order to minimize potential palmitoyl loss during sample preparation. If the reduction and alkylation procedure is required, use tris(2-carboxyethyl)phosphine (TCEP) instead of dithiothreitol (DTT) as the reducing reagent, since DTT is a strong nucleophile that can induce thioester bond cleavage while reduction by TCEP does not lead to appreciable palmitoyl loss. The presence of detergent greatly reduces but does not completely eliminate DTT-induced palmitoyl loss.
10. Keep the digestion time as short as possible to minimize depalmitoylation. It was shown that under digestion conditions described in Subheading 3.4, **step 5**, all cysteine-containing peptides were recovered (there are a total of 11 cysteine residues in RGS4).
11. Keeping the digest in acidic solution for 10 min is necessary to ensure complete degradation of RapiGest so it does not interfere with the subsequent MS analysis.
12. Photographs illustrating the preparation of empty micro columns can be found in ref. [26].
13. It is possible to analyze palmitoyl peptides on a C18-reversed phase analytical column. However, stronger organic solvent containing 10% IPA must be used to elute the hydrophobic palmitoyl peptides from a C18 column. Column clogging is also a tangible problem when a C18 column is used for palmitoyl peptide analysis.
14. HPLC gradient here is optimized for analysis of palmitoyl peptides. The large difference in hydrophobicity between palmitoyl peptides and their unmodified counterparts makes it difficult to analyze both in a single LC-MS run. Derivatization of unmodified peptides with a perfluoroalkyl tag, *N*-[(3-perfluorooctyl)propyl] iodoacetamide, leads to a significant

increase in their hydrophobicity, allowing them to be simultaneously analyzed with palmitoyl peptides for relative palmitoyl quantification [12].

15. Figure 1 shows the CID spectra of several palmitoyl peptides including GHHHHHHMC$_{palm}$K^{3}, ^{4}GLAGLPASC$_{palm}$LR^{14},

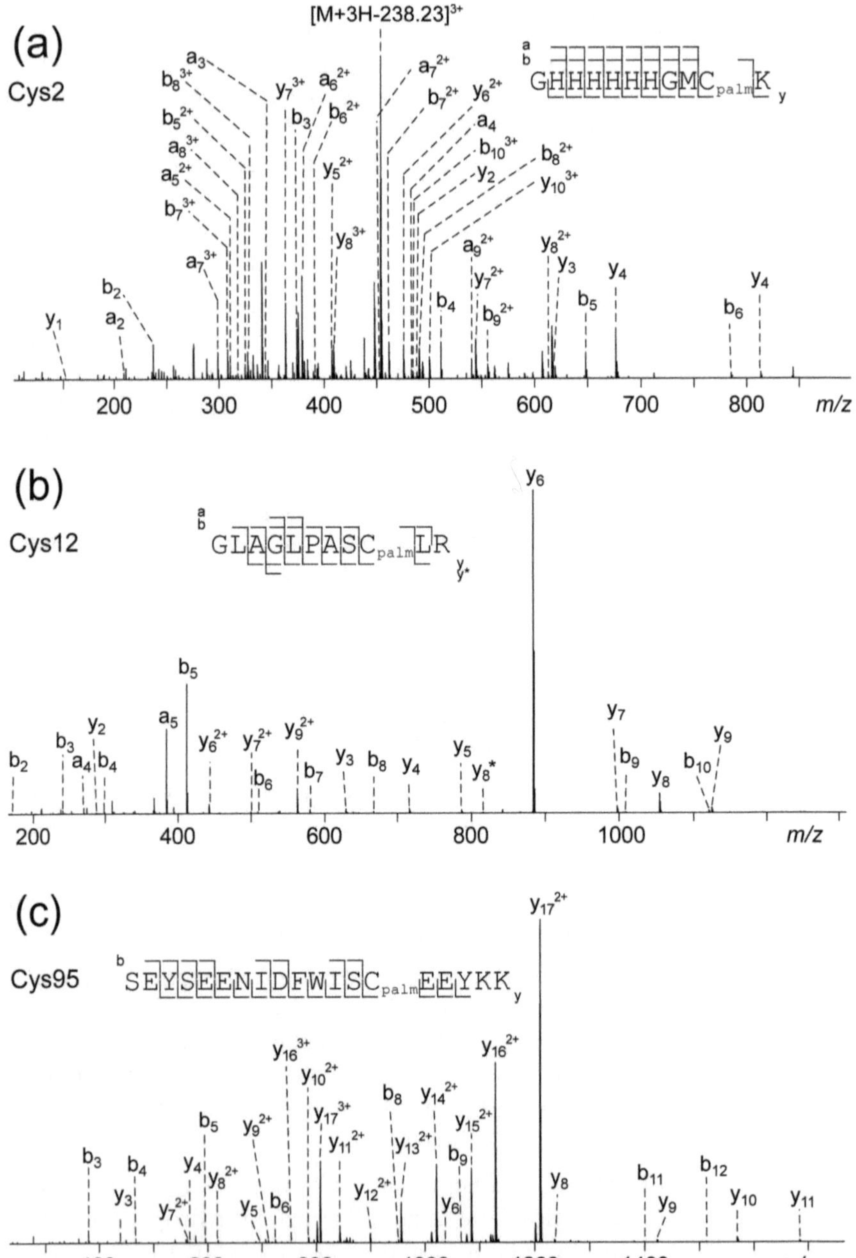

Fig. 1 The CID spectra of GHHHHHHMC$_{palm}$K (**a**), GLAGLPASC$_{palm}$LR (**b**), and SEYSEENIDFWISC$_{palm}$EEYKK (**c**) acquired during an LC-MS/MS analysis of the RGS4-palm-std tryptic digest

and ^{82}SEYSEENIDFWISC$_{palm}$EEYKK^{100}. These three peptides were also detected in His-RGS4 proteins palmitoylated in cells.

16. Although palmitoyl loss does occur in CID, here a sufficient number of fragments with the palmitoyl group still attached are observed, allowing unambiguous determination of the palmitoylation sites. If palmitoyl loss prevents accurate localization of the modification site, alternative fragmentation methods, such as ECD and electron transfer dissociation (ETD), should be used [12].

Acknowledgments

This work is supported by NIH grants P41 RR10999/GM104603, S10 RR020946, and NIH/NHLBI contract HHSN268201000031C. We thank Prof. Elliot M. Ross at the University of Texas Southwestern Medical Center for providing the RGS4 baculovirus. We thank Dr. Minjing Liu for her assistance with sample preparation.

References

1. Resh MD (1999) Fatty acylation of proteins: new insights into membrane targeting of myristoylated and palmitoylated proteins. Biochim Biophys Acta 1451(1):1–16
2. Smotrys JE, Linder ME (2004) Palmitoylation of intracellular signaling proteins: regulation and function. Annu Rev Biochem 73(1):559–587
3. Linder ME, Deschenes RJ (2007) Palmitoylation: policing protein stability and traffic. Nat Rev Mol Cell Biol 8(1):74
4. Resh MD (2016) Fatty acylation of proteins: the long and the short of it. Prog Lipid Res 63:120–131
5. Washburn MP, Wolters D, Yates JR III (2001) Large-scale analysis of the yeast proteome by multidimensional protein identification technology. Nat Biotechnol 19(3):242
6. Aebersold R, Mann M (2003) Mass spectrometry-based proteomics. Nature 422 (6928):198
7. Mann M, Jensen ON (2003) Proteomic analysis of post-translational modifications. Nat Biotechnol 21(3):255
8. Meng F, Forbes AJ, Miller LM, Kelleher NL (2005) Detection and localization of protein modifications by high resolution tandem mass spectrometry. Mass Spectrom Rev 24 (2):126–134
9. Jedrzejewski PT, Lehmann WD (1997) Detection of modified peptides in enzymatic digests by capillary liquid chromatography/electrospray mass spectrometry and a programmable skimmer CID acquisition routine. Anal Chem 69(3):294–301
10. Hoffman MD, Kast J (2006) Mass spectrometric characterization of lipid-modified peptides for the analysis of acylated proteins. J Mass Spectrom 41(2):229–241
11. Guan Z (2002) Identification and localization of the fatty acid modification in ghrelin by electron capture dissociation. J Am Soc Mass Spectrom 13(12):1443–1447
12. Ji Y, Leymarie N, Haeussler DJ, Bachschmid MM, Costello CE, Lin C (2013) Direct detection of *S*-palmitoylation by mass spectrometry. Anal Chem 85(24):11952–11959
13. Drisdel RC, Green WN (2004) Labeling and quantifying sites of protein palmitoylation. Biotechniques 36(2):276–285
14. Roth AF, Wan J, Bailey AO, Sun B, Kuchar JA, Green WN, Phinney BS, Yates JR, Davis NG (2006) Global analysis of protein palmitoylation in yeast. Cell 125(5):1003–1013
15. Wan J, Roth AF, Bailey AO, Davis NG (2007) Palmitoylated proteins: purification and identification. Nat Protoc 2(7):1573

16. Kostiuk MA, Corvi MM, Keller BO, Plummer G, Prescher JA, Hangauer MJ, Bertozzi CR, Rajaiah G, Falck JR, Berthiaume LG (2008) Identification of palmitoylated mitochondrial proteins using a bio-orthogonal azido-palmitate analogue. FASEB J 22(3):721–732
17. Martin BR, Cravatt BF (2009) Large-scale profiling of protein palmitoylation in mammalian cells. Nat Methods 6(2):135
18. Martin BR, Wang C, Adibekian A, Tully SE, Cravatt BF (2012) Global profiling of dynamic protein palmitoylation. Nat Methods 9(1):84
19. Jones ML, Collins MO, Goulding D, Choudhary JS, Rayner JC (2012) Analysis of protein palmitoylation reveals a pervasive role in plasmodium development and pathogenesis. Cell Host Microbe 12(2):246–258
20. Yount JS, Charron G, Hang HC (2012) Bioorthogonal proteomics of 15-hexadecynyloxyacetic acid chemical reporter reveals preferential targeting of fatty acid modified proteins and biosynthetic enzymes. Bioorg Med Chem 20(2):650–654
21. Liang X, Nazarian A, Erdjument-Bromage H, Bornmann W, Tempst P, Resh MD (2001) Heterogeneous fatty acylation of Src family kinases with polyunsaturated fatty acids regulates raft localization and signal transduction. J Biol Chem 276(33):30987–30994
22. Wilson JP, Raghavan AS, Yang Y-Y, Charron G, Hang HC (2011) Proteomic analysis of fatty-acylated proteins in mammalian cells with chemical reporters reveals S-acylation of histone H3 variants. Mol Cell Proteomics 10(3): M110. 001198
23. Ji Y, Bachschmid MM, Costello CE, Lin C (2016) *S*-to *N*-Palmitoyl transfer during proteomic sample preparation. J Am Soc Mass Spectrom 27(4):677–685
24. Tu Y, Popov S, Slaughter C, Ross EM (1999) Palmitoylation of a conserved cysteine in the regulator of G protein signaling (RGS) domain modulates the GTPase-activating activity of RGS4 and RGS10. J Biol Chem 274 (53):38260–38267
25. Ji Y, Liu M, Bachschmid MM, Costello CE, Lin C (2015) Surfactant-induced artifacts during proteomic sample preparation. Anal Chem 87(11):5500–5504. https://doi.org/10.1021/acs.analchem.5b00249
26. Thingholm TE, Larsen MR (2009) The use of titanium dioxide micro-columns to selectively isolate phosphopeptides from proteolytic digests. In: Md G (ed) Phospho-proteomics: methods and protocols. Humana Press, Totowa, NJ, pp 57–66. https://doi.org/10.1007/978-1-60327-834-8_5

Chapter 6

Enrichment of *S*-Palmitoylated Proteins for Mass Spectrometry Analysis

Melanie Cheung See Kit and Brent R. Martin

Abstract

As the 10-year anniversary of their first introduction approaches, alkynyl fatty acids have revolutionized the analysis of *S*-palmitoylation dynamics, acting as functional mimics incorporated into native modification sites in cultured cells. The alkyne functional group provides a robust handle for bioorthogonal Cu(I)-catalyzed azide–alkyne cycloaddition (CuAAC) to reporter-linked azides, forming a stable conjugate for enrichment for mass spectrometry analysis or in-gel fluorescence. Importantly, metabolic labeling enables time-dependent analysis of *S*-palmitoylation dynamics, which can be used to profile incorporation and turnover rates across the proteome. Here we present a protocol for cell labeling, click chemistry conjugation, enrichment, and isobaric tandem mass tag labeling for quantitative mass spectrometry analysis of protein *S*-palmitoylation.

Key words Posttranslational modification, *S*-palmitoylation, Click chemistry, Metabolic labeling, Mass spectrometry, Affinity purification

1 Introduction

Protein *S*-palmitoylation describes the long chain fatty acylation modification of cysteine residues in proteins [1, 2]. zDHHC family protein acyltransferases catalyze acyl group transfer from long chain fatty acyl-CoAs to specific cysteine residues in protein substrates, which can then undergo spontaneous hydrolysis or enzyme-catalyzed depalmitoylation by protein depalmitoylases [3]. Hundreds of membrane-associated proteins require *S*-palmitoylation for proper localization, trafficking, and function. Despite the prevalence of *S*-palmitoylation, proteome-wide enrichment and analysis was confounded for decades by a lack of specific affinity reagents. The classic method for studying *S*-palmitoylation required metabolic addition of ^{3}[H]-labeled fatty acids analogs, immunoprecipitation of select targets, and lengthy exposure times. The introduction of hydroxylamine-switch methods (acyl-biotin exchange [4], acyl-RAC [5], etc.) provided the first approach

Maurine E. Linder (ed.), *Protein Lipidation: Methods and Protocols*, Methods in Molecular Biology, vol. 2009,
https://doi.org/10.1007/978-1-4939-9532-5_6, © Springer Science+Business Media, LLC, part of Springer Nature 2019

to enrich and annotate *S*-acylation, demonstrating the widespread modification of hundreds of proteins in protozoa [6], single cell eukaryotes [7], and mammals [8, 9]. While this method is especially useful for analysis of primary tissues, hydroxylamine-switch methods only measure the steady-state levels of *S*-palmitoylation and thus cannot provide any information on incorporation dynamics or turnover rates.

Alternatively, metabolic labeling with alkynyl fatty acid analogs has greatly simplified analysis of *S*-palmitoylation dynamics [10]. The commercially available alkynyl fatty acid analog 17-octadecynoic acid (17-ODYA) is metabolically incorporated into endogenous *S*-palmitoylation sites, enabling Cu(I)-catalyzed azide–alkyne cycloaddition (click chemistry) to fluorophore-linked azides for in-gel fluorescence or to biotin-azide for enrichment and mass spectrometry analysis [11]. This approach enables simple analysis of probe incorporation rates, as well as pulse–chase analysis of *S*-palmitoylation turnover dynamics [12, 13]. When coupled to quantitative mass spectrometry methods, 17-ODYA labeling and enrichment has identified *S*-palmitoylated proteins with accelerated turnover kinetics, including proteins involved in cell polarity and growth [13, 14]. Through these methods, it is now experimentally feasible to broadly profile *S*-palmitoylation dynamics in cultured yeast [15], protozoa [6, 16], and mammalian cells [1, 10, 11]. Here we present a workflow for multiplexed analysis of *S*-palmitoylation leveraging isobaric TMT quantitation by mass spectrometry. This protocol assumes collaboration with a mass spectrometry facility, able to perform the experiment and assist in any data analysis.

2 Materials

Use ultrapure water (prepared by purifying deionized water, to attain a sensitivity of 18 MΩ-cm at 25 °C) and analytical grade reagents. Use phosphate buffers throughout the protocol. Do not use Tris buffers or EDTA prior to performing the click reaction, since these reagents inhibit the copper(I)-catalyzed click chemistry reaction. Discard waste materials following the appropriate waste disposal regulations.

2.1 Equipment and Supplies

1. Cell culture facility including CO_2 incubator and biosafety cabinets cell culture hoods.
2. Branson Ultrasonics Sonifier S-450A Analog Ultrasonic Cell Disruptor/Homogenizer with double stepped tip (64 to 247 μm amplitude).
3. Beckman Optima L-70 ultracentrifuge with Sorvall F50L-24 × 1.5 mL microrotor.

4. Absorbance spectrophotometer or plate reader to measure 700 nm absorbance measurements for protein quantitation.
5. 15 mL conical centrifuge tubes, 2 mL microcentrifuge tubes, 1.5 mL thick-walled microcentrifuge tubes (Beckman), 1.5 mL screw cap microcentrifuge tubes, mass spectrometry autosampler vials.
6. SDS-PAGE gels and electrophoresis equipment, such as Bio-Rad Mini-PROTEAN tetra cell vertical electrophoresis system with precast 10% SDS-PAGE gels, 4× SDS loading buffer supplemented with β-mercaptoethanol, SDS-PAGE running buffer, and prestained protein molecular weight ladder.
7. Eppendorf 5810R tabletop centrifuge or equivalent with swinging bucket rotor for 15 mL tubes.
8. Promega Vac-Man Vacuum Manifold.
9. Poly-Prep chromatography column, 10 mL volume (Bio-Rad).
10. Oasis HLB μElution plate (2 mg sorbent per well, 30 μm particle size; Waters) and vacuum manifold.
11. Amersham Typhoon 5 Biomolecular Imager for in-gel fluorescence analysis.
12. Thermo Savant SPD1010 SpeedVac Concentrator System.

2.2 Chemicals and Reagents

1. 20 mM 17-ODYA (1000× stock): 5.61 mg/mL in DMSO (Cayman Chemical). Store aliquots at −80 °C.
2. ODYA labeling medium: culture medium, dialyzed serum, any required supplements, and 20 μM 17-ODYA (1×). Sonicate the medium if necessary to dissolve 17-ODYA.
3. 20 mM palmitic acid (1000× stock): 5.13 mg/mL in DMSO. Sonicate the medium if necessary to dissolve the palmitic acid.
4. Palmitic acid labeling medium: culture medium, dialyzed serum, any required supplements, and 20 μM palmitic acid (1×).
5. Lysis Buffer: Dulbecco's phosphate buffered saline (DPBS), 20 μM hexadecylfluorophosphonate (HDFP) (*see* **Note 1**).
6. DC Protein Assay Reagent (Bio-Rad).
7. TBTA solution (1×): Prepare 83 mM Tris((1-benzyl-1H-1,2,3-triazol-4-yl)methyl)amine ligand (TBTA; Click Chemistry Tools) 50× stock solution (44.25 mg/mL) in DMSO. Store at −80 °C for long-term storage. Dilute the 50× stock solution in DMSO and t-butanol following a 1:9:40 ratio respectively. Store at room temperature for extended periods.
8. TCEP solution: 50 mM stock solution (14.35 mg/mL) tris (2-carboxyethyl)phosphine (TCEP-HCl-Sigma) in DPBS and

adjusted to pH 7 using concentrated base (NaOH or KOH). Prepare fresh for each use.

9. Copper sulfate solution: 50 mM stock solution (7.98 mg/mL) Copper (II) sulfate anhydrous ($CuSO_4$) solution in water. Store at room temperature for extended periods.
10. Biotin-PEG3-azide solution: 10 mM stock (4.44 mg/mL) (Click Chemistry) in DMSO, and store at −80 °C.
11. TAMRA-azide solution: 1 mM stock (0.63 mg/mL) (Click Chemistry Tools) in DMSO. Store at −80 °C.
12. Resuspension buffer: 6 M Urea (360 mg/mL) and 2% w/v (20 mg/mL) sodium dodecyl sulfate (SDS), adjusting pH to 7 with HCl. Make fresh before use (*see* **Note 2**).
13. Reducing solution: 200 mM (57.4 mg/mL) TCEP-HCl stock solution. Reconstitute with equimolar potassium hydroxide to neutralize HCl. Confirm with pH indicator strips. Make fresh before use.
14. Alkylating solution: 400 mM (74.0 mg/mL) Iodoacetamide (Sigma) stock solution in water. Make fresh before use. Do not store in direct light.
15. Hydroxylamine solution: 5 M (347 mg/mL) Hydroxylamine hydrochloride (Sigma), neutralized to pH 7 with concentrated NaOH or KOH. Make fresh before use.
16. Chloroform (stored in light-resistant bottle).
17. Acetonitrile.
18. Methanol.
19. Streptavidin–agarose beads (Pierce).
20. Wash buffer A: for streptavidin beads: DPBS with 2 M (120 mg/mL) Urea, 0.2% w/v (2 mg/mL) SDS.
21. Wash buffer B: 2 M urea in DPBS, pH 7.
22. TMTsixplex Isobaric Label Reagent Set (Thermo Fisher).
23. TEAB buffer: 50 mM triethylammonium bicarbonate (TEAB, from 1 M stock solution- Thermo Fisher).
24. Mass spectrometry-grade Trypsin/Lys C (Promega). Reconstitute 20 μg in 100 μL resuspension buffer provided.
25. Acetonitrile (MS grade).
26. Equilibration solution: 0.1% trifluoroacetic acid in water (MS grade).
27. Elution buffer: 70% acetonitrile in water (MS grade).

3 Methods

This procedure performs optimally when carried out in a single day, without freeze–thaw steps. In order to minimize keratin contamination, work in a clean area and frequently change gloves throughout the procedure. Be sure to prepare a sufficient number of replicates ($N = 3$ or more) per condition for confident analysis. Here we describe a comparison between palmitic acid and 17-ODYA treated cells.

3.1 Metabolic Cell Labeling with 17-ODYA

1. Grow cells to desired density in standard culture medium. Plan for at least three biological replicates per condition; depending on the cell type, 15 cm dishes should provide enough protein for one biological replicate.
2. Wash cells gently with warm DPBS (37 °C) and add ODYA Labeling Medium or Palmitic Acid Labeling Medium for the desired time (*see* **Note 3**).
3. Wash cells three times with cold DPBS and harvesting using a cell scraper into a 1.5 mL microcentrifuge tube.
4. Pellet cells by low-speed centrifugation (500 × *g*) for 2 min at 4 °C. Aspirate DPBS and store the cell pellet at −80 °C (stable for several months).

3.2 Preparation of Cell Lysates

1. Add 0.5–1 mL Lysis Buffer per 15 cm plate of confluent cells in the pellet. Sonicate pellet on ice for 2 × 10 s.
2. Transfer cell lysate to a thick-walled ultracentrifuge tube. Balance tube pairs and centrifuge for 45 min at 100,000 × *g* at 4 °C.
3. Carefully remove supernatant (S100) and add 0.5 mL lysis buffer to the pellet (P100). Sonicate the P100 sample 1 × 10 s. The S100 fraction can be saved or discarded since it is not used in the remaining purification (*see* **Note 4**).
4. Quantify protein concentration using the Lowry assay (DC Protein Assay) or other equivalent method. Normalize each sample to 2 mg/mL protein using Lysis Buffer. Aliquot 50 μg of P100 lysate to a 1.5 mL microcentrifuge tube for later in-gel fluorescence analysis.
5. Add 1–2 mg of the P100 sample in 1 mL of Lysis Buffer and split evenly across 2 × 1.5 mL microcentrifuge tubes (0.5 mL each).
6. Add 0.5 mL methanol and 0.188 mL chloroform and vortex. Centrifuge at max speed for 10 min at 4 °C. Carefully remove the aqueous and organic layers to isolate the white protein layer. Add 0.5 mL cold methanol to each tube and briefly sonicate. Recombine the samples and centrifuge at 8000 × *g* for 5 min at 4 °C. Carefully aspirate methanol and add 1 mL DPBS. Sonicate to create a disperse precipitate (*see* **Note 5**).

3.3 Click Reaction for In-Gel Fluorescence

1. Add 1 μL TAMRA-azide solution, 1 μL copper sulfate solution, 1 μL TCEP solution (50 mM stock) and 3 μL TBTA solution to the 50 μL aliquot of 1 mg/mL P100 lysate in DPBS that was set aside. Vortex and let sample sit for 30 min at room temperature. Vortex again and wait for another 30 min.
2. Add reducing 4× SDS sample loading buffer (supplemented with β-mercaptoethanol) to both samples. Avoid boiling to prevent thiol exchange or thioester hydrolysis. Analyze samples using standard SDS-PAGE. When the separation is complete, image the gel using a laser scanning flatbed fluorescence imager to detect TAMRA fluorescence (with approximate excitation and emission wavelengths at 546 and 579 nm respectively).

3.4 Click Reaction and Enrichment

1. Using 1–1.5 mg of P100 lysate, add 58 μL of the Biotin-PEG3-azide solution, 23 μL of the TCEP solution, 23 μL of the copper sulfate solution, and 70 μL of the TBTA solution in 1 mL total volume (*see* **Note 6**). Rotate end over end for 1 h at room temperature.
2. Optional: For additional experimental controls for thioester-dependent labeling, to a separate series of replicates, add 1 M hydroxylamine (pH 7) and heat to 65 °C for 15 min.
3. Centrifuge the sample >8000 × *g* for 10 min at 4 °C. Decant and discard the supernatant. Add 0.5 mL cold DPBS, 0.5 mL cold methanol and sonicate a few seconds to disperse the protein pellet. Add 0.187 mL chloroform, vortex, and centrifuge >8000 × *g* for 10 min. Discard the aqueous and organic layers, saving the insoluble protein interface. Add 1 mL cold methanol, sonicate, and centrifuge 5 min at >8000 × *g*, and repeat. After decanting the residual methanol, add 0.5 mL Resuspension Buffer. Briefly vortex and sonicate if necessary to dissolve the protein pellet.
4. Add 25 μL of the TCEP reducing solution (pH 7), vortex and leave at room temperature to react for 20 min. Then add 25 μL of the iodoacetamide alkylation solution, mix and leave at room temperature for 20 min in the dark.
5. Quantify protein concentration using the Lowry assay (DC Protein Assay) or other equivalent method. Normalize each sample to a common protein concentration (*see* **Note 7**).
6. Transfer the sample to a 15 mL capped conical vial and dilute with 5 mL DPBS. Transfer 75 μL streptavidin–agarose bead slurry to a fresh 15 mL conical vial and wash twice with Wash Buffer A. Add to sample and rotate end over end for 90 min.
7. Place samples in a table top centrifuge and pellet beads at 500 × *g* for 3 min. Decant the supernatant. Add 10 mL of Wash Buffer A, turn vials upside down several times, and centrifuge at 500 × *g* for 3 min. Decant the supernatant and transfer beads in DBPS to individual Poly-Prep chromatography columns.

8. Attach each chromatography column to a multiport vacuum manifold, and add 3 × 10 mL of Wash Buffer 1, followed by 3 × 10 mL Wash Buffer B, each time allowing the buffer to nearly dry the column bed. Alternatively, the streptavidin beads can be sequentially washed with four cycles of centrifugation (500 × *g* × 3 min) in 10 mL Wash Buffer A and four cycles of centrifugation with 10 mL Wash Buffer B, carefully decanting the supernatant after each wash.

3.5 Digestion and Desalting

1. Transfer beads to screw cap microcentrifuge tubes in Wash Buffer B. Carefully remove any residual Wash Buffer and add 100 μL TEAB Buffer, supplemented with 6 μL Trypsin/LysC solution. Shake at 225 rpm at 37 °C for at least 4 h or overnight (*see* **Note 8**).
2. The next day, centrifuge samples at 500 × *g* for 1 min. Transfer the supernatant to a new microcentrifuge tube. Wash the beads with an additional 50 μL TEAB buffer and centrifuge again. Collect and combine the supernatants.
3. Resuspend each TMT label reagent (0.2 mg) in 50 μL anhydrous acetonitrile and mix for 5 min. (*see* **Note 9**). Add 45 μL of each TMT label reagent to separate samples (i.e., palmitic acid control and 17-ODYA labeled; $N = 3$) and leave for 1 h to incubate at room temperature.
4. Add 1.2 μL of 5 M Hydroxylamine (pH 7) to each sample and wait for 15 min to allow reaction to quench. Add 1.5 μL of 50% trifluoroacetic acid for a final concentration of about 0.8%. Combine labeled samples in one tube before desalting peptides.
5. Desalt using an Oasis PRIME HLB μElution Plates (Waters), or equivalent solid phase extraction column. Condition resin with 200 μL acetonitrile, followed by 200 μL equilibration buffer. Load peptides onto the sorbent at low speed. Wash three times with 400 μL equilibration buffer, followed by another three washes with 400 μL water to remove salts. Elute peptides with 75 μL of elution buffer (70% acetonitrile in water). Add an additional 75 μL elution buffer and pool the eluents.
6. Dry the desalted elutions using a SpeedVac (Thermo), transfer peptides to mass spectrometry autosampler vials and store at −80 °C.
7. Before mass spectrometry analysis, resuspend samples in MS-grade water containing 3% acetonitrile and 0.1% formic acid.
8. Contact an appropriate mass spectrometry core facility or collaborator for mass spectrometry analysis, preferably using a quadrupole Orbitrap (Q-Exactive series) or Tribrid Orbitrap (Fusion series) instrument.

4 Notes

1. HDFP is not commercially available, but can be substituted with other acyl fluorophosphonates, such as methyl arachidonyl fluorophosphonate. Hexadecylsulfonyl fluoride can also be used to block depalmitoylase activity.
2. It is important to measure and neutralize the pH of all buffers, since basic conditions can promote thioester hydrolysis. Be aware, 6 M urea can shift the solution to a more basic pH, which should be equilibrated to neutral pH before use.
3. The labeling time for steady-state analysis typically requires four or more hours. Shorter time points can be used to measure differences in labeling rates between samples. Pulse–chase experiments require sufficient time for probe metabolic incorporation (>90 min) prior to addition of excess palmitic acid (chase). Different cell lines have different labeling kinetics, so it is important to use in-gel fluorescence analysis to confirm labeling conditions. Also, cells labeled with palmitic acid provide a control for nonspecific enrichment from streptavidin beads.
4. The S100 fraction has no detectable 17-ODYA labeling in-gel fluorescence analysis, and thus does not add any additional information to the mass spectrometry analysis. It is not essential to perform this fractionation step, but it concentrates the membrane proteome, removes soluble thioesterase enzymes, and limits nonspecific enrichment of common false positives.
5. The initial chloroform–methanol precipitation is critical to increase reproducibility by removing unincorporated probe prior to click chemistry conjugation. In addition, the efficiency of the click chemistry reaction is not affected when the protein is present as an insoluble slurry.
6. Typical click chemistry reactions for proteomics analysis use 100 μM biotin-azide. We find that increasing the biotin-azide concentration fivefold significantly improves the reaction efficiency with alkynyl fatty acid conjugated proteins. We have not explored other commercially available copper ligands, which may further improve the reaction efficiency.
7. Following click chemistry, the chloroform–methanol precipitation and methanol washes can lead to variable levels of protein loss. It is critical to include this additional protein normalization for TMT-based mass spectrometry quantitation.
8. Agitation is used to maintain the bead slurry during digestion. The length of the digestion has not been thoroughly optimized, but longer incubations may affect thioester stability (which does not affect the analysis of tryptic peptides).

9. The TMT reagents can be diluted severalfold and used over multiple separate experiments. Experiments should be designed to match experimental groups ($N = 3$) within the single multiplexed mass spectrometry experiment.

Acknowledgments

Support for this work was provided by the National Institutes of Health grant DP2 GM114848 and the University of Michigan.

References

1. Tom CT, Martin BR (2013) Fat chance! Getting a grip on a slippery modification. ACS Chem Biol 8:46–57
2. Jiang H, Zhang X, Chen X, Aramsangtienchai P, Tong Z, Lin H (2018) Protein lipidation: occurrence, mechanisms, biological functions, and enabling technologies. Chem Rev 118:919–988
3. Won SJ, Cheung See Kit M, Martin BR (2018) Protein depalmitoylases. Crit Rev Biochem Mol Biol 53:83–98
4. Wan J, Roth AF, Bailey AO, Davis NG (2007) Palmitoylated proteins: purification and identification. Nat Protoc 2:1573–1584
5. Forrester MT, Hess DT, Thompson JW, Hultman R, Moseley MA, Stamler JS, Casey PJ (2011) Site-specific analysis of protein S-acylation by resin-assisted capture. J Lipid Res 52:393–398
6. Jones ML, Collins MO, Goulding D, Choudhary JS, Rayner JC (2012) Analysis of protein palmitoylation reveals a pervasive role in Plasmodium development and pathogenesis. Cell Host Microbe 12:246–258
7. Roth AF, Wan J, Bailey AO, Sun B, Kuchar JA, Green WN, Phinney BS, Yates JR 3rd, Davis NG (2006) Global analysis of protein palmitoylation in yeast. Cell 125:1003–1013
8. Kang R, Wan J, Arstikaitis P, Takahashi H, Huang K, Bailey AO, Thompson JX, Roth AF, Drisdel RC, Mastro R, Green WN, Yates JR 3rd, Davis NG, El-Husseini A (2008) Neural palmitoyl-proteomics reveals dynamic synaptic palmitoylation. Nature 456:904–909
9. Collins MO, Woodley KT, Choudhary JS (2017) Global, site-specific analysis of neuronal protein S-acylation. Sci Rep 7:4683
10. Gao X, Hannoush RN (2018) A decade of click chemistry in protein palmitoylation: impact on discovery and new biology. Cell Chem Biol 15:236. https://doi.org/10.1016/j.chembiol.2017.12.002
11. Martin BR, Cravatt BF (2009) Large-scale profiling of protein palmitoylation in mammalian cells. Nat Methods 6:135–138
12. Zhang MM, Tsou LK, Charron G, Raghavan AS, Hang HC (2010) Tandem fluorescence imaging of dynamic S-acylation and protein turnover. Proc Natl Acad Sci U S A 107:8627–8632
13. Martin BR, Wang C, Adibekian A, Tully SE, Cravatt BF (2011) Global profiling of dynamic protein palmitoylation. Nat Methods 9:84–89
14. Hernandez JL, Davda D, Cheung See Kit M, Majmudar JD, Won SJ, Gang M, Pasupuleti SC, Choi AI, Bartkowiak CM, Martin BR (2017) APT2 inhibition restores scribble localization and S-palmitoylation in snail-transformed cells. Cell Chem Biol 24:87–97
15. Zhang MM, Wu PY, Kelly FD, Nurse P, Hang HC (2013) Quantitative control of protein S-palmitoylation regulates meiotic entry in fission yeast. PLoS Biol 11(7):e1001597. https://doi.org/10.1371/journal.pbio.1001597
16. Foe IT, Child MA, Majmudar JD, Krishnamurthy S, van der Linden WA, Ward GE, Martin BR, Bogyo M (2015) Global analysis of palmitoylated proteins in toxoplasma gondii. Cell Host Microbe 18:501–511

Part II

Reversible Fatty Acylation

Chapter 7

Systematic Screening of Depalmitoylating Enzymes and Evaluation of Their Activities by the Acyl-PEGyl Exchange Gel-Shift (APEGS) Assay

Takashi Kanadome, Norihiko Yokoi, Yuko Fukata, and Masaki Fukata

Abstract

Palmitoylation is a reversible posttranslational lipid modification of proteins involved in a wide range of cellular functions. More than a thousand proteins are estimated to be palmitoylated. In neurons, PSD-95, a major postsynaptic scaffold protein, requires palmitoylation for its specific accumulation at the synapse and dynamically cycles between palmitoylated and depalmitoylated states. Although palmitoylating enzymes of PSD-95 have been well characterized, little is known about the depalmitoylating enzymes (e.g., thioesterases for palmitoylated PSD-95). An elegant pharmacological analysis has suggested that subsets of α/β hydrolase domain (ABHD)-containing proteins of the metabolic serine hydrolase superfamily involve thioesterases for palmitoylated proteins. Here, we describe a systematic method to screen the ABHD serine hydrolase genes, which unveiled ABHD17 as the depalmitoylating enzyme for PSD-95. Furthermore, we introduce the acyl-PEGyl exchange gel-shift (APEGS) method that enables quantification of palmitoylation levels/stoichiometries on proteins in various biological samples and can be used to monitor the dynamic depalmitoylation process of proteins.

Key words Posttranslational modification, Palmitoylation, Depalmitoylation, PSD-95, 2-bromopalmitate, Serine hydrolase, ABHD genes, ABHD17, APEGS, Thioesterase

1 Introduction

Posttranslational modifications of proteins such as phosphorylation, ubiquitination, glycosylation, and lipidation play critical roles in regulating the localization, stability, and protein-protein interaction of various proteins. Palmitoylation, one type of lipidation, occurs between specific cysteine residues of proteins and palmitic acid through a labile thioester linkage, and confers hydrophobicity to proteins, which facilitates the interaction of modified proteins with lipid bilayers [1–3].

PSD-95, a postsynaptic scaffold protein that plays a pivotal role in maturation of excitatory synapses [4], is a representative palmitoylated protein in neurons [5]. Palmitoylation of PSD-95 occurs

Maurine E. Linder (ed.), *Protein Lipidation: Methods and Protocols*, Methods in Molecular Biology, vol. 2009, https://doi.org/10.1007/978-1-4939-9532-5_7,

at its N-terminal two cysteine residues [5] and is essential for its specific targeting to the postsynaptic density (PSD) [6], a specialized membrane region at the postsynapse. Importantly, treatment of neurons with 2-bromopalmitate (2-BP), an inhibitor of protein palmitoylation [7], causes the time-dependent decrease of palmitoylated PSD-95 and delocalizes PSD-95 from the PSD [8–10], indicating that PSD-95 undergoes continuous palmitoylation cycles between (re)palmitoylation and depalmitoylation. In 2004, we isolated 23 mouse ZDHHC palmitoylating enzymes and established a systematic screening method for the palmitoyltransferase activity to a specific substrate protein [11, 12]. Among 23 ZDHHC proteins, synaptic ZDHHC2 mediates local palmitoylation of PSD-95 at the synapse [9, 10]. However, depalmitoylating enzymes for PSD-95 have remained unknown although APT/Lypla and PPT were reported as classic depalmitoylating enzymes [13–15].

A global proteomic analysis combined with a lipid hydrolase-selective inhibitor, hexadecylfluorophosphonate (HDFP) confirmed the enzyme-mediated depalmitoylation on several palmitoyl proteins such as H-Ras, N-Ras, and MPP6 [16]. Because serine hydrolases targeted by HDFP (e.g., APT/Lypla and PPT) mostly possess the α/β hydrolase domain (ABHD) as a catalytic domain, it was speculated that some HDFP-sensitive serine hydrolases and the related ABHD proteins mediate depalmitoylation. To identify the physiological depalmitoylating enzymes for particular substrates such as PSD-95, we isolated mouse or rat ABHD proteins (Fig. 1a) and systematically assessed their depalmitoylating activity to PSD-95 by the metabolic labeling method with [^{3}H]palmitic acid in transfected HEK293T or COS7 cells (Fig. 1b). We found that ABHD17 members (17A, 17B, and 17C) robustly reduced PSD-95 palmitoylation levels and showed the strongest activity among those tested [17].

To further examine the relevance of ABHD17 as the PSD-95 depalmitoylating enzyme, we developed a method named acyl-PEGyl exchange gel-shift (APEGS) [17] (also known as PEG switch [18] or acyl-PEG exchange (APE) [19, 20]) for specific labeling of palmitoylated proteins obtained from cells with defined molecular mass tags, the polyethylene glycol (PEG) polymers. The APEGS assay consists of four chemical reactions: (1) break of disulfide bonds with a reducing reagent, tris(2-carboxyethyl) phosphine (TCEP), (2) blockade of free, nonmodified cysteine thiols with *N*-ethylmaleimide (NEM), (3) specific cleavage of palmitoyl-cysteine thioester linkages with hydroxylamine (NH_2OH), and (4) labeling of newly exposed cysteine thiols with the maleimide-conjugated PEG (mPEG) of defined molecular mass (e.g., 5 k or 2 k), which causes the mobility shift of palmitoylated (now PEGylated) proteins on SDS-PAGE (Fig. 2a). Western

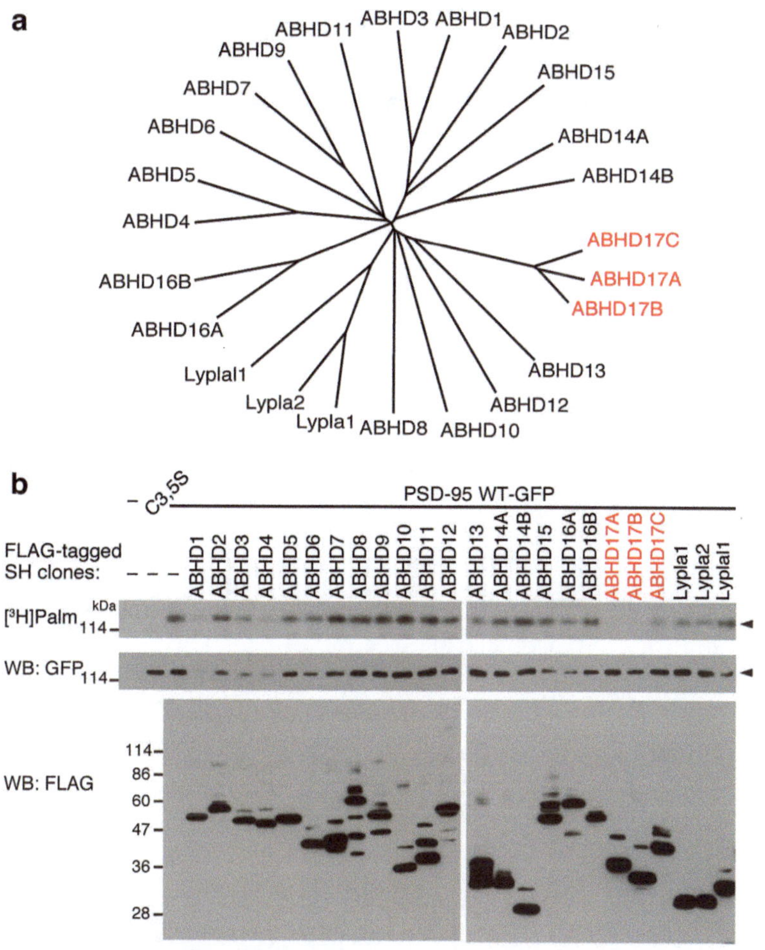

Fig. 1 ABHD17 proteins are depalmitoylating enzymes for PSD-95. (**a**) The phylogenetic tree of the mouse ABHD protein family. (**b**) Screening of the serine hydrolase library for PSD-95 depalmitoylating enzymes. HEK293T cells cotransfected with PSD-95-GFP and FLAG-tagged serine hydrolase (SH) clones were metabolically labeled with [^{3}H]palmitic acid for 4 h. The labeled cells were lysed and subjected to SDS-PAGE, followed by fluorography (upper panel) and Western blot analysis using the indicated antibodies. Arrowheads indicate the position of PSD-95-GFP. Reproduced from ref. [17] with permission from the Society for Neuroscience

blotting with antibodies against proteins of interest visualizes distinct palmitoylated states (i.e., non-, mono-, di-, tri-, and more) of the protein (Fig. 2a), thereby providing the quantitative information about the number of palmitoylated cysteines and their occupancy ratio (i.e., levels/stoichiometries of palmitoylation). This assay detected the mobility shifts of PSD-95 that depend on the number of palmitoylation sites (Fig. 2b), confirming that PSD-95

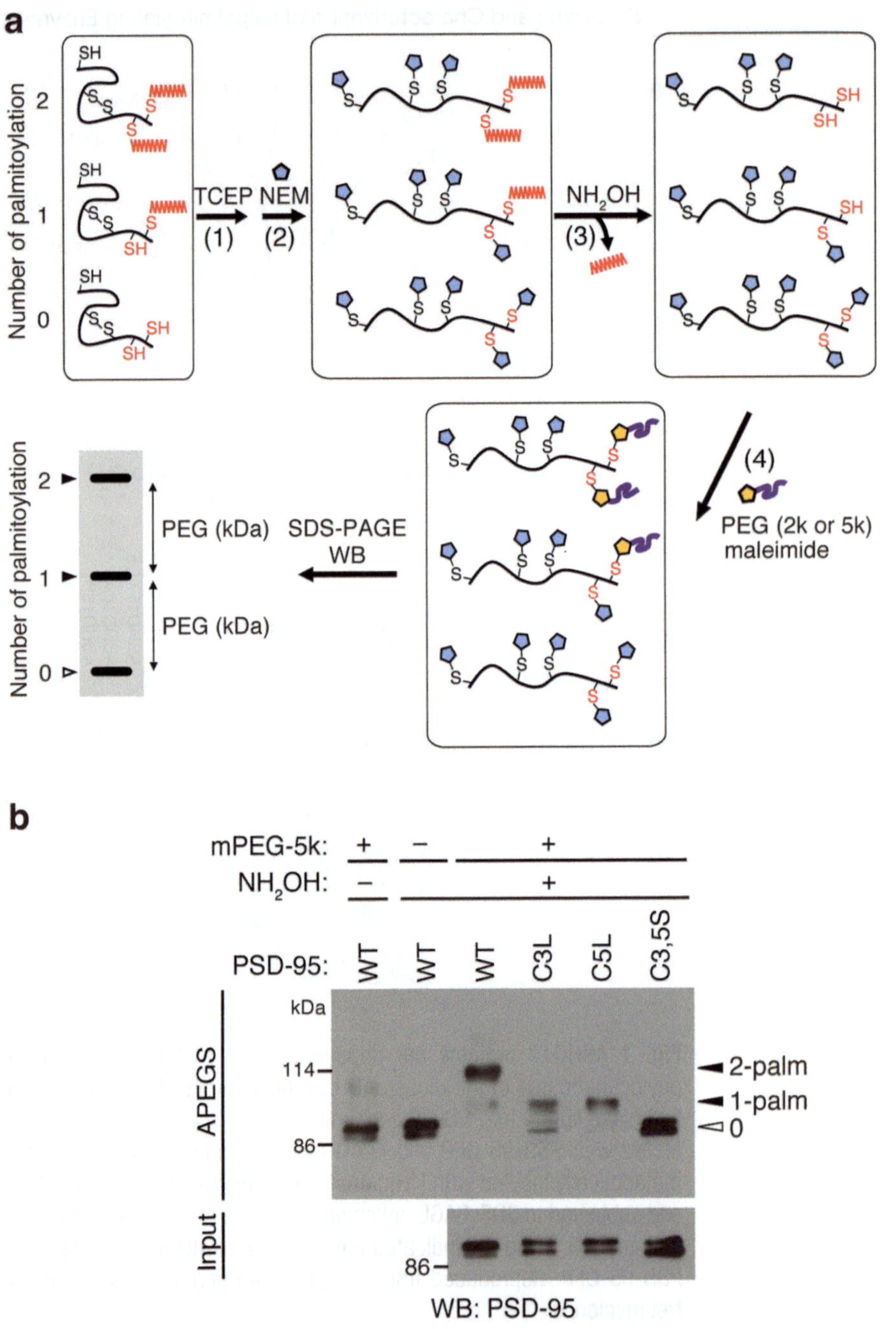

Fig. 2 The APEGS assay to examine palmitoylation stoichiometries of proteins. (**a**) Schematics of the APEGS assay. The APEGS assay consists of four steps: (1) reduction of disulfide bonds by TCEP, (2) blockage of free cysteine residues by NEM, (3) cleavage of palmitoyl-thioester linkages by hydroxylamine (NH_2OH) and (4) labeling of newly exposed cysteine residues with maleimide-PEG (mPEG) (2 or 5 kDa). The subsequent SDS-PAGE and Western blot analysis reveal palmitoylation levels and states of proteins. (**b**) The APEGS assay detects mobility shifts of PSD-95 in a palmitoylation-number dependent manner. HEK293T cells cotransfected with the ZDHHC3 palmitoylating enzyme and either wild type (*WT*), monopalmitoylated mutants (*C3L* and *C5L*) or a nonpalmitoylated mutant (*C3,5S*) of PSD-95 were subjected to the APEGS assay. PSD-95 is palmitoylated fully at two cysteine residues in cells. Reproduced from ref. [17] with permission from the Society for Neuroscience

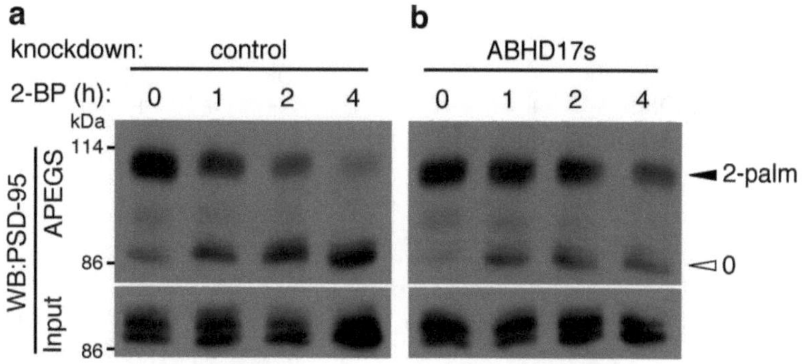

Fig. 3 ABHD17 mediates PSD-95 depalmitoylation in neurons. (**a**) Cultured hippocampal neurons (14 days in vitro) were treated with 2-BP for the indicated periods. The APEGS assay reveals a high level of palmitoylation (80 ± 4% at 0 h) and rapid depalmitoylation kinetics (palmitoylation half-life, 2.4 ± 0.2 h) of PSD-95 in neurons. Closed and open arrowheads represent the positions of palmitoylated and nonpalmitoylated proteins, respectively. (**b**) Knockdown of ABHD17A, B and C increased the PSD-95 palmitoylation level (89 ± 3% at 0 h; $p < 0.01$, compared to control in (**a**) and delayed depalmitoylation kinetics of PSD-95 (4.8 ± 1.3 h; $p < 0.05$, compared to control in (**a**). Reproduced from ref. [17] with permission from the Society for Neuroscience

palmitoylation occurs at maximally two cysteine residues when expressed in HEK293T cells. When this assay was applied to cultured neurons, it was revealed that endogenous PSD-95 is palmitoylated at a very high level (~80%) and exists predominantly in the dipalmitoylated state in neurons (Fig. 3a). Furthermore, dynamic depalmitoylation events on endogenous PSD-95 can be monitored: hippocampal neurons treated with 2-BP showed the time-dependent decrease of the dipalmitoylated fraction and reciprocal increase of the nonpalmitoylated fraction (Fig. 3a). The observed depalmitoylation kinetics was significantly delayed when the expression of ABHD17A, B, and C was knocked down (Fig. 3b).

Finally, the APEGS assay is also applicable to tissue samples for measurement of in vivo palmitoylation stoichiometries, which cannot be measured by current acyl-biotinyl exchange (ABE) chemistry and metabolic labeling methods with [^{3}H]palmitic acid or clickable probes. For example, the APEGS assay using brain tissue extracts enables the profiling of in vivo palmitoylation levels and states of various neuronal proteins: GluA1, GluA2, GluN2A, mGluR5, $G\alpha_q$, and vGluT1 (Fig. 4). The methodology in this chapter enables us to identify physiological depalmitoylating enzymes for various palmitoylated substrates and quantitatively evaluate their activities [17].

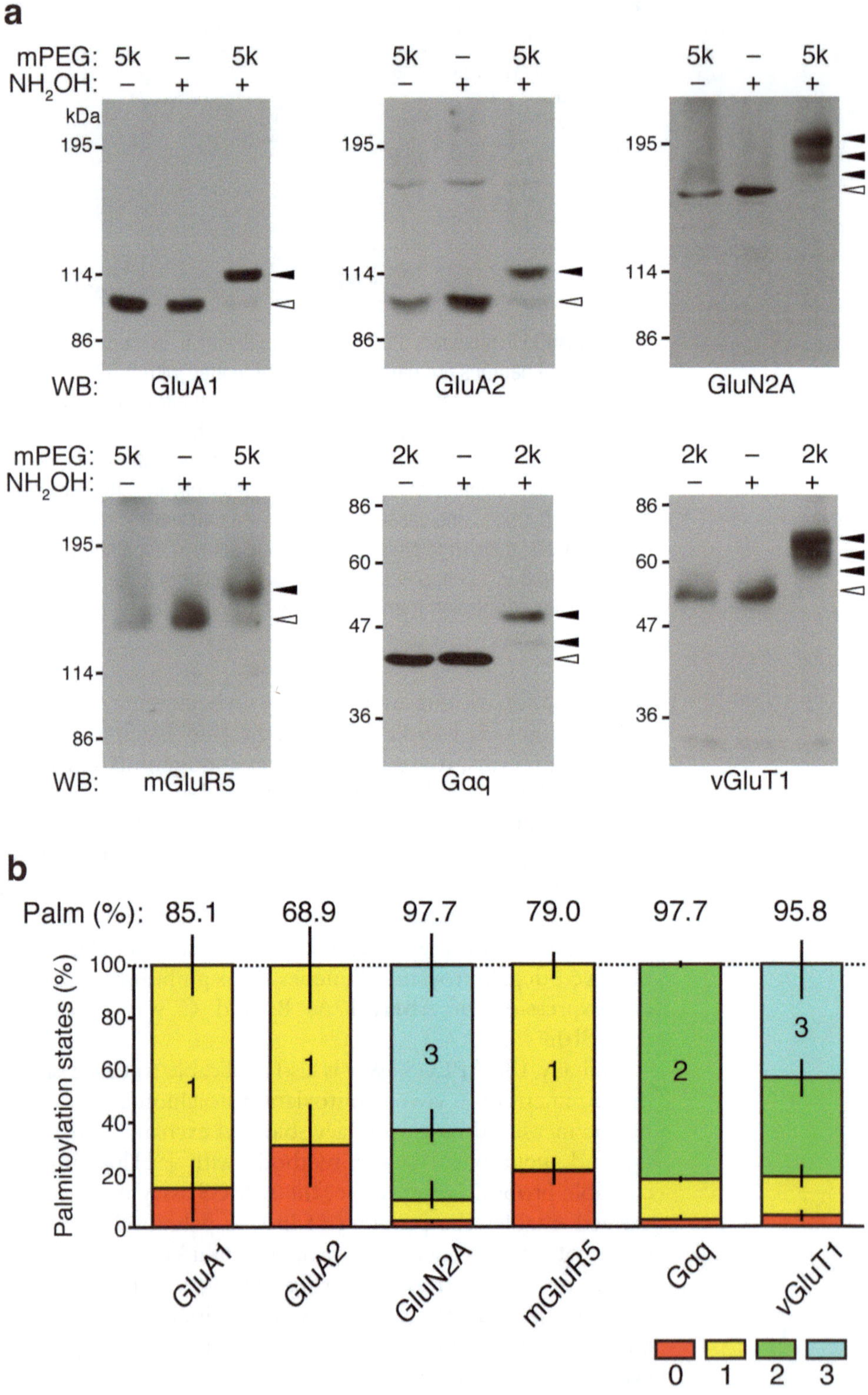

Fig. 4 Profiling of in vivo palmitoylation stoichiometries. (**a**) Adult rat cerebrum extract was subjected to the APEGS assay with mPEG-5k or 2k, SDS-PAGE, and Western blot analysis using the indicated antibodies.

2 Materials

2.1 Cell Culture, Transfection, and Metabolic Labeling with [^{3}H] Palmitic Acid

1. HEK293T cells (or COS7 cells) (ATCC).
2. 10% FBS/DMEM: Dulbecco's modified Eagle's medium (DMEM) supplemented with 10% (v/v) fetal bovine serum (FBS) (*see* **Note 1**).
3. Transfection reagent (*see* **Note 1**).
4. 20% FBS/DMEM: DMEM supplemented with 20% (v/v) FBS.
5. [^{3}H]palmitic acid: Dry [^{3}H]palmitic acid (PerkinElmer, 45 Ci/mmol, 5 mCi/mL in ethanol) under reduced pressure using a concentrator such as SpeedVac and reconstitute it in small volume of 100% ethanol (e.g., ~50 mCi/mL) (*see* **Note 2**). Store at −20 °C.
6. Preincubation medium: Serum-free DMEM supplemented with 5 mg/mL fatty-acid free bovine serum albumin (BSA), filter-sterilized through a vacuum filtration system (0.22 μm).

2.2 SDS-PAGE and Fluorography

1. Phosphate-buffered saline (PBS): 137 mM NaCl, 2.7 mM KCl, 10 mM Na_2HPO_4, 1.8 mM KH_2PO_4 (pH 7.4).
2. SDS-PAGE sample buffer: 62.5 mM Tris–HCl, pH 6.8, 10% (v/v) glycerol, 2% (w/v) SDS, 0.005% (w/v) bromophenol blue (BPB) with 10 mM dithiothreitol (DTT) (for fluorography, *see* Subheading 3.3, **steps 1** and **2**) or 2% (v/v) 2-mercaptoethanol (for Western blot analysis).
3. SDS–polyacrylamide gel: Resolving gels (6–10% nongradient gel, depending on the size of substrate proteins) with the stacking gel (3% acrylamide).
4. Fixing solution: 25% (v/v) isopropanol and 10% (v/v) acetic acid.
5. Amplify fluorographic solution (GE Healthcare) or equivalent: 1 M sodium salicylate and 15% (v/v) methanol.
6. Fuji Medical X-ray film Super RX or equivalent.

2.3 APEGS Assay

1. Buffer A: PBS with 4% (w/v) SDS and 5 mM EDTA.
2. Buffer B: PBS with 4% (w/v) SDS, 5 mM EDTA and 8.9 M Urea.

Fig. 4 (continued) Closed and open arrowheads represent the positions of palmitoylated and nonpalmitoylated proteins, respectively. (**b**) Palmitoylation stoichiometries (Palm %) are determined by measuring the relative band intensities of various palmitoylation states (non-, mono-, di-, tripalmitoylated). Under the condition, 90.1 ± 5.3% of PSD-95 is dipalmitoylated (not shown) ref. [17]

3. Homogenate buffer: 20 mM Tris–HCl, pH 8.0, 2 mM EDTA, 0.32 M sucrose.
4. Protease inhibitors (PIs): phenylmethylsulfonyl fluoride (PMSF, 1000×), 50 mg/mL in ethanol; leupeptin (1000×), 10 mg/mL in ultrapure water; and Pepstatin (1000×), 10 mg/mL in DMSO.
5. Bicinchoninic acid (BCA) protein assay kit.
6. 0.5 M Tris(2-carboxyethyl)phosphine (TCEP) solution, neutral pH (e.g., Bond-Breaker TCEP Solution, Thermo Fisher Scientific) (*see* **Note** 3).
7. 2 M *N*-ethylmaleimide (NEM): Dissolve 250 mg of NEM in approximately 0.8 mL of ethanol and make up to 1 mL with ethanol. Prepare fresh.
8. Methanol.
9. Chloroform, dispensed in a glass bottle.
10. Buffer H: 1.33 M hydroxylamine (NH_2OH), pH 7.0, 5 mM EDTA, 0.2% (w/v) Triton X-100. Dissolve 463 mg of hydroxylamine hydrochloride in 2 mL of ultrapure water and adjust pH to 7.0 with NaOH. Add 0.1 mL of 250 mM EDTA and 0.1 mL of 10% (w/v) Triton X-100. Make up with ultrapure water to 5 mL. Prepare fresh.
11. Buffer T: 1.33 M Tris–HCl, pH 7.0, 5 mM EDTA, 0.2% (w/v) Triton X-100. Dissolve 8.05 g of Tris in approximately 40 mL of ultrapure water and adjust pH to 7.0 with HCl. Add 1 mL of 250 mM EDTA and make up with ultrapure water to 50 mL. This solution can be stored at room temperature. Take 5 mL of the solution and add 0.1 mL of 10% (w/v) Triton X-100.
12. 200 mM (400 mg/mL) maleimide-conjugated PEG-2k (mPEG-2k): Transfer 1 g of mPEG-2k (SUNBRIGHT ME-020MA, NOF corporation) to a 5-mL centrifuge tube and add 1 mL of nitrogen-purged ultrapure water (*see* **Note 4**). Collect all the contents of the tube by centrifugation at 6800 × *g* for 3 min. Bring up to 2.5 mL with nitrogen-purged ultrapure water and mix well by vigorous inversion. After centrifugation at 6800 × *g* for 3 min, store in small aliquots (200 μL in 1.5-mL tube) at −70 °C (*see* **Note 5**).
13. 100 mM mPEG-5k (500 mg/mL): As described in Subheading 2.3, **step 12** for 200 mM mPEG-2k, dissolve 1 g of mPEG-5k (SUNBRIGHT ME-050MA, NOF corporation) in nitrogen-purged ultrapure water and bring up to a final volume of 2 mL.

3 Methods

Carry out all experiments at room temperature (or 25 °C) and use all solutions and microcentrifuges for centrifugation at room temperature, unless otherwise indicated.

3.1 Transfection

1. Seed HEK293T (or COS7) cells (2×10^5 HEK293T cells/well, 12-well plate) and culture in 1 mL of 10% FBS/DMEM without antibiotics at 37 °C for 16–18 h.
2. Transfect using Lipofectamine PLUS reagent according to the manufacturer's instructions as described in **steps 3–8**.
3. Dilute the DNA mixtures (1 μg) in 50 μL of DMEM (or Opti-MEM from Thermo Fisher Scientific). The DNA mixture includes 0.5 μg each of the plasmids for the substrate (e.g., PSD-95-GFP) and the individual serine hydrolase clone [17].
4. Add 3 μL of Lipofectamine PLUS reagent and mix gently. Incubate for 15 min.
5. Combine the diluted DNA mixture with diluted Lipofectamine (2 μL in 50 μL of DMEM or Opti-MEM). Incubate the mixtures for 15 min.
6. Add 400 μL of DMEM or Opti-MEM to each tube containing the complexes. Total volume will be 500 μL.
7. Remove the culture medium from cells and add 500 μL of the complexes onto the cells.
8. Incubate cells at 37 °C for 4 h and then add 500 μL of 20% FBS/DMEM (final concentration of FBS in the culture medium is 10%). Culture the cells for 18–20 h at 37 °C before metabolic labeling.

3.2 Metabolic Labeling

1. Replace the culture medium with 1 mL of the preincubation medium and incubate the cells for 30 min at 37 °C.
2. Replace the preincubation medium with 0.5 mL of the preincubation medium containing 0.25–0.5 mCi/mL [^{3}H]palmitic acid and incubate the cells for 4 h at 37 °C (*see* **Note 2**).

3.3 SDS-PAGE, Fluorography, and Western Blot Analysis

1. Remove the labeling medium and wash the cells with 1 mL of PBS. Add 500 μL of SDS-PAGE sample buffer supplemented with 10 mM DTT to each well of the tissue culture plates (*see* **Note 6**).
2. Pipet the cell suspension up and down in the well to help cell lysis, and then transfer it to a 1.5-mL tube. Heat the sample at 90 °C for 2 min (*see* **Note 7**).
3. For fluorography, resolve proteins by SDS-PAGE. Fix the gel in fixing solution for 30 min, then treat it with fluorographic

solution for 30 min. Dry the gel using a vacuum gel dryer system. Expose the wrapped dried gel to an X-ray film at −70 °C for 48–72 h (without an intensifying screen) and develop the film using a film processer (*see* **Note 8**).

4. For Western blot analysis to verify the expression of the substrate protein and serine hydrolases (*see* **Note 8**), appropriately dilute the aliquot of samples prepared for the fluorography with SDS-PAGE sample buffer containing 2% (v/v) 2-mercaptoethanol. Heat the samples at 100 °C for 5 min. Process the samples for Western blotting. Detect chemiluminescent signals by exposing the membrane to an X-ray film. Alternatively, use a chemiluminescent image analyzer equipped with a cooled CCD camera.

5. Scan the fluorography image and measure the signal intensity of radiolabeled bands corresponding to the substrate protein with ImageJ software. To compare the relative palmitoylation levels across the samples, normalize signal intensities of the radiolabeled substrate to the amount of the substrate protein expressed in each sample detected by Western blotting (*see* **Notes 9** and **10**).

3.4 APEGS Assay

1. Wash cultured cells [e.g., 3 × 10^5 cells/well, 6-well plate (*see* **Note 11**)] with 2–3 mL of PBS twice. Add 1 mL of buffer A with protease inhibitors (PIs). Swirl the plate until the cell layer dissolves and transfer the cell suspension to a 1.5-mL tube by pipetting (for tissue samples, *see* **Note 12**).

2. Ultrasonicate briefly to completely solubilize proteins. Clear the cell lysate by centrifugation at maximum speed in a microcentrifuge for 15 min and transfer the supernatant to a 1.5-mL tube. Remove an aliquot (50 μL) of the supernatant as the input for Western blot analysis.

3. Measure the protein concentration by the BCA protein assay according to the manufacture's protocol. Dilute the cell lysate with buffer A containing PIs in a new 1.5-mL tube to get ~0.4 mg/mL of protein concentration in 500 μL (150–200 μg of protein) (*see* **Note 13**).

4. To break disulfide bonds, add 25 μL of 0.5 M TCEP (final 25 mM) to the 500 μL of sample. Incubate for 1 h at 55 °C in a water bath (*see* **Note 14**).

5. To alkylate and block free cysteine residues, centrifuge briefly and add 12.5 μL of 2 M NEM (final 50 mM). Rotate for 3 h. Centrifuge briefly and transfer the sample to a new 5-mL centrifuge tube made of polypropylene (*see* **Note 15**).

6. To remove excess TCEP and NEM, perform the first methanol–chloroform precipitation as described in **steps 7–12** (*see* **Note 16**).

7. Add 2.0 mL (4 vol.) of methanol and shake vigorously.
8. Add 1.0 mL (2 vol.) of chloroform and shake vigorously.
9. Add 1.5 mL (3 vol.) of ultrapure water and shake vigorously.
10. Centrifuge the sample at 6800 × *g* for 10 min (e.g., using Beckman Avanti HP26XP JS5.3 swinging bucket rotor) and carefully remove the upper layer over the protein pellet disk, which is observed between the two phases.
11. Add 1.5 mL of methanol and mix gently (be careful not to break the pellet into small pieces). Centrifuge the sample at 6800 × *g* for 10 min to completely precipitate proteins. After removal of the supernatant, rinse the pellet and inner wall of the tube with 1 mL of methanol twice.
12. Remove the supernatant completely (*see* **Note 17**).
13. Suspend the pellet with 125 μL of buffer A supplemented with 10 μg/mL pepstatin and incubate at 37 °C until the pellet is completely dissolved (for ~10 min). Centrifuge and transfer the sample to a new 5-mL centrifuge tube.
14. Add 375 μL (3 vol.) of freshly prepared buffer H (final 1 M NH_2OH) or buffer T (final 1 M Tris–HCl) (total volume is 500 μL). After mixing and centrifugation, incubate at 37 °C for 1 h (do not rotate the tube).
15. To remove NH_2OH, perform the second methanol–chloroform precipitation (processing samples treated with buffer H and T in parallel), repeating **steps 7–12** (*see* **Note 17**).
16. Suspend the pellet in 150 μL of buffer A supplemented with 10 μg/mL pepstatin. Incubate at 37 °C until the pellet is completely dissolved (~10 min). Centrifuge and transfer the sample to a new 1.5-mL tube.
17. Centrifuge at maximum speed in a microcentrifuge for 15 min and transfer the supernatant to a new 1.5-mL tube.
18. Measure the protein concentration by the BCA protein assay. Dilute the sample with buffer A containing 10 μg/mL pepstatin to yield 0.5–0.75 mg/mL lysate in 100–140 μL (usually 50–75 μg of protein for the next step) (*see* **Note 18**).
19. Add 0.5 M TCEP (final 10 mM) in 100–140 μL of the sample.
20. Add 200 mM mPEG-2k (final 20 mM) in 100–140 μL of the sample (use NEM as a negative control to see the mPEG-dependent molecular shift). Then, suspend the content completely by pipetting (*see* **Note 19**). After centrifugation, incubate for 1 h with shaking (e.g., at 25 °C with Eppendorf Thermomixer Comfort). In the case of mPEG-5k, add 100 mM mPEG-5k (final 20 mM).
21. To remove uncoupled mPEG, perform the third methanol-–chloroform precipitation as described in **steps 22–26**.

22. Add 4 vol. (400 μL for 100 μL sample) of methanol and shake vigorously.
23. Add 2 vol. (200 μL for 100 μL sample) of chloroform and shake vigorously.
24. Add 3 vol. (300 μL for 100 μL sample) of ultrapure water and shake vigorously.
25. Centrifuge the sample at maximum speed in a microcentrifuge for 10 min (e.g., Eppendorf 5417R F45-30-11 angle rotor) and remove the upper layer carefully.
26. Add 1 mL of methanol and mix gently (be careful not to break the pellet into small pieces). Centrifuge the sample at maximum speed in a microcentrifuge for 10 min and remove the supernatant. Rinse with 1 mL of methanol. After centrifugation at 20,000 × *g* for 5 min, remove the supernatant completely (*see* **Note 17**).
27. Add 60 μL (depending on the expected sample concentration) of SDS-PAGE sample buffer without 2-mercaptoethanol and suspend well by tapping or pipetting. Incubate for 20 min at 37 °C with shaking until the pellet is completely dissolved.
28. Measure the protein concentration by the BCA protein assay to adjust the protein amount for SDS-PAGE.
29. For SDS-PAGE, add 2-mercaptoethanol (final 2%) and heat for 3 min at 100 °C. Dilute the sample appropriately to apply 10–50 μg protein (depending on the protein of interest) for SDS-PAGE.
30. Perform SDS-PAGE (*see* **Note 20**) and Western blot analysis with specific antibodies against the protein of interest (*see* **Note 21**).

4 Notes

1. Efficient transfection, especially efficient cotransfection of plasmids for the substrate and individual serine hydrolases, is required for detection of the serine hydrolase-dependent effects. We usually use Lipofectamine and PLUS reagents (Thermo Fisher Scientific), but other reagents are also applicable. Antibiotics in the medium before/during/after transfection sometimes reduce transfection efficiency and cell viability.
2. Final concentration of ethanol (derived from [^{3}H]palmitic acid stock solution) in the labeling medium must be less than 1%.
3. TCEP reduces disulfide bonds but does not break thioester bonds. If making up the TCEP solution from powder, be aware that the pH will be ~2.5 and should be brought to neutral pH with 1 M NaOH or 1 M KOH.

4. We recommend dissolving PEG powder in degassed water (e.g., nitrogen gas-purged water) because PEGs are susceptible to oxidative degradation.
5. The centrifugation step is to collect the viscous PEG stuck on the walls of the tube. Freezing of mPEG solution causes its swelling. Do not fill the plastic tube completely with mPEG solution for storage at −70 °C.
6. In this protocol, the substrate protein, PSD-95-GFP, is not immunoprecipitated before detection of radiolabeled PSD-95-GFP. Because the levels of its expression and palmitoylation are very high in transfected HEK293T cells, radiolabeled PSD-95 GFP is readily distinguishable from other radiolabeled proteins on the fluorograph even if whole cell lysates are analyzed. (We usually include a palmitoylation-deficient mutant of PSD-95 as a palmitoyl-negative control). If enrichment of the protein of interest by immunoprecipitation is necessary, use 1% SDS-containing buffer for the cell lysis/protein extraction so that palmitoylated proteins can be solubilized. The cell lysate should then be diluted with Triton X-100-containing buffer to prevent SDS denaturation of the antibody added for the immunoprecipitation [12].
7. Prolonged heating of the sample should be avoided as thioester bonds are very labile in the presence of reducing reagents such as DTT. As freezing and thawing of samples might cause palmitoylated proteins to aggregate, perform SDS-PAGE as soon as the samples are prepared.
8. Some serine hydrolases transfected (e.g., ABHD17A, B and C) are also detected as radiolabeled proteins on the fluorograph when whole-cell lysates are analyzed [17], indicating that they are palmitoylated in the cell. Because our cloned serine hydrolase genes are tagged with FLAG, the molecular weight (the position) of individual serine hydrolases can be verified by Western blotting with anti-FLAG antibody (Fig. 1b).
9. Some serine hydrolases (e.g., ABHD1, 3, and 4) affect the expression of substrate protein in transfected HEK293T or COS7 cells [17].
10. Once candidate depalmitoylating enzymes are obtained by this screening method that is based on the enzyme overexpression, the candidate should be validated by other approaches to check (1) expression of the candidate enzyme in the cell/tissue where the substrate is expressed and (2) the effect of knockdown/knockout of the candidate enzymes on the overall stoichiometry or depalmitoylation kinetics of the (endogenous) substrate protein (Fig. 3) [17].
11. Cells (e.g., HEK293T cells) cultured in a 6-well plate at a density of 3×10^5 cells/well contain approximately

500–1000 μg of proteins. The protein quantity is cell-type dependent. The cell culture format can be scaled-down to a 12-well plate, as 150–200 μg of proteins will be used for one assay point of the APEGS.

12. The APEGS assay is applicable to tissue samples. For example, homogenize a mouse brain tissue (approximately 0.5 g) in 2.5 mL (5 vol.) of homogenate buffer with 100 μg/mL PMSF using Potter-Elvehjem homogenizer. Dilute the brain homogenate (200 μg/point) with buffer A and B (both with PIs) to adjust the protein concentration to 0.4 mg/mL (500 μL/point) and the urea concentration to 8 M. Subsequent procedures are common to those for cultured cells unless otherwise specified (*see* **Note 14**).
13. To insure that the reactions of TCEP and NEM (and also of the subsequent NH_2OH and mPEGs) go to completion, adjust the protein concentration, volume and amount of proteins to 0.4–0.5 mg/mL, 500 μL and ~200 μg, respectively. The maximum amount of protein used should be ~250 μg. In addition, there are some other restrictions: (1) the maximum volume of starting lysate is 500 μL. This volume is well suited for the centrifuge tube used for the later steps of methanol–chloroform precipitation (e.g., 5-mL tubes with Beckman Avanti HP26XP, JS5.3 rotor). If a 1.5-mL tube is used for the methanol–chloroform precipitation, the starting volume should be further reduced (*see* **Note 18**), and (2) if the protein amount of starting lysate is very low (<50 μg), the protein disk/precipitate obtained by methanol–chloroform precipitation will be very fragile, likely resulting in reduced protein recovery.
14. For tissue samples treated in urea-containing buffer, incubate at room temperature for 1 h during TCEP reaction. Do not heat the samples to avoid protein carbamylation.
15. A 15-mL tube can be used for methanol–chloroform precipitation, but the recovery of proteins may be lower.
16. Methanol–chloroform precipitation is a critical procedure to prevent the carryover of NEM to the subsequent reactions. However, this procedure requires many steps and often results in inconsistent sample loss and reduced assay sensitivity. A recently reported method, chemical scavenging of NEM by 2,3-dimethyl-1,3-butadiene, is an alternative approach [21].
17. Precipitated proteins can be stored at −20 °C before the next step (before dissolving the precipitate).
18. The maximum volume for the PEGylation step is 140 μL. This volume is suitable for methanol–chloroform precipitation in a 1.5-mL microcentrifuge tube, using a microcentrifuge. To obtain complete PEGylation, the concentrations of mPEG and protein amount in the reaction are critical. Optimization

of the mPEG concentration may be required because the effective concentration or purity of mPEG reagents may vary from vendor to vendor.

19. mPEG-2k is used for proteins smaller than 60 kDa, while mPEG-5k is used for proteins larger than 60 kDa. As the mPEG stock solution is highly viscous, care should be taken for complete mixing with the sample. We recommend using a wide-bore tip to pipette mPEG solutions.
20. We typically use: 14% polyacrylamide gels for 20–30 kDa proteins coupled with mPEG-2k; 10–12% (nongradient) polyacrylamide gels for 30–60 kDa proteins coupled with mPEG-2k; 8% polyacrylamide gels for 60–90 kDa proteins coupled with mPEG-5k; and 6% polyacrylamide gels for >90 kDa proteins coupled with mPEG-5k. If the protein of interest is expected to be palmitoylated at multiple cysteine sites, a lower percentage polyacrylamide gel than described above may be more suitable. We have observed the palmitoylation-dependent gel shifts for proteins with a broad molecular weight range between 20 kDa (e.g., H-Ras) and 170 kDa (e.g., GluN2A).
21. Transfer efficiency of PEGylated proteins for Western blotting can sometimes be lower due to uncertain reasons, possibly strong hydrophilicity or large molecular weight. We recommend using EzFastBlot HMW transfer buffer (ATTO). Its detailed composition is not known, but it gives better results with a semidry transfer blotter.

Acknowledgments

We thank previous lab members Atsushi Sekiya and Tatsuro Murakami for their scientific contribution. This work was supported by the Ministry of Education, Culture, Sports, Science and Technology (Grant numbers 17K14969 to N.Y.; 15H04279 to Y.F; 16H01371, 16K14560, 17H03678, 17H05709, 18H04873 to M.F.); the Ministry of Health, Labour and Welfare (Intramural Research Grant [H27-7] for Neurological and Psychiatric Disorders to Y.F.); Takeda Science Foundation to Y.F and M.F.; The Japan Epilepsy Research Foundation and The Kato Memorial Trust for Nambyo Research to Y.F.; and The Naito Foundation and The Hori Sciences and Arts Foundation to M.F.

References

1. Chamberlain LH, Shipston MJ (2015) The physiology of protein S-acylation. Physiol Rev 95:341–376
2. Fukata Y, Fukata M (2010) Protein palmitoylation in neuronal development and synaptic plasticity. Nat Rev Neurosci 11:161–175

3. Linder ME, Deschenes RJ (2007) Palmitoylation: policing protein stability and traffic. Nat Rev Mol Cell Biol 8:74–84
4. El-Husseini AE, Schnell E, Chetkovich DM et al (2000) PSD-95 involvement in maturation of excitatory synapses. Science 290:1364–1368
5. Topinka JR, Bredt DS (1998) N-terminal palmitoylation of PSD-95 regulates association with cell membranes and interaction with K^+ channel, Kv1.4. Neuron 20:125–134
6. Craven SE, El-Husseini AE, Bredt DS (1999) Synaptic targeting of the postsynaptic density protein PSD-95 mediated by lipid and protein motifs. Neuron 22:497–509
7. Webb Y, Hermida-Matsumoto L, Resh MD (2000) Inhibition of protein palmitoylation, raft localization, and T cell signaling by 2-bromopalmitate and polyunsaturated fatty acids. J Biol Chem 275:261–270
8. El-Husseini AE, Schnell E, Dakoji S et al (2002) Synaptic strength regulated by palmitate cycling on PSD-95. Cell 108:849–863
9. Fukata Y, Dimitrov A, Boncompain G et al (2013) Local palmitoylation cycles define activity-regulated postsynaptic subdomains. J Cell Biol 202:145–161
10. Noritake J, Fukata Y, Iwanaga T et al (2009) Mobile DHHC palmitoylating enzyme mediates activity-sensitive synaptic targeting of PSD-95. J Cell Biol 186:147–160
11. Fukata M, Fukata Y, Adesnik H et al (2004) Identification of PSD-95 palmitoylating enzymes. Neuron 44:987–996
12. Fukata Y, Iwanaga T, Fukata M (2006) Systematic screening for palmitoyl transferase activity of the DHHC protein family in mammalian cells. Methods 40:177–182
13. Camp LA, Hofmann SL (1993) Purification and properties of a palmitoyl-protein thioesterase that cleaves palmitate from H-Ras. J Biol Chem 268:22566–22574
14. Duncan JA, Gilman AG (1998) A cytoplasmic acyl-protein thioesterase that removes palmitate from G protein alpha subunits and p21 (RAS). J Biol Chem 273:15830–15837
15. Yeh DC, Duncan JA, Yamashita S et al (1999) Depalmitoylation of endothelial nitric-oxide synthase by acyl-protein thioesterase 1 is potentiated by Ca^{2+}-calmodulin. J Biol Chem 274:33148–33154
16. Martin BR, Wang C, Adibekian A et al (2012) Global profiling of dynamic protein palmitoylation. Nat Methods 9:84–89
17. Yokoi N, Fukata Y, Sekiya A et al (2016) Identification of PSD-95 depalmitoylating enzymes. J Neurosci 36:6431–6444
18. Howie J, Reilly L, Fraser NJ et al (2014) Substrate recognition by the cell surface palmitoyl transferase DHHC5. Proc Natl Acad Sci U S A 111:17534–17539
19. Percher A, Ramakrishnan S, Thinon E et al (2016) Mass-tag labeling reveals site-specific and endogenous levels of protein S-fatty acylation. Proc Natl Acad Sci U S A 113:4302–4307
20. Percher A, Thinon E, Hang H (2017) Mass-tag labeling using acyl-PEG exchange for the determination of endogenous protein S-fatty acylation. Curr Protoc Protein Sci 89:14.17.11–14.17.11
21. Hurst CH, Turnbull D, Plain F et al (2017) Maleimide scavenging enhances determination of protein S-palmitoylation state in acyl-exchange methods. BioTechniques 62:69–75

Chapter 8

Measuring *S*-Depalmitoylation Activity In Vitro and In Live Cells with Fluorescent Probes

Rahul S. Kathayat and Bryan C. Dickinson

Abstract

S-palmitoylation is a reversible lipid posttranslational modification (PTM) that can mediate protein localization, trafficking, interaction with membranes, and a host of other biophysical characteristics. Over the past decade, a suite of chemoproteomic strategies have uncovered the breadth of *S*-palmitoylation, revealing widespread susceptibility to modification by this PTM throughout the human proteome. A focal point of research toward understanding the role of *S*-palmitoylation in varied cellular processes has focused on understanding how "writer" and "eraser" proteins function together to control the levels of *S*-palmitoylation of target proteins. The spatial and temporal regulation of *S*-palmitoylation by its "erasers"—acyl protein thioesterases (APTs)—is not fully understood. Tools which enable monitoring of the activity levels of the APTs in real-time in live cells illuminate how spatial control of these enzymes redecorate the lipidation state of the local proteome. To this end, we have developed fluorescence-based *de*palmitoylation *p*robes (DPPs), which report *S*-depalmitoylase activity in live cells. Using DPPs, we have demonstrated that *S*-depalmitoylase activity changes in response to growth factor stimulation, unveiling potential regulation of cell growth and metabolism by APTs. Additionally, we recently discovered APTs in mitochondria using targeted DPPs, indicating new roles for *S*-depalmitoylation in this critical cellular compartment. Here, we present detailed protocols on how to carry out in vitro *S*-depalmitoylase activity assays and live cell fluorescence imaging employing the growing DPP toolbox.

Key words Protein lipidation, *S*-Palmitoylation, Acyl protein thioesterases, Depalmitoylation, Live-cell fluorescence imaging

1 Introduction

S-palmitoylation is an abundant lipid posttranslational modification (PTM) that arises from the attachment of a C16 palmitate to the thiol of a cysteine residue through a thioester bond [1–3]. More than 10% of human proteome is reported to be susceptible to modification by *S*-palmitoylation [4], which makes it one of the most prevalent PTMs. Although thioesters are chemically stable in biological conditions, thioesters can be removed biochemically, making palmitoylation one of the few dynamic lipid PTMs [5, 6]. Installation of the palmitate group (*S*-palmitoylation) is carried out

Maurine E. Linder (ed.), *Protein Lipidation: Methods and Protocols*, Methods in Molecular Biology, vol. 2009,
https://doi.org/10.1007/978-1-4939-9532-5_8,

in humans by 23 "DHHC" motif-containing palmitoyl acyl transferases (PATs), which are primarily associated with Golgi and ER membranes [2, 7–9]. Removal of the palmitate (*S*-depalmitoylation) is mediated by acyl protein thioesterases (APTs), which includes PPT1, APT1, APT2, and the ABHD17 family of proteins (A, B, and C) [10–14]. PPT1 is a lysosomal protein, while APT1, APT2, and ABHD17A/B/C are generally thought to be cytosolic. However, we recently discovered that APT1 is also highly enriched in mitochondria [15].

The palmitoylation status of the proteome is not static, but instead can change in response to various input signals, such as growth factors [16, 17]. Measuring the palmitoylation status of the proteome can be achieved through a variety of proteomic methods such as acyl-biotin exchange (ABE) [18, 19], acyl-RAC [20], mass-shift assays [21], and metabolically labeling cells with clickable lipid analogs [22–24]. These techniques, however, report on changes to the palmitoylation level of target proteins and do not reveal the mechanisms of those changes through alterations in the activities or localization of the PATs or APTs. Additionally, traditional proteomic methods do not generally reveal spatial information about how the activities of regulatory machinery result in local proteomic consequences. Indeed, we recently discovered both growth factor stimulation [25] and lipid stress [26] results in rapid alterations in the activities of the APTs. Therefore, probing the activities of the APTs, in real time and in live cells, especially with spatial information, is critical to understanding the role(s) of the APTs in regulating cellular signaling.

The activities of the APTs can be monitored using activity-based protein profiling (ABPP) techniques employing general serine hydrolase probes [27] or APT-targeted probes [28]. Additionally, fluorescent substrates, including small molecules [29] and fluorescent proteins [30, 31] can be monitored for changes in cellular localization due to *S*-palmitoylation. Recently, we initiated a research program focused on developing new chemical tools that would allow us to probe the dynamic activity levels of the APTs in live cells. To accomplish this, we developed fluorescent probes that respond to *S*-deacylation of a peptide substrate with enhanced fluorescence (Fig. 1) [15, 25, 26]. The resultant family of depalmitoylation probes (DPPs; Fig. 2a–c) can monitor changes in APT activity levels by live cell fluorescence microscopy or flow cytometry. Although outside the scope of this chapter, we also recently developed ratiometric depalmitoylation probes (RDPs), which allow for better quantification of APT activity levels, including in complex primary tissue samples [32].

Among the first-generation DPPs, DPP-2, reports on general cytosolic activity (Figs. 2a, 3 and 4), while DPP-3 is more selective for APT1 over APT2 [25]. These probes use a surrogate C-8 lipid

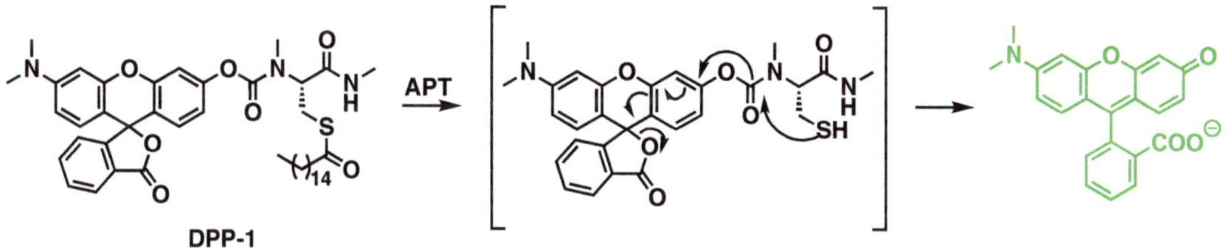

Fig. 1 Schematic of mechanism of DPPs, which "*turn-on*" when processed by acyl protein thioesterases (APTs). A peptide substrate featuring an acylated cysteine residue is tethered to a profluorescent molecule by a biologically stable carbamate linker. The activity of an APT on the peptide, which removes the thioester and unmasks the thiol of the cysteine, results in a cascade of rapid intramolecular cyclization, lactone opening, and release of a fluorescent product

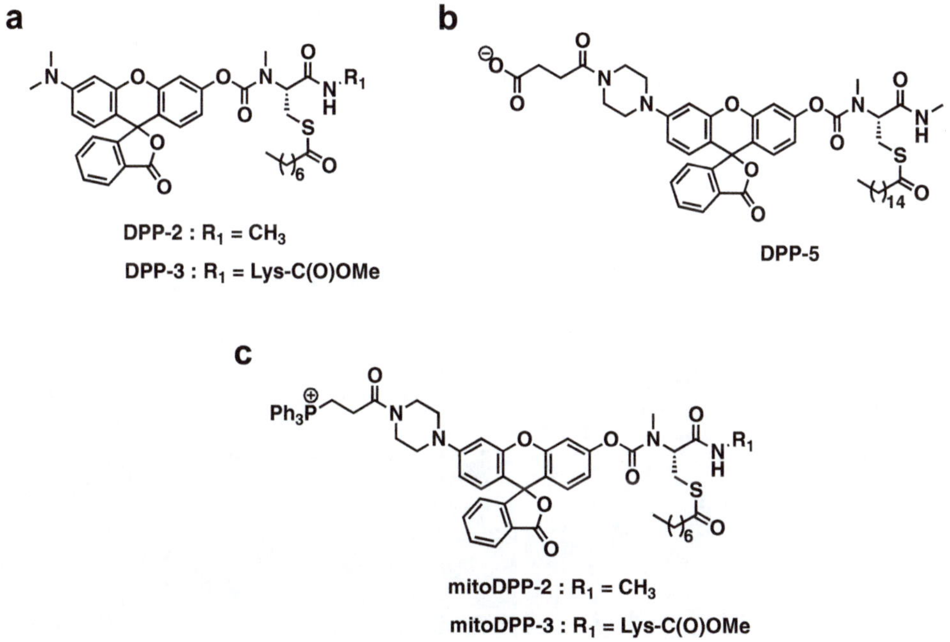

Fig. 2 Structures of a subset of the delpamitoylation probes (DPPs) for APT activity. (**a**) First-generation untargeted *S*-deacylase probe DPP-2 and more APT1-targeted deacylase probe DPP-3. (**b**) Second-generation *S*-depalmitoylase probe with enhanced water solubility, DPP-5. (**c**) Mitochondrial-targeted derivatives of DPP-2 and DPP-3, mitoDPP-2 and mitoDPP-3

substrate to increase cell uptake. DPP-1 used a natural palmitate lipid, but due to solubility issues is only applicable to in vitro assays. DPP-5, however, features an additional water solubilizing group on the scaffold (Fig. 2b), which allows for the use of natural palmitate group and measurement of depalmitoylation (rather than peptide *S*-deacylation) [26]. Finally, mitoDPP-2 and mitoDPP-3 feature a triphenylphosphonium group (Fig. 2c), which shuttles the probes

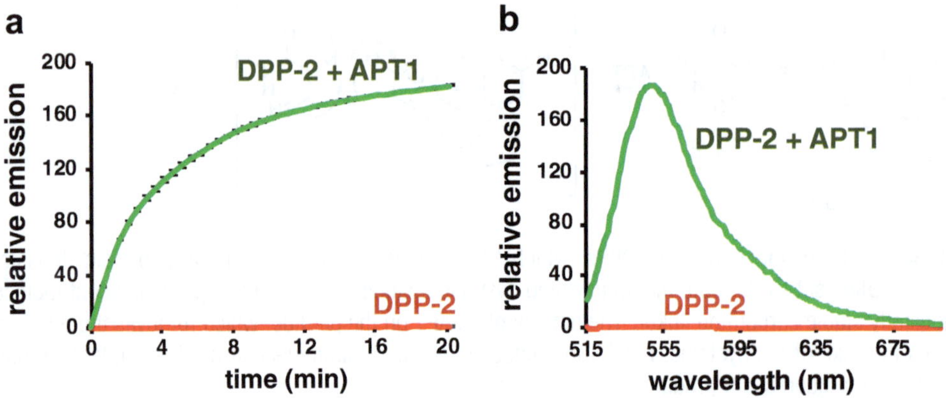

Fig. 3 A representative in vitro biochemical assay of DPP-2 with recombinant APT1. (**a**) Normalized in vitro fluorescence assays of 5 μM DPP-2, incubated with or without 50 nM purified APT1 (λ_{ex} 490/9 nm, λ_{em} 545/20 nm) in HEPES Buffer (20 mM, pH 7.4, 150 mM NaCl, 0.1% Triton X-100). Error bars are standard error ($n = 3$). (**b**) Normalized fluorescence emission spectrum of 5 μM DPP-2, incubated with or without 50 nM APT1 for 20 min ($\lambda_{ex} = 485$ nm). Spectra are plotted as average of $n = 3$

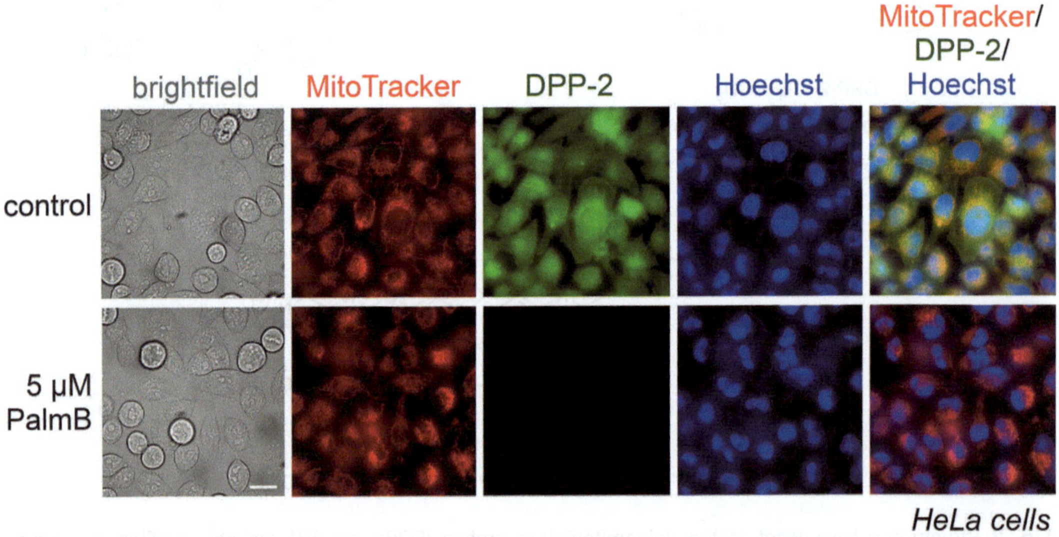

Fig. 4 HeLa cells treated for 30 min with 1 μM Hoechst 33342, 100 nM MitoTracker Deep Red, and either DMSO or 5 μM PalmB (pan APT inhibitor), washed, loaded with 1 μM DPP-2 for 15 min, and then analyzed by epifluorescence microscopy. Representative images for brightfield, MitoTracker, DPP-2, Hoechst 33342 nuclear stain, and an overlay of MitoTracker, DPP-2, and Hoechst 33342 are shown for each set of conditions. 25 μm scale bar shown

to the mitochondria and reports on APT activity selectively within this compartment [15]. Here, we will describe the general use of the DPP family of probes to measure APT activities both in vitro and in live biological samples, using cytosolic and mitochondrial activity measurements as example experiments.

2 Materials

1. 384-well black flat bottom plate.
2. Infinite M200 Pro plate reader (Tecan).
3. Poly-D-lysine (30–70 KDa).
4. Sterile water.
5. 8-well imaging dish.
6. MitoTracker Deep Red FM (MitoTracker).
7. Hoechst 33342.
8. Palmostatin B (PalmB).
9. Live Cell Imaging Solution.
10. HEPES Buffer: 20 mM HEPES, 150 mM NaCl and 0.1% Triton X-100 at pH 7.4.
11. 5 mM DPP-1 in DMSO, stored at −80 °C.
12. 1 and 5 mM DPP-2 in DMSO, stored at −80 °C.
13. 1 mM DPP-5 in DMSO, stored at −80 °C.
14. 0.5 mM mitoDPP-2 in DMSO, stored at −80 °C.
15. 100 μM MitoTracker Deep Red in DMSO, stored at −80 °C.
16. 1 mM Hoechst 33342 in DMSO, stored at −80 °C.
17. 1 μg/μL poly-D-lysine in sterile H_2O, stored at −20 °C.
18. HeLa cells from ATCC, freeze at an early passage (passage 6) in individual aliquots. Use cells for fewer than 25 passages for all experiments.
19. Growth medium: DMEM glutamax, 10% FBS, 1% Pencillin/ Streptomycin antibiotic mixture.

2.1 Fluorescence Microscopy

1. Lieca DMi8 epifluorescence micrscope with Leica LASX software.
2. Hamamatsu Orca-Flash 4.0 camera.
3. 63× oil objective (N/A 1.4).
4. Images were taken in four channels: brigthfield, Hoechst 33342 (ET 402/15×, Quad-S, ET 455/50m), MitoTracker Deep Red (ET 645/30×, Quad-S, ET 705/72m), and DPPs/ mitoDPPs (YFP filter cube 1525306).

3 Methods

Detailed synthetic procedures for the preparation of DPP-1, DPP-2, DPP-5 and mitoDPP-2 can be accessed from previous reports [15, 25, 26]. This chapter will focus on optimal utilization of these chemical tools to report activity of *S*-depalmitoylases in in vitro biochemical and live cell imaging experiments.

3.1 In Vitro S-Depalmitoylation Assays with Purified APTs

We present here a number of steps to study biochemical activity of APTs in vitro using DPPs. As a representative case, the following steps involve a fluorescence plate reader assay to monitor the in vitro reaction kinetics of recombinant APT1 with DPP-2 (*see* **Notes 1–6**).

1. Warm HEPES Buffer at 37 °C.
2. Add 50 μL of 7.5 μM DPP-2 in HEPES Buffer to a 384-well plate.
3. Place the 384-well plate in plate reader at 37 °C for 3–5 min.
4. Add 25 μL of either HEPES Buffer alone or HEPES Buffer containing 150 nM recombinant APT1 (*see* **Notes 1–3**) using a multichannel pipette. The final concentration of DPP-2 and APT1 are 5 μM and 50 nM, respectively.
5. To obtain a kinetic curve, measure fluorescence intensities every 30 s until the fluorescence emission saturates. Use the following parameters to run a kinetic assay using the Infinite M200 Pro plate reader: λ_{ex} 490/9 nm, λ_{em} 545/20 nm, Gain 70, No. of flashes 25, Integration time 100 μs, and *Z*-position 20,000 μm. These parameters need to be optimized for other plate readers.
6. Take the average of fluorescence emission for three experimental replicates for every time interval.
7. Normalize the average fluorescence emission of both control ± APT1 to the average emission for control at $t = 0$.
8. Plot kinetic curves from normalized emission (Fig. 3a) to reveal the enzymatic activity of APT1. To obtain kinetic parameters, vary the concentration of probe, generally from approximately 50 nM–20 μM and repeat this protocol to obtain initial rates (*see* ref. [26]).
9. To compare an end point fluorescence emission spectra between control ± APT1, at the end of the kinetic run in **step 5**, obtain emission spectra. We use the following parameters to obtain emission spectra: λ_{ex} 485 nm, λ_{em} 510–700 nm, Gain 100, No. of flashes 25, Integration time 100 μs and *Z*-position 20,000 μm. Normalize average fluorescence intensities for three biological replicates at all wavelengths for both control ± APT1 to maximum fluorescence intensity for control (~543 nm).
10. Plot normalized emission to wavelength (Fig. 3b) to demonstrate APT1 activity.

3.2 Poly-D-Lysine Coating of Imaging Dishes

Detachment of cells from the surface of imaging dish during a number of washing steps can be a problem depending on cell lines. As a precautionary measure, we use poly-D-lysine to coat our dishes to improve cell adherence to the imaging surface (*see* **Note 7**).

1. Perform all following steps in a biosafety cabinet to maintain sterility.
2. Add 50 μL of poly-D-lysine stock solution (1 μg/μL) to 5 mL of sterile water for a final concentration of 0.01 μg/μL.
3. Add 400 μL of 0.01 μg/μL of poly-D-lysine to each well of the 8-well imaging dish at room temperature.
4. After 2 h, remove the solution from each well of the dish. The treated dish is now ready for plating cells.

3.3 Epifluorescence Imaging of DPP-2/DPP-5/MitoDPP-2 with PalmB

1. Plate HeLa cells (50,000–55,000/well) in 450 μL of growth media into an 8-well dish (*see* **Note 8**).
2. After 24–28 h, remove the growth media and wash the cells with 400 μL of DMEM glutamax.
3. Add 400 μL of DMEM glutamax containing 1 μM Hoechst 33342, 100 nM MitoTracker Deep Red and 5 μM PalmB/equivalent DMSO, or other experimental variable (*see* **Notes 9** and **10**).
4. Incubate the cells at 37 °C with a supply of 5% CO_2.
5. After 30 min of incubation, wash the cells with 400 μL of Live Cell Imaging Solution and replace by 1 μM DPP-2/1 μM DPP-5/500 nM mitoDPP-2 in 400 μL of Live Cell Imaging Solution (*see* **Note 11**). To minimize experimental variability due to pipetting error, we recommend making one master stock of DPP and then splitting it between the wells in an experiment (*see* **Notes 12** and **13**).
6. After 15 min of incubation at 37 °C with a supply of 5% CO_2, acquire images on an inverted epifluorescence microscope (*see* **Notes 14–17**). We recommend using the MitoTracker Deep Red channel for focusing the objective lens on cells to minimize photobleaching of the DPP signal.

3.4 Details of Quantification of Images

Images are quantified using Fiji (Image J, Wayne Rasband, NIH), which is available free of cost. Fiji is a very versatile software and provides different ways to quantify and process images for publication purposes. We recommend using the Fiji online resources for in-depth training on the software. Nevertheless, in this section we will briefly discuss the simplest way to process images, which we have found provides consistent results.

1. Use "*free hand selection*" tool from Fiji tool bar to select region of interest in the brightfield image, which comprises healthy cells that are on the same plane. Ideally, this step is researcher "blinded," where the images from various experimental points are scrambled to avoid bias in the analysis.

2. Open the ROI manager window,which is available under "*Analyze*" and then "*Tool*" option on the menu bar.
3. Add the region of interest selected in **step 1** into ROI manager window.
4. Gate the selected region of interest from **step 1** in corresponding MitoTracker image by clicking first on MitoTracker image and then on the code for the region of interest in ROI window.
5. Press "*m*" on the keyboard to get the mean fluorescence intensity from MitoTracker image.
6. Repeat **steps 4** and **5** for the corresponding image for DPP-2/DPP-5/mitoDPP-2.
7. Use mean fluorescence intensities for all the images to calculate average fluorescence intensity and standard errors for both MitoTracker channel and DPP-2/DPP-5/mitoDPP-2 channel.
8. Normalize average fluorescence intensity for both MitoTracker and DPPs/mitoDPPs for various set of conditions to the control experiment to show relative effect of external perturbations on mitochondrial health and APTs activity, respectively.

4 Notes

1. For purification of recombinant APT1 and APT2 *see* ref. [25].
2. The purified APT1 and APT2 (not included here) should be aliquoted and stored at −80 °C. Avoid repeated freeze-thaws; it is preferable to use the purified enzyme once after thawing.
3. Additionally, we found that the activity of both APT1 and APT2 decrease with time, even when stored at −80 °C. Thus, for optimal resuts, purify APT1 and APT2 right before use.
4. The core fluorophore constructs of DPP-5 and mitoDPP-2 are slightly different from that of the first generation depalmitoylation probe, DPP-2 (Fig. 2). APT activity on DPP-5 and mitoDPP-2 release fluorescent products that are brighter than the product from DPP-2. Therefore in vitro assays can be run on lower concentrations of probes *see* refs. [15, 26].
5. The in vitro depalmitoylation assay presented here can be extended to assays with cell lysates or Immunoprecipitated proteins.
6. To demonstrate enzymatic depalmitoylation by cell lysates or immunoprecipitated proteins, PalmB is recommended as a control for in vitro APT activity, as in our hands PalmB is a robust pan-active APT inhibitor that can be used to distinguish

between enzymatic and nonenzymatic *S*-depalmitoylation activities. We recommend using 5–20 μM PalmB in incubations with cell lysates or immunoprecipitated proteins for 20 min before addition of DPPs/mitoDPPs.

7. For some cells, poly-D-lysine may be toxic, in which case lower concentrations or other adherents can be used.
8. Depending on growth conditions and cell health, the cell number and resultant density of cells for imaging may need to be optimized.
9. Treatment with PalmB and APT1/APT2 specific inhibitors ML348/ML349 [27] (not included in this work), respectively, should always be done in plain DMEM glutamax, as we have seen that activity of these inhibitors is reduced by the presence of fetal bovine serum.
10. For experiments just requiring MitoTracker and Hoechst staining (shRNA/siRNA silencing, genetic overexpression, etc.), growth medium can be used for incubation of cells with these cell markers.
11. To avoid detachment of cells from the imaging surface during the imaging experiments all the washing and addition of various probes should be done very gently. The addition of solutions should preferably be done along the sides of imaging dishes.
12. There are many steps which can contribute to technical variability in an experiment (e.g., pipetting errors, different stocks of probes, nonuniform intensity of light source, especially at the end of its life span). Therefore, to minimize false results from these techincal issuses we recommend running experiments with the control conditions in parallel.
13. Since the kinetics of DPPs and mitoDPPs are very fast, for experiments where one might expect only subtle changes in activity of APTs due to external stimulation (EGF treatment, starvation, palmitate stress, etc.), multichannel pipettes are recommended to add probes solution simultaneously in both control and treatment conditions.
14. We generally do imaging after 15 min of incubation (which includes 10–12 min of incubation at 37 °C and roughly 3–5 min to get the dishes ready on microscope for imaging) with probes. Nevertheless, one can try to obtain images from earlier time points if the experiment demands. We have found that the experimental error from earlier time points is generally higher.
15. We recommend imaging experiments in sets (treatment vs. control) to obtain the highest quality data, which minimizes image acquisition time, resulting in lower

experimental error. Alternatively, image acquisition on a microscope capable of autofocussing and processing many samples in a short time is recommended.

16. For acquiring images, we recommend taking one or two images per well from all of the samples. Then, collect a second set of images from all of the samples, ultimately collecting three to five images from each well. Collecting images in this way helps to decrease bias due to timing.
17. In our experience, it takes 7–10 min to acquire 5–6 high quality images for a set of samples. Therefore, for fairly representing images, we recommend users deploy the entire set of images for quantification purposes, but when selecting images to display, use images that were aquired at nearly the same time.

Acknowledgments

This work was supported by the University of Chicago, the National Institute of General Medical Sciences (R35 GM119840) of the National Institutes of Health, and a Research Fellowship from the Alfred P. Sloan Foundation.

References

1. Blanc M, David F, Abrami L et al (2015) SwissPalm: protein palmitoylation database. F1000Res 4:261
2. Linder ME, Deschenes RJ (2007) Palmitoylation: policing protein stability and traffic. Nat Rev Mol Cell Biol 8:74–84
3. Lanyon-Hogg T, Faronato M, Serwa RA, Tate EW (2017) Dynamic protein acylation: new substrates, mechanisms, and drug targets. Trends Biochem Sci 42:566–581
4. Sanders SS, Martin DD, Butland SL et al (2015) Curation of the mammalian palmitoylome indicates a pivotal role for palmitoylation in diseases and disorders of the nervous system and cancers. PLoS Comput Biol 11(8): e1004405. https://doi.org/10.1371/journal.pcbi.1004405
5. Hernandez JL, Majmudar JD, Martin BR (2013) Profiling and inhibiting reversible palmitoylation. Curr Opin Chem Biol 17:20–26
6. Peng T, Thinon E, Hang HC (2016) Proteomic analysis of fatty-acylated proteins. Curr Opin Chem Biol 30:77–86
7. Gottlieb CD, Linder ME (2017) Structure and function of DHHC protein S-acyltransferases. Biochem Soc Trans 45:923–928
8. Rana MS, Kumar P, Lee CJ et al (2018) Fatty acyl recognition and transfer by an integral membrane S-acyltransferase. Science 359 (6372):eaao6326
9. Zheng B, DeRan M, Li X et al (2013) 2-Bromopalmitate analogues as activity-based probes to explore palmitoyl acyltransferases. J Am Chem Soc 135:7082–7085
10. Duncan JA, Gilman AG (2002) Characterization of *Saccharomyces cerevisiae* acyl-protein thioesterase 1, the enzyme responsible for G protein alpha subunit deacylation in vivo. J Biol Chem 277:31740–31752
11. Lin DT, Conibear E (2015) ABHD17 proteins are novel protein depalmitoylases that regulate N-Ras palmitate turnover and subcellular localization. elife 4:e11306
12. Toyoda T, Sugimoto H, Yamashita S (1999) Sequence, expression in Escherichia coli, and characterization of lysophospholipase II. Biochim Biophys Acta 1437:182–193
13. Verkruyse LA, Hofmann SL (1996) Lysosomal targeting of palmitoyl-protein thioesterase. J Biol Chem 271:15831–15836
14. PMID 27307232, https://www.ncbi.nlm.nih.gov/pubmed/?term=27307232

15. Kathayat RS Cao Y, Elvira PD et al (2018) Active and dynamic mitochondrial S-depalmitoylation revealed by targeted fluorescent probes. Nat Commun 9:334
16. El-Husseini Ael D, Schnell E, Dakoji S et al (2002) Synaptic strength regulated by palmitate cycling on PSD-95. Cell 108:849–863
17. Ponimaskin E, Dityateva G, Ruonala MO et al (2008) Fibroblast growth factor-regulated palmitoylation of the neural cell adhesion molecule determines neuronal morphogenesis. J Neurosci 28:8897–8907
18. Drisdel RC, Green WN (2004) Labeling and quantifying sites of protein palmitoylation. BioTechniques 36:276–285
19. Wan J, Roth AF, Bailey AO, Davis NG (2007) Palmitoylated proteins: purification and identification. Nat Protoc 2:1573–1584
20. Forrester MT, Hess DT, Thompson JW et al (2011) Site-specific analysis of protein S-acylation by resin-assisted capture. J Lipid Res 52:393–398
21. Percher A, Ramakrishnan S, Thinon E et al (2016) Mass-tag labeling reveals site-specific and endogenous levels of protein S-fatty acylation. Proc Natl Acad Sci U S A 113:4302–4307
22. Charron G, Wilson J, Hang HC (2009) Chemical tools for understanding protein lipidation in eukaryotes. Curr Opin Chem Biol 13:382–391
23. Martin BR, Wang C, Adibekian A, Tully SE, Cravatt BF (2011) Global profiling of dynamic protein palmitoylation. Nat Methods 9:84–89
24. Yap MC, Kostiuk MA, Martin DD et al (2010) Rapid and selective detection of fatty acylated proteins using omega-alkynyl-fatty acids and click chemistry. J Lipid Res 51:1566–1580
25. Kathayat RS, Elvira PD, Dickinson BC (2017) A fluorescent probe for cysteine depalmitoylation reveals dynamic APT signaling. Nat Chem Biol 13:150–152
26. Qiu T, Kathayat RS, Cao Y, Beck MW, Dickinson BC (2018) A fluorescent probe with improved water solubility permits the analysis of protein S-depalmitoylation activity in live cells. Biochemistry 57:221–225
27. Adibekian A, Martin BR, Chang JW et al (2012) Confirming target engagement for reversible inhibitors in vivo by kinetically tuned activity-based probes. J Am Chem Soc 134:10345–10348
28. Garland M, Schulze CJ, Foe IT et al (2018) Development of an activity-based probe for acyl-protein thioesterases. PLoS One 13: e0190255
29. Creaser SP, Peterson BR (2002) Sensitive and rapid analysis of protein palmitoylation with a synthetic cell-permeable mimic of SRC oncoproteins. J Am Chem Soc 124:2444–2445
30. Dekker FJ, Rocks O, Vartak N (2010) Small-molecule inhibition of APT1 affects Ras localization and signaling. Nat Chem Biol 6:449
31. Görmer K, Bürger M, Kruijtzer JA et al (2012) Chemical-biological exploration of the limits of the Ras de- and repalmitoylating machinery. Chembiochem 13:1017–1023
32. Beck M, Kathayat RS, Cham CM, Chang EB, Dickinson BC (2017) Michael addition-based probes for ratiometric fluorescence imaging of protein S-depalmitoylases in live cells and tissues. Chem Sci 8:7588–7592

Chapter 9

Dynamic Radiolabeling of *S*-Palmitoylated Proteins

Laurence Abrami, Robin A. Denhardt-Eriksson, Vassily Hatzimanikatis, and F. Gisou van der Goot

Abstract

Proteins can be radiolabeled either during synthesis, typically using ^{35}S-cysteine/methionine (^{35}S-Cys/Met), or after synthesis, by adding a radiolabeled posttranslational modification. Here we describe how protein *S*-palmitoylation, and its dynamics, can be monitored by ^{3}H-palmitate labeling and how the importance of *S*-palmitoylation in protein biogenesis and turnover can be investigated using ^{35}S-Cys/Met pulse–chase metabolic labeling. Proteins frequently have multiple palmitoylation sites. The importance thereof on the design and interpretation of metabolic labeling experiments is discussed.

Key words Pulse–chase, Radiolabeling, Metabolic labeling, *S*-palmitoylation, Turnover, Data-driven mathematical modeling, Protein *S*-acyltransferase (PAT)

1 Introduction

Protein function can be regulated by a plethora of posttranslational modifications, one of which is *S*-acylation, often referred to as *S*-palmitoylation, which consists of the addition of an acyl chain (C14 to C18) to a cysteine via a thioester bond [1–3]. This modification is mediated in the cytosol by members of the DHHC *S*-acyltransferase family of polytopic transmembrane enzymes, of which there are 23 in the human genome [4]. As opposed to other protein lipid modifications, such as myristoylation, *S*-acylation can be reversed by thioesterases, of which five have so far been identified [5, 6].

S-palmitoylation can affect protein function in different manners [1, 2]. For example, it can promote biogenesis of membrane proteins in the endoplasmic reticulum [7], affect trafficking [8], localization to specific membrane domains [3], or protein turnover and function [9].

Laurence Abrami and Robin A. Denhardt-Eriksson contributed equally to this work.

Maurine E. Linder (ed.), *Protein Lipidation: Methods and Protocols*, Methods in Molecular Biology, vol. 2009, https://doi.org/10.1007/978-1-4939-9532-5_9,

Unraveling the different consequences of *S*-palmitoylation generally requires a combination of approaches including functional assays, mutation of *S*-palmitoylation sites, and localization by microscopy. Probing the dynamics of *S*-palmitoylation, and the consequence on protein biogenesis and turnover requires metabolic labeling. Here we describe the methods and the wealth of information metabolic labeling experiments may provide when coupled to mathematical modeling. This chapter provides protocols for the measurement of palmitate incorporation, and palmitate loss as well as a metabolic labeling method to investigate the effects of *S*-palmitoylation on protein biogenesis and turnover. We also describe how the obtained data may be interpreted, raising awareness as to how easy it is to misinterpret metabolic labeling data.

1.1 Species, Mutants and Computational Representation of S-Palmitoylation

Most proteins that undergo *S*-palmitoylation can get modified on more than one cysteine residue and the consequences of acyl chain attachment may differ from one site to the other both in terms of dynamics and function. Moreover, acylation of one site may affect palmitoylation or depalmitoylation rates of the other. Thus, in-depth understanding of the *S*-palmitoylation of a given protein requires taking into account the specific *S*-palmitoylation status of the protein, leading to the notion of different "species." If a protein has *n* *S*-palmitoylation sites, it may exist in 2^n different *S*-palmitoylation species. The cases of 2 and 3 sites are shown in (Fig. 1). Complexity may further increase if the protein of interest can be found in multiple locations (for example, the plasma membrane and the Golgi apparatus), if a protein can form higher order structures (dimers or more), or if other posttranslational modifications also influence the studied parameters.

In this chapter we will use the following nomenclature: the protein of interest is termed WT (Fig. 1). If it has two demonstrated *S*-palmitoylation sites, the WT^{00} species corresponds to the nonpalmitoylated protein. WT^{10} is the protein modified on the first

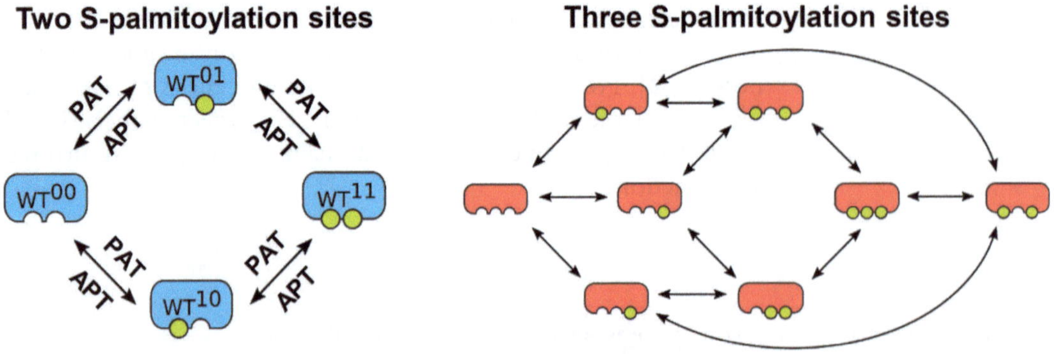

Fig. 1 Schematic representations of the palmitoylation systems of a protein with two *S*-palmitoylation sites (left) or three *S*-palmitoylation sites (right). PAT: protein *S*-acyltransferase; APT: acyl protein thioesterase

site, WT^{11} modified on both sites, and so on. The importance of a given site can be probed by mutagenesis. If for example the second site is mutated to serine (the most similar residue to cysteine), this will lead to the CS mutant, which can only occur as the CS^{00} and CS^{10} species. WT would be CC. For a three-site protein, the CSC mutant would still be able to occupy four states, CSC^{000}, CSC^{100}, CSC^{101}, CSC^{001}. The aim of this nomenclature is to clearly differentiate site occupancy from mutation.

Conversion of one species into the other is mediated by protein *S*-acyltransferases (PATs) and acyl protein thioesterases (APTs) (Fig. 1). Each of these species can undergo degradation with its own specific kinetics. As an illustrative example, we will use calnexin, a chaperone of the endoplasmic reticulum, which is a transmembrane protein and has two cysteines at the cytoplasmic face of its single transmembrane domain that can be modified by palmitoylation [10, 11].

1.2 ^{3}H-Palmitate Incorporation as a Proof of Protein Palmitoylation

If a protein is suspected to undergo *S*-palmitoylation, several assays are available to test this hypothesis. The simplest is to use the so-called Acyl-RAC or ABE capture assays [12, 13]. These "hydroxylamine-switch" [6] methods allow the capture of proteins that contain a thioester bond and detect them using a western blot. While very useful, these assays may lead to both false positives and negatives and are not quantitative. Thus, complementary evidence is required such as metabolic incorporation of clickable [14] or ^{3}H-labeled [15, 16] palmitate, as described below, and the loss of the label upon hydroxylamine treatment of the sample to confirm thioester linkage.

It is important to keep in mind that any palmitate metabolic labeling method will only "see" proteins that have not yet undergone acylation, that is, at least one palmitoylation site needs to be free for labeling to occur. If 5% of a given protein is palmitoylated in the cells, 95% is potentially available for palmitoylation during the labeling period. However, if 95% is stably palmitoylated, only 5% is left to be palmitoylated. This will lead to low palmitate incorporation, especially if a protein is of low abundance or palmitoylation is slow. Therefore, we recommend combining hydroxylamine-switch capture methods, which monitor the steady state palmitoylated pool, with metabolic labeling, which monitors the pool undergoing palmitoylation.

S-palmitoylation can also be evaluated using a variant of the Acyl-RAC method in which the capture step is replaced by a PEGylation step leading to a shift in mass that can be monitored by western blot [17–19]. This method not only reveals the percentage of the protein population that is *S*-palmitoylated at steady state but also, if multiple *S*-palmitoylation sites are present, the percentage of the population that has 1, 2, 3, or more sites occupied.

1.3 Kinetics of Palmitoylation and Depalmitoylation

^{3}H-Palmitate incorporation can also be performed as a function of time [20] typically over a 1–7 h time frame (Fig. 2). Palmitate incorporation may continuously rise as a function of time, plateau, or decline after an initial increase due to the contributions of palmitoylation, depalmitoylation, protein synthesis, and degradation. Such experiments can also be done in the presence of protein synthesis inhibitors, such as cycloheximide, or inhibitors of enzymatic depalmitoylation such as palmostatin B [21], ML348, or ML349 [22], to evaluate the contributions of protein synthesis and enzymatic depalmitoylation to the ^{3}H-palmitate incorporation.

Finally, depalmitoylation can be monitored by performing a ^{3}H-palmitate pulse–chase experiment (Fig. 3). Rapid ^{3}H-palmitate loss is indicative of enzymatic depalmitoylation, while very slow

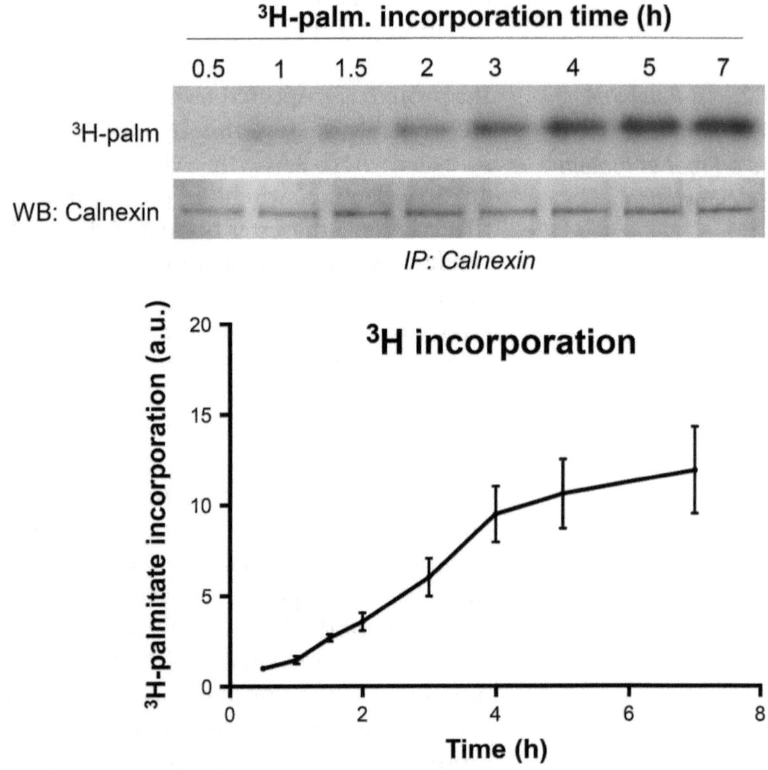

Fig. 2 ^{3}H-Palmitate incorporation into calnexin. RPE1 cells were labeled for different times at 37 °C with ^{3}H-palmitic acid. Protein synthesis before and during labeling was inhibited by addition of 10 μg/mL cycloheximide. Cells were washed, lysed, and immunoprecipitated for calnexin and subjected to SDS-PAGE. Each sample was divided in two: one for autoradiography and the fluorogram for ^{3}H-palmitic acid (1 week exposure) analyzed using the Typhoon; the other was analyzed by western blot (anti-calnexin). After verifying by western blot that equal amounts of protein were immunoprecipitated at each time point, the curve was plotted using only fluorogram data. Error bars correspond to standard deviations ($n = 4$)

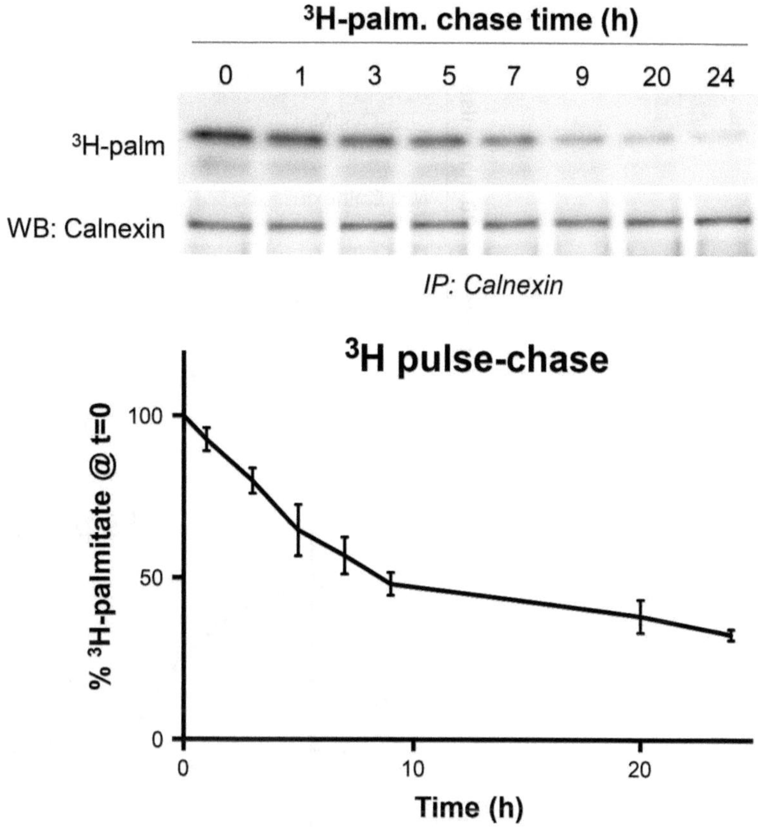

Fig. 3 ^{3}H-Palmitate pulse-chase of calnexin. RPE1 cells were labeled for 2 h at 37 °C with ^{3}H-palmitic acid. After different times of chase with complete medium, cells were washed, lysed, and immunoprecipitated for calnexin and subjected to SDS-PAGE. Samples were analyzed as in Fig. 2. Intensity values were normalized fixing 100% for time = 0, just after the pulse and washes and before the chase. Error bars correspond to standard deviations ($n = 4$)

hydrolysis may occur through intrinsic hydrolysis. Importantly, various pulse times should be tested since this may drastically affect the shape of the ^{3}H-palmitate decay curve (*see* Subheading 2).

1.4 Impact of S-Palmitoylation on Protein Biogenesis, Trafficking and Turnover

Since the earliest studies on the consequences of protein *S*-palmitoylation, it was found that it may affect protein turnover. Two approaches are commonly used to monitor turnover. The first is blocking protein synthesis and following the abundance of a protein by western blot, so called cycloheximide chase. In such experiments, the steady state population present at $t = 0$ is followed as a function of time. The second method is metabolic labeling using a label that is incorporated into the protein during synthesis, such as ^{35}S-Cys/Met [23] (Fig. 4) or other nonnatural amino acids. Labeling is typically performed during a rather short period, 20 min, and then the protein is followed as a function of time. The chase time

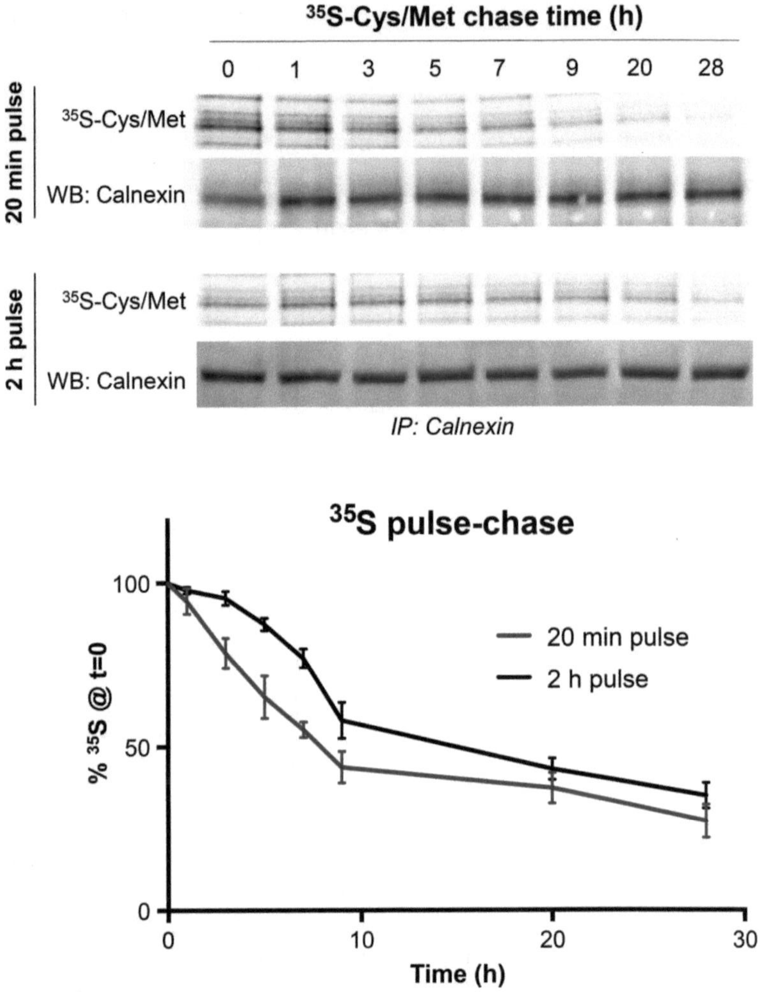

Fig. 4 Decay profile of endogenous calnexin observed by ^{35}S-Cys/Met pulse-chase. RPE1 cells were labeled for 20 min ($n = 4$) or for 2 h ($n = 3$) at 37 °C with ^{35}S-Cys/Met. After different times of chase with culture medium supplemented with cold methionine/cysteine, cells were washed, lysed, and immunoprecipitated for calnexin and subjected to SDS-PAGE. Each sample was divided in two: one for autoradiography and the fluorogram for ^{35}S-Cys/Met (16 h exposure) analyzed with Typhoon; the other for analysis by western blot (anti-calnexin). After verifying by western blot that equal amounts of protein were immunoprecipitated at each time point, the curves were plotted using only fluorogram data. Intensity values were normalized fixing 100% for time = 0, just after the pulse and washes and before the chase. Error bars correspond to standard deviations

may cover protein folding/biogenesis, trafficking if relevant, the life span of the protein, all the way to its degradation by the proteasome or in lysosomes, the only two degradation stations in the cell. The curves that can be generated using these two types of approaches, cycloheximide chase and metabolic labeling, both show a decay with time. However, for a given protein, the shape of the curve may vary greatly between the two methods and lead to apparent half-lives ($t_{1/2}$) that differ by many folds. This is because fully folded mature proteins are monitored in a cycloheximide chase, whereas metabolic labeling monitors proteins from "birth" to "death." If one takes an extreme example, a protein that folds very inefficiently but once folded is very stable, the apparent half-life obtained by metabolic labeling will be short, because most of the protein fails to fold and is degraded by the quality control machineries of the cell at short times after the pulse, but the apparent half-life obtained by cycloheximide chase will be long. Here we will refer to turnover time as the "true" half-life of a mature protein.

^{35}S-Cys/Met pulse–chase experiments can thus be used to evaluate any parameter that may affect protein degradation during the life of a protein, that is, during biogenesis, trafficking, or at its site of action. Such a parameter could be *S*-palmitoylation, and if there are multiple palmitoylation sites, it could be site-specific palmitoylation.

2 Experimental Design and Simulations

^{3}H-Palmitate incorporation and pulse–chase experiments can be performed on the wild type protein and mutants of palmitoylation sites, that is, single cysteine mutants, double, or more, depending on the number of sites. ^{3}H-Palmitate incorporation should initially be tested over a period of 4–5 h. If a plateau is reached, then the incorporation period should be kept shorter than that time. If no plateau or decline is observed, then later times can be tested, but we recommend not exceeding 8 h because of potential depletion of the label and the fact that incorporation is performed in the absence of serum. The optimal labeling time might vary from one protein to another.

To validate that a protein is *S*-palmitoylated, we recommend showing that the palmitate label is lost upon hydroxylamine treatment (which breaks thioester bonds) rather than showing inhibition with 2-bromopalmitate. This compound is a nonspecific inhibitor which may even modify your protein of interest and thus compete for ^{3}H-palmitate labeling [24].

To assay the turnover of palmitate, ^{3}H-palmitate pulse–chase experiments can be initiated with a 2 h pulse (to obtain sufficient signal). If the ^{3}H-palmitate decay signal is rapid, then we recommend also performing experiments with longer pulses such as 4 or 6 h. If the longer pulse leads to slower apparent depalmitoylation, it will be indicative that different species have different palmitoylation and/or depalmitoylation rates. If the ^{3}H-palmitate decay from a 2 h pulse is slow, shorter pulse times can be tested.

To monitor protein turnover, a pulse–chase experiment is performed with ^{35}S-Cys/Met. Labeling can be initiated with a 20-min pulse. We strongly advise repeating the experiment with longer pulses, for example 2 h or even overnight (Fig. 4). The importance of testing different pulse times is nicely illustrated by calnexin [10].

Calnexin has two neighboring sites that undergo palmitoylation extremely slowly, but once one site is modified, palmitoylation of the second site is rapid [10]. The ^{3}H-palmitate incorporation curve indeed shows a gradual increase over a 7 h period, with no plateau, even in the presence of protein synthesis inhibitors (Fig. 2). Depalmitoylation of calnexin is rapid if a single site is modified but is essentially irreversible when both sites are palmitoylated. The half-life of the nonpalmitoylated form, WT^{00}, is 5 h, and that of the dually palmitoylated, WT^{11}, is close to 50 h [10]. As a consequence, calnexin in a cell is mostly dually palmitoylated.

Because only a small population of calnexin is available for palmitoylation (since most of it already is modified), and since palmitoylation is very slow, when labeling cells for 2 h with ^{3}H-palmitate, only a minute proportion of the calnexin population undergoes palmitoylation. Of this ^{3}H-palmitate-labeled population, half, according to our mathematical modeling [10], is in the WT^{10} form (which undergoes fast depalmitoylation), the other half is in the WT^{11} (which does not get depalmitoylated). The observed ^{3}H-palmitate decay curve is a linear combination of the two decay behaviors (Fig. 3). Longer pulse times (4 h or more) would change the distribution of the starting population, with more and more WT^{11}, thus slowing the apparent ^{3}H-palmitate decay. Thus, the pulse time of any metabolic labeling experiment influences the species distribution of the population at time 0. If the property the label reveals differs significantly between species, the curve will be significantly dependent on pulse time.

Similarly, ^{35}S-Cys/Met pulse–chase results for calnexin greatly depend on the length of the pulse period (Fig. 4). Since palmitoylation is slow, the pool of calnexin synthesized during a 20-min pulse has not acquired any palmitate, thus the initial part of the decay curve will reflect the stability of WT^{00}. During the time course of the experiment, more and more calnexin will acquire palmitate, which slows down its decay. If a 2- or 4-h pulse with ^{35}S-Cys/Met is performed, the WT^{11} species will be significantly populated at $t = 0$ of the chase and decay will appear significantly slower (Fig. 4).

Performing the types of experiments described above will indicate how complex the studied system is. If a protein has multiple palmitoylation sites, or locations, or higher-order assembly states, it is likely that the detailed understanding of the system will require mathematical modeling [9, 10]. To illustrate the impact of pulse length, Fig. 5 shows simulations for a protein that undergoes slow *S*-palmitoylation and for which *S*-palmitoylation leads to

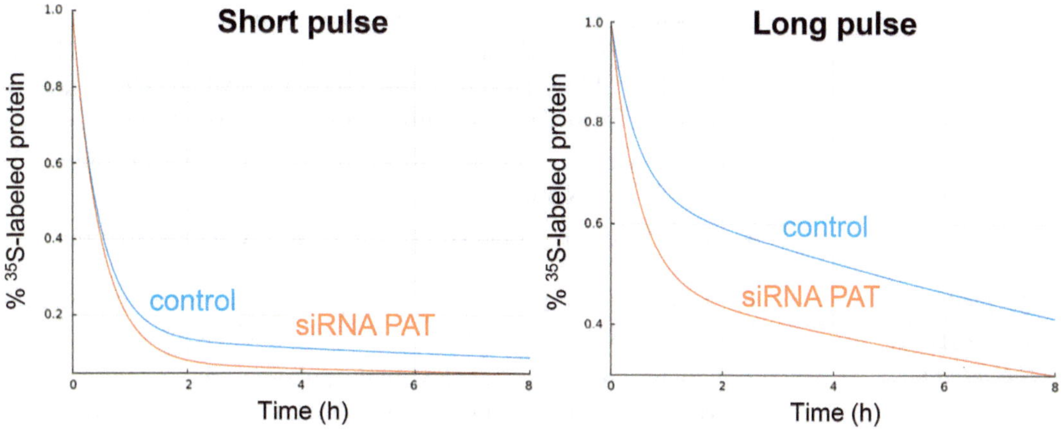

Fig. 5 Simulations of ^{35}S-Cys/Met pulse-chase experiments. Simulation of a simple model of a protein with slow *S*-palmitoylation kinetics and for which *S*-palmitoylation leads to stabilization of the protein. The blue line represents the decay curve for the wild-type protein in control conditions; the orange line represents the decay curve for the protein when its palmitoylation is inhibited by depleting its protein *S*-acyltransferase (PAT) with siRNA

stabilization. When the pulse is short, siRNA silencing of the protein *S*-acyltransferase (PAT) appears to have little impact, which would suggest that *S*-palmitoylation does not affect turnover. However, simulation with a longer pulse time reveals that turnover is faster in the absence of *S*-palmitoylation. This is because during a short pulse, too few molecules have the time to undergo *S*-palmitoylation, so the presence or absence of the PAT makes little or no difference. If the pulse is longer, *S*-palmitoylation has time to occur and thus a difference in the turnover rate is observed in the absence of the PAT.

3 Materials

3.1 Cell Culture

1. Human epithelial cells immortalized: RPE1 cells (ATCC #CRL-4000).
2. Culture Medium: Dulbecco's Modified Eagle Medium (DMEM) GlutaMAX supplement, 10% fetal bovine serum (FBS), 1% penicillin–streptomycin (10,000 U/mL).
3. Trypsin–EDTA 0.005% phenol red.
4. Fugene® 6 transfection reagent diluted in reduced serum medium Opti-MEM medium.

3.2 Media for ^{3}H-Palmitic Acid Incorporation

1. Starvation medium: Glasgow's Minimum Essential Medium with L-Glutamine (GMEM), 10 mM HEPES, pH 7.4. For 5 L, mix GMEM powder with 50 mL 1 M HEPES in water, adjust pH with NaOH and filter using 0.45 μM filter in a tissue culture hood (*see* **Note 1**).

2. Labeling medium: GMEM, 10 mM HEPES, pH 7.4, 200 μCi/mL ^{3}H-palmitic acid {9,10-^{3}H(N)} $CH_3(CH_2)_5CH_2CH_2(CH_2)COOH$, 5 μM, (Bioactif #ART0129-25); 1:50 dilution from a 10 mCi/mL stock in ethanol, 250 μM.
3. Chase Medium: same as Culture Medium (Subheading 3.1, **step 2**) (*see* **Note 2**).
4. Wash Medium: Dulbecco's PBS without calcium/magnesium.

3.3 Media for ^{35}S-Cys/Met Labeling of Cells

1. Starvation medium: DMEM high glucose without L-Glutamine, without L-Methionine, without L-Cysteine.
2. Labeling medium: DMEM high glucose without L-glutamine, without L-methionine, without L-cysteine, 50 μCi/mL ^{35}S-methionine-cysteine. 1:200 dilution from a 10 mCi/mL stock (*see* **Note 3**).
3. Chase medium: Culture medium (Subheading 3.1, **step 2**), 0.015% methionine, 0.05% cysteine (*see* **Note 4**). Filter 15 g/l L-methionine and 50 g/l L-cysteine dissolved in PBS; dilute 1:100 in Culture Medium (Subheading 3.1, **step 2**).
4. Wash medium: Dulbecco's PBS without calcium/magnesium.

3.4 Immunoprecipitation

1. Immunoprecipitation buffer: 500 mM Tris–HCl, pH 7.4, 20 mM EDTA, 10 mM NaF, 30 mM sodium pyrophosphate decahydrate, 1 mM orthovanadate, 2 mM benzamidine, 1 mM PMSF, complete™ protease inhibitor cocktail tablets (Roche), 0.5% NP40.
2. Protein G Sepharose 4 Fast Flow beads.
3. Calnexin monoclonal antibody, clone C8.B6 (Millipore #MAB3126).

3.5 SDS-PAGE

1. 2× Laemmli buffer 2×: 120 mM Tris–HCl, pH 6.8, 20% glycerol, 4% SDS, 0.02% bromophenol blue, 10% beta-mercaptoethanol.
2. SDS-PAGE: 4–20% acrylamide gel, Novex Tris-Glycine Mini gels Wedge well, 10-well.
3. Running buffer SDS-PAGE: 1.44% glycine, 0.1% SDS in 50 mM Tris. Do not adjust the pH.

3.6 Immunoblotting

1. Nitrocellulose membrane.
2. Iblot system with mini transfer stack (Novex #IB301002).
3. Immunoblotting blocking solution: 5% blotting-grade blocker nonfat dry milk in PBS, 0.05% Tween® 20.
4. Washing solution: 0.05% Tween® 20 in PBS.
5. Primary antibody solution: dilution of anti-calnexin antibody 1:2000 in Immunoblotting blocking solution.

6. Secondary antibody solution: dilution of sheep anti-mouse IgG-HRP linked antibody 1:3000 in Immunoblotting blocking solution.
7. Chemiluminescence substrate solution: Super signal west Dura Extended Duration Substrate (Pierce #34076), Mix part A and B 1:1 and prepare just before use.
8. Chemiluminescence detection system—Fusion FX Imaging system (VILBER).

3.7 Fluorography

1. Fixation solution: isopropanol–water–acetic acid (25:65:10).
2. Enhancer solution: Amersham Amplify Fluorography Reagent (*see* **Note 5**).
3. Gel dryer system under vacuum (*see* **Note 6**).
4. X-ray hyper film MP.
5. Screen for a phosphorimager.
6. Typhoon 5 biomolecular Imager (GE Healthcare).

4 Methods

4.1 Cell Culture and Transfection

1. Cultivate RPE1 cells in Culture Medium. Trypsinize and dilute 1:10 twice a week.
2. Seed trypsinized RPE1 cells, 10^6 cells in a 60-mm tissue culture dish.
3. The next day, transfect RPE1 cells with plasmids codifying for calnexin-HA WT and mutants using Fugene® 6. For one 60-mm dish, mix 6 μg of plasmid DNA with 12 μL of Fugene® diluted in 200 μL of Opti-MEM. Let stand for 30 min.
4. During the incubation, replace the medium on the cells with 5 mL fresh Culture Medium.
5. After the 30 min incubation period, add the DNA transfection mix to the cells.
6. Perform radioactive labeling experiments 24 h after transfection. If labeling cells that are not transfected, perform radioactive labeling of endogenous proteins on cells 24 h after trypsinization.

4.2 Steady-State ^{3}H-Palmitic Acid Incorporation

1. Prepare starvation medium, 3 × 5 mL per 60-mm dish.
2. Wash each dish twice with 5 mL starvation medium.
3. Add 5 mL starvation medium in each dish for 1 h at 37 °C in a water bath or cell culture incubator (*see* **Note 7**). During the last 30 min, add 10 μg/mL cycloheximide to block protein synthesis (*see* **Note 8**).

4. Prepare labeling medium, 3 mL per 60-mm dish. Keep 10 μg/mL cycloheximide during the labeling period (*see* **Note 8**).
5. Replace the starvation medium with 3 mL labeling medium on each dish and incubate for 2 h at 37 °C in water bath or incubator. The time of incubation can be increased to 7 h maximum to allow more incorporation of radioactivity (*see* **Note 9**).
6. Following the incubation with ^{3}H-palmitic acid, place cells on ice. Remove medium and wash three times with ice-cold PBS. Collect washes and radioactive media as radioactive waste.
7. Lyse cells with 300 μL of Immunoprecipitation buffer per dish. Place dishes on ice on an orbital shaking platform for 20 min.
8. Collect cells into a 1.5 mL microcentrifuge tube.
9. Spin lysate for 3 min at 2500 × *g* at 4 °C in a tabletop centrifuge.
10. Transfer supernatant to a new 1.5 mL microcentrifuge tube.
11. Continue with immunoprecipitation and SDS-PAGE (Subheading 4.5).

4.3 Depalmitoylation: ^{3}H-Palmitic Acid Pulse–Chase

1. Prepare starvation medium, 3 × 5 mL per 60-mm dish.
2. Wash each dish twice with 5 mL starvation medium.
3. Add 5 mL starvation medium to each dish for 1 h at 37 °C in a water bath or in a cell culture incubator.
4. Prepare labeling medium, 3 mL per 60-mm dish.
5. Remove starvation medium and replace with 3 mL labeling medium and incubate for 2 h at 37 °C in water bath or incubator (*see* **Note 10**).
6. Remove radioactive medium and wash three times with chase medium prewarmed at 37 °C (*see* **Note 2**). Collect washes and radioactive media as radioactive waste.
7. Incubate cells in CO_2 incubator for different times of chase (e.g., 0, 1, 2, 4, 6, 8, 20, 24 h).
8. After each chase incubation time, place cells on ice. Remove medium and wash three times with ice-cold PBS.
9. Lyse cells with 300 μL of Immunoprecipitation buffer per dish, placing the dish on ice on an orbital shaking platform for 20 min.
10. Collect cells into a 1.5 mL microcentrifuge tube.
11. Spin lysate for 3 min at 2500 × *g* at 4 °C in a tabletop centrifuge.

12. Transfer the supernatant to a 1.5 mL micro centrifuge tube. Keep all the supernatants on ice or at −20 °C until the last time point is collected and processed.
13. Continue with immunoprecipitation and SDS-PAGE (Subheading 4.5).

4.4 Pulse–Chase Labeling of Cells with ^{35}S-Cysteine/Methionine

1. Prepare starvation medium, 3 × 5 mL per 60-mm dish.
2. Wash each dish twice with 5 mL starvation medium.
3. Add 5 mL starvation in each dish and incubate for 30 min at 37 °C in a cell culture incubator.
4. Prepare labeling medium, 3 mL per 60-mm dish.
5. Remove starvation medium and replace with 3 mL labeling medium. Incubate for 20 min at 37 °C in a cell culture incubator (*see* **Note 10**).
6. Remove radioactive medium and wash three times with chase medium prewarmed at 37 °C. Collect washes and radioactive media as radioactive waste.
7. Incubate cells in CO_2 incubator for different times of chase (e.g., 0, 1, 2, 4, 6, 8, 20, 24 h).
8. After each time point, place cells on ice. Remove medium and wash three times with ice-cold PBS.
9. Lyse cells with 300 μL of Immunoprecipitation buffer per dish, placing the dish on ice on an orbital shaking platform for 20 min.
10. Collect cells into a 1.5 mL micro centrifuge tube.
11. Spin lysate for 3 min at 2500 × *g* at 4 °C in a tabletop centrifuge.
12. Transfer supernatant to a 1.5 mL microcentrifuge tube. Keep all the supernatants on ice or at −20 °C until the last time point is collected and processed.
13. Continue with immunoprecipitation and SDS-PAGE (Subheading 4.5).

4.5 Immunoprecipitation and SDS-PAGE

1. Prepare Protein G beads for preclearing lysates (50 μL/sample) and immunoprecipitations (50 μL/sample) by washing three times with immunoprecipitation buffer.
2. Preclear the lysates by adding 50 μL of Protein G beads to each sample. Rotate for 30 min at 4 °C (*see* **Note 11**).
3. Spin lysate and beads for 3 min at 2500 × *g* at 4 °C in a tabletop centrifuge.
4. Transfer the supernatant to a new 1.5 mL microcentrifuge tube.
5. Add 1 μL (1 μg) of anti-calnexin antibody and 50 μL of prewashed Protein G beads to the supernatants. Rotate overnight at 4 °C.

6. Wash the beads three times at room temperature with immunoprecipitation buffer.
7. Suspend the beads in 40 μL 2× Laemmli buffer. Vortex, heat for 5 min at 95 °C, vortex, and spin to pellet the beads keeping the supernatant.
8. Divide the samples, performing SDS-PAGE on duplicate gels: one gel for western blotting and one gel to dry to expose for fluorography.

4.6 Western Blots

1. Transfer proteins from the gel to a nitrocellulose membrane using the iBlot semi dry transfer system, 15 V for 15 min.
2. Block membranes with immunoblotting blocking solution 15 min.
3. Incubate with primary antibodies (anti-calnexin 1:2000 in immunoblotting blocking solution) for 2 h at room temperature or overnight at 4 °C.
4. Wash three times with washing solution.
5. Incubate 1 h at room temperature with anti-mouse HRP 1:3000 in immunoblotting blocking solution.
6. Wash 6 × 10 min with washing solution.
7. Incubate membrane 5 min with 4 mL chemiluminescence substrate solution and visualize and quantify calnexin signal using a Fusion FX Imaging system.

4.7 Fluorography and Quantitation

1. Incubate the gel 30 min in fixation solution and then 30 min in enhancer solution.
2. Place the gel on Whatman paper and dry gels for 2 h on gel dryer system (*see* **Note 6**).
3. In dark room, place X-ray film or phosphor screen into a cassette containing the dried gel. Store the cassette at −80 °C (*see* **Note 12**).
4. For ^{3}H-palmitic acid labeling, expose the screen to the dried radioactive gel for several weeks; for ^{35}S-Cys/Met incorporation, expose the screen for 1 day, depending on the abundance of the protein of interest.
5. Allow the cassette to thaw at room temperature before developing the film or scanning the phosphor screen.
6. For ^{3}H-palmitic acid and ^{35}S-Cys/Met incorporation, obtain values for quantification by scanning the phosphor screen with a Typhoon 5 biomolecular Imager.
7. To prepare images of ^{35}S-Cys/Met-labeled proteins, use the scans from the phosphor-imager.
8. To prepare images of ^{3}H-palmitate labeled proteins, use the X-ray film (*see* **Note 13**).

5 Notes

1. The starvation medium for ^{3}H-palmitate pulse chase experiments lacks serum, which is a source of fatty acids.
2. Chase medium for ^{3}H-palmitate experiments is culture medium. The serum in culture medium is the source of exogenous fatty acids to replenish palmitate pools with nonradioactive fatty acid.
3. To increase labeling efficiency of newly synthesized proteins, it is recommended to use two labeled amino acids ^{35}S-Cys/Met, since proteins have few methionines and cysteines.
4. The methionine and cysteine added to the chase medium are nonradioactive amino acids.
5. The use of Amplify fluorography reagent reduces the amount of exposure time for the detection of ^{35}S- and ^{3}H-labeled proteins.
6. To avoid cracking of the gel, do not dry under vacuum more than 2 h.
7. Control the bath or incubator temperature and adjust to 37 °C with precision because ^{3}H-palmitic acid incorporation is temperature dependent.
8. For ^{3}H-palmitic acid incorporation experiments, we recommend blocking protein synthesis before and during the labeling to avoid interference from nascent proteins.
9. Only nonpalmitoylated proteins are available for ^{3}H-palmitate labeling. Thus, a protein that is extensively and stably palmitoylated may lead to a low ^{3}H-palmitate incorporation signal after a 2 h labeling time, which could be misleading.
10. Pulse–chase decay curves, whether ^{35}S-Cys/Met or ^{3}H-palmitate, may vary significantly depending on the time of the pulse. *See* Subheading 2 on experimental design for discussion of this issue.
11. Incubation of the cell lysates with Protein G beads is done to remove proteins that nonspecifically bind to the beads.
12. The cold will increase the signal by increasing the sensitivity of the film.
13. In fluorography, detection of the radiolabeled proteins is indirect. ^{3}H-palmitate labeled proteins excite the fluorophore (enhancer solution, Amplify), which emits light and exposes the X-ray film.

Acknowledgments

We thank Maria-Eugenia Zaballa for her help in finalizing the manuscript. This work benefited from funding from the European Research Council under the European Union's Seventh Framework Programme (FP/2007-2013)/ERC Grant Agreement no. 340260—PalmERa. This work was also supported by grants from the Swiss National Science Foundation (to G.v.d.G), and the Swiss SystemsX.ch initiative evaluated by the Swiss National Science Foundation (LipidX) (to G.v.d.G).

References

1. Blaskovic S, Blanc M, van der GFG (2013) What does *S*-palmitoylation do to membrane proteins? FEBS J 280:2766–2774
2. Chamberlain LH, Shipston MJ (2015) The physiology of protein *S*-acylation. Physiol Rev 95:341–376
3. Fukata Y, Murakami T, Yokoi N et al (2016) Local palmitoylation cycles and specialized membrane domain organization. Curr Top Membr 77:97–141
4. Gottlieb CD, Linder ME (2017) Structure and function of DHHC protein *S*-acyltransferases. Biochem Soc Trans 45:923–928
5. Lin DTS, Conibear E (2015) ABHD17 proteins are novel protein depalmitoylases that regulate N-Ras palmitate turnover and subcellular localization. elife 4:e11306
6. Won SJ, Cheung See Kit M, Martin BR (2018) Protein depalmitoylases. Crit Rev Biochem Mol Biol 53:83–98
7. Perrody E, Abrami L, Feldman M et al (2016) Ubiquitin-dependent folding of the Wnt signaling coreceptor LRP6. elife 5:e19083
8. Daniotti JL, Pedro MP, Valdez Taubas J (2017) The role of *S*-acylation in protein trafficking. Traffic 18:699–710
9. Abrami L, Dallavilla T, Sandoz PA et al (2017) Identification and dynamics of the human ZDHHC16-ZDHHC6 palmitoylation cascade. elife 6:e27826
10. Dallavilla T, Abrami L, Sandoz PA et al (2016) Model-driven understanding of palmitoylation dynamics: regulated acylation of the endoplasmic reticulum chaperone calnexin. PLoS Comput Biol 12:e1004774
11. Lakkaraju AK, Abrami L, Lemmin T et al (2012) Palmitoylated calnexin is a key component of the ribosome–translocon complex. EMBO J 31:1823–1835
12. Forrester MT, Hess DT, Thompson JW et al (2011) Site-specific analysis of protein *S*-acylation by resin-assisted capture. J Lipid Res 52:393–398
13. Hurst CH, Turnbull D, Plain F et al (2017) Maleimide scavenging enhances determination of protein *S*-palmitoylation state in acyl-exchange methods. BioTechniques 62:69–75
14. Gao X, Hannoush RN (2017) A decade of click chemistry in protein palmitoylation: impact on discovery and new biology. Cell Chem Biol 25:236–246
15. Hancock JF (1995) [24] Prenylation and palmitoylation analysis. In: Methods in enzymology. Elsevier, Amsterdam, pp 237–245
16. O'Brien PJ, Zatz M (1984) Acylation of bovine rhodopsin by [3H]palmitic acid. J Biol Chem 259:5054–5057
17. Howie J, Reilly L, Fraser NJ et al (2014) Substrate recognition by the cell surface palmitoyl transferase DHHC5. Proc Natl Acad Sci 111:17534–17539
18. Percher A, Ramakrishnan S, Thinon E et al (2016) Mass-tag labeling reveals site-specific and endogenous levels of protein S-fatty acylation. Proc Natl Acad Sci 113:4302–4307
19. Yokoi N, Fukata Y, Sekiya A et al (2016) Identification of PSD-95 depalmitoylating enzymes. J Neurosci 36:6431–6444
20. O'Dowd BF, Hnatowich M, Caron MG et al (1989) Palmitoylation of the human beta

2-adrenergic receptor. Mutation of Cys341 in the carboxyl tail leads to an uncoupled nonpalmitoylated form of the receptor. J Biol Chem 264:7564–7569

21. Dekker FJ, Rocks O, Vartak N et al (2010) Small-molecule inhibition of APT1 affects Ras localization and signaling. Nat Chem Biol 6:449–456
22. Won SJ, Davda D, Labby KJ et al (2016) Molecular mechanism for isoform-selective inhibition of acyl protein thioesterases 1 and 2 (APT1 and APT2). ACS Chem Biol 11:3374–3382
23. Esposito AM, Kinzy TG (2014) In vivo [35 S]-methionine incorporation. In: Methods in enzymology. Elsevier, Amsterdam, pp 55–64
24. Davda D, El Azzouny MA, Tom CTMB et al (2013) Profiling targets of the irreversible palmitoylation inhibitor 2-bromopalmitate. ACS Chem Biol 8:1912–1917

Chapter 10

Fluorogenic Assays for the Defatty-Acylase Activity of Sirtuins

Jun Young Hong, Ji Cao, and Hening Lin

Abstract

Sirtuins are type III histone deacetylases (HDAC) that uses nicotinamide adenine dinucleotide as cosubstrate. Dysfunction of sirtuins is implicated in wide varieties of human diseases. As such, there has been increased interest in the development of small molecule to modulate sirtuin activities. Besides deacetylase activity, recent studies suggest SIRT1, 2, 3, and 6 efficiently remove fatty acyl groups on lysine. In vitro sirtuin enzymatic activity assays established so far are mainly based on the deacetylation activity. Here, we describe a fluorogenic assay for monitoring defatty-acylase activity of SIRT1, 2, 3 and 6 using peptide substrates. This assay can be utilized to evaluate sirtuin modulators in high-throughput manners.

Key words Fluorogenic assay, Enzymatic assay, SIRT1, SIRT2, SIRT3, SIRT6, Defatty-acylase

1 Introduction

Sirtuins (SIRTs), which are evolutionarily conserved in all domains of life, are the type III histone deacetylases (HDACs). Seven sirtuins have been identified in mammals and can be classified into four classes. SIRT1, 2, and 3 belong to class I. SIRT4 is in class II, SIRT5 is in class III, while SIRT6 and 7 are both in class IV [1, 2]. Sirtuins are known to influence a wide range of physiological processes, such as aging, transcription, stress response, and inflammation [2, 3]. Initially, sirtuins were known to act as nicotinamide adenine dinucleotide (NAD^+)-dependent protein lysine deacetylases. In recent years, however, it was discovered that SIRT1, 2, 3, and 6 can remove long-chain fatty acyl groups, like myristoyl and palmitoyl, from protein lysine residues with high catalytic efficiency [4–7]. Of importance, by removing fatty acyl groups, SIRT2 positively affects the transforming ability of KRas4A [8], while SIRT6 functions as a tumor suppressor by disrupting the plasma membrane localization of RRas2 [9]. These findings highlight the emerging key roles of sirtuin-regulated lysine fatty acylation in regulating cell functions.

Maurine E. Linder (ed.), *Protein Lipidation: Methods and Protocols*, Methods in Molecular Biology, vol. 2009,
https://doi.org/10.1007/978-1-4939-9532-5_10,

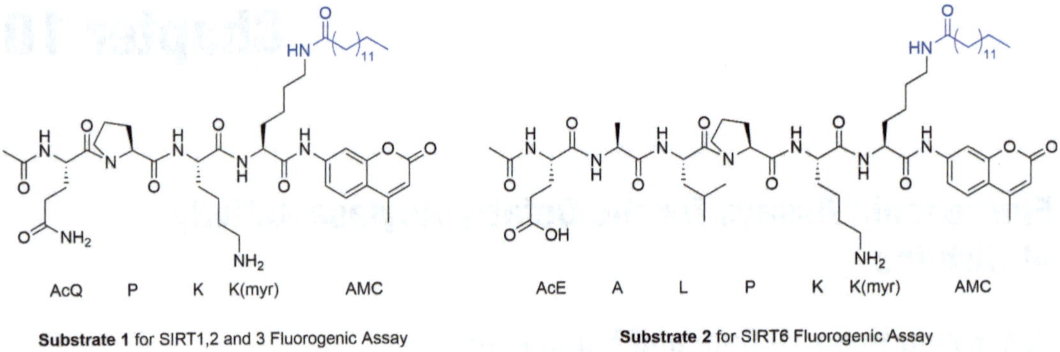

Fig. 1 Fluorogenic assay substrates

Because of sirtuins' extensive regulatory activities and their pivotal role in human diseases, there has been increased interest in developing activators and inhibitors of sirtuins for various potential therapeutic applications [10, 11]. Since deacetylation was the initial accepted activity of sirtuins, established high-throughput assays for screening sirtuin modulators are mainly based on their deacetylation activity. Very few methods have been developed to assess defatty-acylation activity. It is possible that the same compound may exhibit different abilities to inhibit the deacetylation and defatty-acylation activities of sirtuins. Thus, screening assays based on the defatty-acylation activities of sirtuin can be useful. Here, we describe a high-throughput assay protocol for screening small-molecule modulators of the defatty-acylase activity of SIRT1, 2, 3 and 6 [12, 13].

The fluorogenic substrates used in this assay are designed based on sirtuins' natural substrate peptides, p53 for SIRT1, 2, and 3 and tumor necrosis factor-α (TNF-α) for SIRT6, with myristoylated lysine and 7-amino-4-methylcoumarin moiety (AMC) at the C-termini of the peptide chains. Our group has previously demonstrated that these substrates give stronger fluorescence read out than other substrates tried [12, 13]. Substrate 1 (Fig. 1) is used for SIRT1, 2 and 3, while substrate 2 is used for SIRT6. Using NAD^+ as the cosubstrate, the sirtuins remove the myristoyl group on the lysine residues in the peptides, which allows trypsin to cleave off the AMC group. Once released, the AMC group is fluorescent (excitation/emission maximal wavelengths: 355/460 nm). The strength of the fluorescent signal portrays the sirtuin activity. In principle, a stronger signal reflects a higher activity of the sirtuin. This assay can be performed in 96-well plates, which allows testing multiple compounds at different concentrations at the same time. If a small molecule inhibits the sirtuin effectively, the well treated with the small molecule will exhibit a substantially lower fluorescent signal than vehicle-treated wells. Based on the results, the concentration of the inhibitor and the fluorescent signal can be plotted to calculate the IC_{50} or EC_{50} values (concentrations needed to reduce or activate the enzyme by 50%).

2 Materials

2.1 Expression and Purification of Sirtuins

1. SIRT1 cDNA (full-length) in pET28a vector with N-terminal His6 tag.
2. SIRT2 cDNA (aa 38–256) in pET28a vector with N-terminal His6-SUMO tag.
3. SIRT3 cDNA (aa 102–399) in pET28a vector with N-terminal His6 tag.
4. SIRT6 cDNA (full-length) in pET28a vector with an N-terminal His6 tag.
5. Competent *E. coli* BL21 cells (use for SIRT1, SIRT2, and SIRT3).
6. Competent *E. coli* BL21 pRARE2 cells (use for SIRT6).
7. Culture Medium: Luria Broth, 20 μg/mL chloramphenicol, 50 μg/mL kanamycin.
8. 1 M isopropyl β-D-1-thiogalactopyranoside (IPTG) in water.
9. TNGP buffer: 20 mM Tris–HCl (pH 8.0), 500 mM NaCl, 10% glycerol, and 1 mM phenylmethane sulfonyl fluoride (PMSF).
10. TN buffer: 20 mM Tris–HCl (pH 8.0) and 500 mM NaCl.
11. Imidazole Elution Buffers for step gradient: TN buffer with 20 mM, 50 mM, 100 mM, 200 mM, or 500 mM imidazole.
12. Nickel column.
13. Dialysis tubing with 14,000 Molecular Weight Cutoff.
14. Superdex 75 gel filtration column.
15. 50% glycerol.
16. ULP1 (SUMO protease).

2.2 Fluorogenic Assay

1. 96-well plate with solid black bottom (*see* **Note 1**).
2. Fluorescence plate reader (*see* **Note 2**).
3. 10× trypsin–EDTA (0.5%), no phenol red.
4. Plate incubator.

2.3 Fluorogenic Assay for the Demyristoylase Activity of SIRT1, 2, and 3

1. Reaction buffer: 50 mM Tris–HCl (pH 8.0), 150 mM NaCl, 2.7 mM KCl, 1 mM $MgCl_2$ (*see* **Note 3**).
2. Enzyme: 20 μM of SIRT1, SIRT2, or SIRT3 in 20 mM Tris–HCl (pH 8.0), 500 mM NaCl, and 10% glycerol (*see* **Note 4**).
3. NAD^+: stock solution of 25 mM in water (*see* **Note 5**).
4. Quench buffer: 10× trypsin–EDTA (0.5%) without phenol red, 4 mM nicotinamide (*see* **Note 6**).
5. Substrate 1: 0.5 mM in DMSO (*see* **Note** 7).

6. Activator/Inhibitor: various concentrations dissolved in vehicle solvent (e.g., DMSO).
7. Vehicle control: vehicle solvent that the activator/inhibitor is dissolved in.

2.4 Fluorogenic Assay for the Demyristoylase Activity of SIRT6

1. Reaction Buffer: 20 mM Tris–HCl (pH 8.0), 1 mM dithiothreitol (DTT), 1 mg/mL of bovine serum albumin (BSA) (*see* **Note 8**).
2. Enzyme: 20 μM of SIRT6 in 20 mM Tris–HCl (pH 7.2), 500 mM NaCl, and 10% glycerol.
3. NAD^+: solution of 25 mM in water (*see* **Note 5**).
4. Quench buffer: 10× trypsin–EDTA (0.5%) with 4 mM nicotinamide without phenol red (*see* **Note 6**).
5. Substrate 2: 0.5 mM in DMSO (*see* **Note 7**).
6. Activator/Inhibitor: varied concentrations in vehicle solvent (e.g., DMSO).
7. Vehicle control: vehicle solvent that the activator/inhibitor is dissolved in.

3 Method

3.1 Purification of SIRT1, SIRT3, or SIRT6

1. Transform the SIRT1 or SIRT3 plasmid into *E. coli* BL21 cells; transform the SIRT6 plasmid into *E. coli* BL21 cells—pRARE2 strain.
2. Pick a colony from an agar plate, and grow in 25 mL Luria Broth with corresponding bacterial resistance antibiotic overnight at 37 °C.
3. Inoculate 2 L culture medium with the overnight culture.
4. Incubate the cells at 37 °C for 2–4 h.
5. When the optical density at 600 nm reaches ~0.6, add 40 μL 1 M IPTG to the culture (final concentration, 20 μM) to induce expression of sirtuins.
6. Incubate the cells for an additional 16–20 h at 15 °C and 200 rpm.
7. Collect the cells by centrifugation at 4 °C, 8000 rpm (11,295 × *g*) for 5 min using Beckman Coulter JA-10 rotor.
8. Suspend the cell pellet in 20 mL of TNGP buffer and lyse using a cell disrupter.
9. Remove cell debris by centrifuging at 4 °C, 20,000 rpm (48,254 × *g*) for 30 min using Beckman Coulter JA-20 rotor.
10. Apply the supernatant containing the sirtuin protein to a nickel column equilibrated with TN buffer.

11. Use a step gradient to elute the sirtuin protein from the nickel resin. Add 1.5 mL of elution buffers containing increasing concentrations (20 mM, 50 mM, 100 mM, 200 mM, 500 mM) of imidazole sequentially; collect as individual fractions.
12. Resolve aliquots of the elutions by SDS-PAGE and stain with Coomassie Blue to identify fractions containing the sirtuin protein.
13. Pool the desired fractions and dialyze at 4 °C for 16–20 h in TN buffer to remove imidazole.
14. If the protein still contains other impurities, further purify by gel filtration using a Superdex 75 gel filtration column equilibrated in TN buffer.
15. After the purification, concentrate the protein using a centrifugal concentrator.
16. Add 50% glycerol to the enzyme solution to a final concentration of 10%. The desired final enzyme concentration is 20 μM.
17. Store the purified protein at −80 °C in aliquots.

3.2 Purification of SIRT2

1. Transform the SIRT2 plasmid into *E. coli* BL21 cells.
2. Follow **steps 2–11** in Subheading 3.1.
3. Pool fractions with SIRT2 into dialysis tubing, adding 0.1 mL of ULP1, the SUMO protease that will cleave the His6-SUMO tag on SIRT2.
4. Dialyze the mixture at 4 °C for 16–20 h in TN buffer.
5. If desired, further purify SIRT2 by gel filtration on a Superdex 75 column.
6. Follow **steps 14–16** in Subheading 3.1.

3.3 Fluorogenic Assay for the Demyristoylase Activity of SIRT1, 2, and 3 (See Note 9)

1. Thaw the purified SIRT1, 2, or 3 on ice.
2. To 48.4 μL of the reaction buffer, add 2.4 μL of 25 mM NAD^+, and 6 μL of 20 μM sirtuin enzyme. Place the plate on ice (*see* **Note 10**).
3. Add 2 μL of activator/inhibitor or vehicle control to the reaction buffer, mix well by pipetting up and down, and incubate the plate at 37 °C for 15 min (*see* **Note 11**).
4. Add 1.2 μL of 0.5 mM fluorogenic substrate 1 to the well and mix. The total volume of the reaction system is 60 μL and the final concentration of the substrate is 10 μM (*see* **Note 12**).
5. Incubate the plate at 37 °C for 45 min.
6. Add 60 μL of the quench buffer to the well and mix by pipetting up and down a few times.

7. Incubate the plate at 37 °C for 30 min.
8. Measure the fluorescence at excitation of 355 nm and emission of 460 nm.

3.4 Fluorogenic Assay for the Demyristoylase Activity of SIRT6 (See Note 13)

1. Thaw the purified SIRT6 on ice.
2. To 51.4 μL of the reaction buffer, add 2.4 μL of 25 mM NAD^+ (*see* **Note 10**).
3. Add 3 μL of 20 μM SIRT6 enzyme to the well.
4. Add 2 μL of activator/inhibitor or vehicle control to the reaction buffer, mix, and incubate the plate at 37 °C using the dry block incubator for 15 min (*see* **Note 11**).
5. Add 1.2 μL of 0.5 mM fluorogenic substrate 2 to the well and mix (*see* **Note 12**).
6. Incubate the plate at 37 °C for 2 h.
7. Add 60 μL of the quench buffer and mix.
8. Incubate the plate at 37 °C for 30 min.
9. Measure the fluorescence at excitation of 355 nm and emission of 460 nm.

4 Notes

1. We have been using a Costar assay plate, 96 well, flat bottom, nontreated nonsterile, black polystyrene plate from Corning (Cat#3916).
2. The plate reader should be equipped to measure the fluorescence which excites at 355 nm and emits at 460 nm.
3. Stock solutions for the reaction buffer: 1 M Tris–HCl (pH 8.0) (20× solution), 1.5 M NaCl (10× solution), 2.7 M KCl (1000× solution), and 1 M $MgCl_2$ (1000× solution).
4. Avoid repetitive freeze–thaw of the enzyme. Each time the enzyme is thawed, it may lose activity slightly. Aliquot the enzyme into smaller portions and store at −80 °C for future usage.
5. Prepare NAD^+ stock solution fresh every time.
6. Nicotinamide, a universal sirtuin inhibitor, will stop the enzyme activity during the trypsin cleavage.
7. The fluorogenic substrates are prepared by solid-state peptide synthesis, and AMC is coupled to the C-terminal of the peptides using isobutyl chloroformate and N-methylmorpholine in dichloromethane. For more detailed synthesis procedures, please refer to Chiang et al. (Org. Biolmol. Chem., 2016, **14**, 2186) for substrate 1 and Hu et al. (Org. Biomol. Chem., 2013, **11**, 5213) for substrate 2.

8. Prepare a stock solution of the reaction buffer. To 41.8 μL of water, add 1.2 μL of 1 M Tris–HCl (pH 8.0), 2.4 μL of 25 mM DTT, and 6 μL of 10 mg/mL BSA in H_2O. This can be scaled up depending on how many reactions are needed (each reaction volume is 60 μL).
9. In the assay, SIRT1, 2, or 3 is incubated with fluorogenic substrate 1, NAD^+, and a small molecule modulator or vehicle control. If the small molecule is a mechanism-based inhibitor, the enzyme is preincubated with NAD^+ and the inhibitor before the addition of the fluorogenic substrate. After 15 min of preincubation, the fluorogenic substrate is added to the reaction system, and the reaction is stopped by adding the quench buffer, which contains trypsin and nicotinamide. Trypsin will cleave off the AMC group from the demyristoylated peptide, producing a fluorescent signal. An excessive amount of nicotinamide, which inhibits the sirtuins, is added to quench the activity of the enzyme [14]. The stronger the fluorescent signal, the stronger the sirtuin activity. Thus, if the activator is working efficiently, the signal will be significantly stronger than that of the control group.
10. A master mix of the reaction components multiplied by the number of the reaction tubes can be utilized. The components include everything except the inhibitor and the fluorogenic peptide substrate, which will be added separately. This way, one can test multiple compounds at multiple concentrations simultaneously.
11. This preincubation can be skipped, if the tested modulator is not mechanism-based. For the vehicle control sample, add the vehicle solution instead of the modulator.
12. Using a multichannel pipettor throughout the assay can expedite the process.
13. The assay principle for SIRT6 is the same as that described in **Note 9**.

References

1. Imai S, Guarente L (2014) NAD+ and sirtuins in aging and disease. Trends Cell Biol 24 (8):464–471. https://doi.org/10.1016/j.tcb.2014.04.002
2. Hu J, Jing H, Lin H (2014) Sirtuin inhibitors as anticancer agents. Future Med Chem 6 (8):945–966. https://doi.org/10.4155/fmc.14.44
3. Guarente L (2011) Sirtuins, aging, and medicine. N Engl J Med 364(23):2235–2244. https://doi.org/10.1056/NEJMra1100831
4. Feldman JL, Baeza J, Denu JM (2013) Activation of the protein deacetylase SIRT6 by long-chain fatty acids and widespread deacylation by mammalian sirtuins. J Biol Chem 288 (43):31350–31356. https://doi.org/10.1074/jbc.C113.511261
5. He B, Hu J, Zhang X, Lin H (2014) Thiomyristoyl peptides as cell-permeable Sirt6 inhibitors. Org Biomol Chem 12(38):7498–7502. https://doi.org/10.1039/c4ob00860j
6. Teng Y-B, Jing H, Aramsangtienchai P, He B, Khan S, Hu J, Lin H, Hao Q (2015) Efficient

demyristoylase activity of SIRT2 revealed by kinetic and structural studies. Sci Rep 5:8529. https://doi.org/10.1038/srep08529

7. Zhang X, Khan S, Jiang H, Antonyak MA, Chen X, Spiegelman NA, Shrimp JH, Cerione RA, Lin H (2016) Identifying the functional contribution of the defatty-acylase activity of SIRT6. Nat Chem Biol 12(8):614–620. https://doi.org/10.1038/nchembio.2106
8. Jing H, Zhang X, Wisner SA, Chen X, Spiegelman NA, Linder ME, Lin H (2017) SIRT2 and lysine fatty acylation regulate the transforming activity of K-Ras4a. eLife 6. https://doi.org/10.7554/eLife.32436
9. Zhang X, Spiegelman NA, Nelson OD, Jing H, Lin H (2017) SIRT6 regulates Ras-related protein R-Ras2 by lysine defatty-acylation. eLife 6: e25158. https://doi.org/10.7554/eLife.25158
10. Outeiro TF, Kontopoulos E, Altmann SM, Kufareva I, Strathearn KE, Amore AM, Volk CB, Maxwell MM, Rochet J-C, McLean PJ, Young AB, Abagyan R, Feany MB, Hyman BT, Kazantsev AG (2007) Sirtuin 2 inhibitors rescue alpha-synuclein-mediated toxicity in models of Parkinson's disease. Science 317 (5837):516–519. https://doi.org/10.1126/science.1143780
11. Jing H, Hu J, He B, Negrón Abril YL, Stupinski J, Weiser K, Carbonaro M, Chiang Y-L, Southard T, Giannakakou P, Weiss RS, Lin H (2016) A SIRT2-selective inhibitor promotes c-Myc oncoprotein degradation and exhibits broad anticancer activity. Cancer Cell 29(3):297–310. https://doi.org/10.1016/j.ccell.2016.02.007
12. Chiang YL, Lin H (2016) An improved fluorogenic assay for SIRT1, SIRT2, and SIRT3. Org Biomol Chem 14(7):2186–2190. https://doi.org/10.1039/c5ob02609a
13. Hu J, He B, Bhargava S, Lin H (2013) A fluorogenic assay for screening Sirt6 modulators. Org Biomol Chem 11(32):5213–5216. https://doi.org/10.1039/c3ob41138a
14. Avalos JL, Bever KM, Wolberger C (2005) Mechanism of sirtuin inhibition by nicotinamide: altering the NAD(+) cosubstrate specificity of a Sir2 enzyme. Mol Cell 17 (6):855–868. https://doi.org/10.1016/j.molcel.2005.02.022

Chapter 11

Global Profiling of Sirtuin Deacylase Substrates Using a Chemical Proteomic Strategy and Validation by Fluorescent Labeling

Shuai Zhang, Nicole A. Spiegelman, and Hening Lin

Abstract

Protein fatty-acylation is an important posttranslational modification (PTM) and has been associated with many fundamental biological processes. Sirtuins, the nicotinamide adenine dinucleotide (NAD)-dependent class of histone deacetylases have been reported to possess lysine defatty-acylase activity. Comprehensive substrate profiling of sirtuins will help to establish the function of both protein lysine fatty acylation and its regulation by sirtuins. Here, we describe a chemical proteomic strategy to globally profile sirtuin defatty-acylation substrates and a fluorescent labeling method to validate sirtuin substrates.

Key words Protein fatty acylation, Sirtuin substrates, Proteomic profiling, SILAC, Fluorescent labeling

1 Introduction

Most of our current understanding of protein lipidation comes from studies of *N*-terminal glycine myristoylation (*N*-myristoylation), cysteine palmitoylation (*s*-palmitoylation), cysteine prenylation, and modifications by GPI anchors [1]. However, recently it has been reported that NAD^+-dependent sirtuins (SIRT) 1, 2, 3, 6, and 7, as well as the zinc-dependent histone deacetylase 8 (HDAC8) possess lysine defatty-acylase activity, suggesting that protein lysine fatty acylation is an understudied, but important PTM [2–5]. To date, the only mammalian proteins that have been identified to have lysine fatty acylation are tumor necrosis factor alpha (TNF-α), interleukin-1 alpha, lens integral membrane

Shuai Zhang and Nicole A. Spiegelman contributed equally to this work.

Maurine E. Linder (ed.), *Protein Lipidation: Methods and Protocols*, Methods in Molecular Biology, vol. 2009, https://doi.org/10.1007/978-1-4939-9532-5_11,

protein aquaporin-0, Ras Related Protein 2 (R-Ras2), K-Ras4a and H-Ras, and several members of the Rac family of small GTPases [3, 6–11]. Among these, TNF-α, R-Ras2, and K-Ras4a are all sirtuin defatty-acylation substrates [3, 6, 7]. Profiling the defatty-acylation substrates of sirtuins will not only help to elucidate the physiological function of sirtuins, but will also help to understand the role of protein lysine fatty acylation.

Chemical proteomic technology has allowed the global profiling of fatty-acylated proteins, especially *N*-myristoylated and *S*-palmitoylated proteins, in cells and even animals [12, 13]. Fatty acid analogs functionalized with either an alkyne or azide group, such as Alk14 (Fig. 1A), can be metabolically incorporated into fatty-acylated proteins [14]. Utilizing bioorthogonal click chemistry, a biotin tag can be introduced to Alk14-labeled proteins for selective enrichment of fatty-acylated proteins on streptavidin beads. Treatment of the immobilized proteins with hydroxylamine, which cleaves fatty acids linked to cysteine, further enriches proteins that are fatty acylated at lysine residues, that can then be identified through proteomic studies. To specifically identify sirtuin defatty-acylase substrates, we have used a quantitative proteomics technique, SILAC (stable isotope labeling with amino acids in cell culture), to identify fatty-acylated proteins in sirtuin wild type (WT) and knock down or knock out (KD/KO) cells (Fig. 1a) [7]. If a protein is a substrate for the defatty-acylase activity of the sirtuin, then it should have more Alk14 labeling in the KD or KO cells. The use of SILAC mitigates sample preparation variation, such as click chemistry reaction efficiency, producing more reliable quantitative results. SILAC proteomics using alkyne fatty acid analogues can also be applied to cells treated with small molecules that inhibit defatty-acylase activity of sirtuins as an alternative approach to identifying substrates.

To validate candidate substrates identified using the proteomics method described above, we compare the fatty acylation level of a specific protein in sirtuin WT and KD/KO cells by in-gel fluorescence. The protein of interest is expressed or overexpressed in WT and sirtuin KD/KO cells treated with the Alk14 probe and then isolated by immunoprecipitation. An azide-tagged fluorophore can be attached by click chemistry and the sirtuin-dependent fatty acylation can be detected through in-gel fluorescence (Fig. 1b). Confirmation that the labeled protein is fatty acylated at lysine is assessed by hydroxylamine resistance of the in-gel fluorescence.

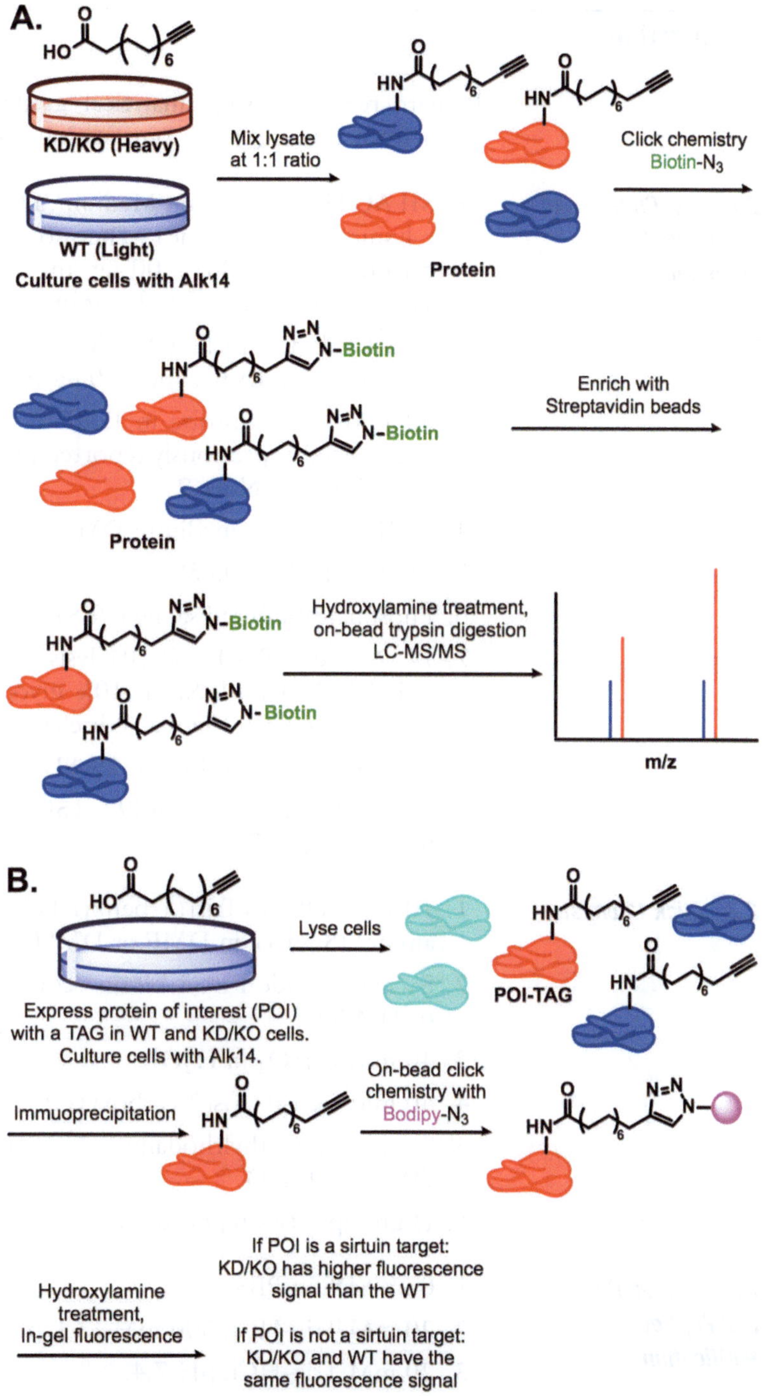

Fig. 1 Workflows for (**a**) profiling of sirtuin deacylation substrates and (**b**) fluorescent labeling of fatty-acylated proteins

2 Materials

Prepare reagents using analytical grade reagents and ultrapure water from a MilliQ water purification system.

2.1 Cell Culture and Alkyne Probe Treatment

1. SILAC Heavy Labeling Medium: DMEM for SILAC (without L-lysine and L-arginine) (Thermo Fisher), 10% dialyzed fetal bovine serum (FBS), 100 μg/mL [$^{13}C_6$, $^{15}N_2$]-L-lysine, and 100 μg/mL [$^{13}C_6$, $^{15}N_4$]-L-arginine (Sigma) (*see* **Note 1**).
2. SILAC Light Labeling Medium: DMEM, 10% dialyzed FBS, 100 μg/mL L-lysine, and 100 μg/mL L-arginine.
3. 50 mM Alkyne probe in DMF or DMSO: Alk12 or Alk14 synthesized as previously reported [15], or purchased (Caymen Chemical) (*see* **Note 2**).
4. DMEM Culture medium: DMEM, 10% FBS.
5. Trypsin–EDTA (0.05%).
6. Phosphate-buffered saline (PBS).
7. 1% Nonidet P-40 (NP40) lysis buffer: 25 mM Tris–HCl, pH 7.4, 150 mM NaCl, 10% glycerol, 1% NP-40, protease inhibitor cocktail (Sigma Aldrich) (add just before use).
8. SDS Solubilization Buffer: 2% SDS and 50 μM EDTA in PBS.
9. Brij Buffer: 1% (w/v) Brij97, 150 mM NaCl, 50 mM triethanolamine (pH 7.4).

2.2 Click Chemistry

1. 10 mM TBTA (Tris[(1-benzyl-1H-1,2,3-triazol-4-yl)methyl] amine) (Sigma), in DMF or DMSO.
2. 5 mM Azide-PEG3-biotin conjugate (Sigma) in DMF or DMSO.
3. 40 mM $CuSO_4$ in H_2O.
4. 40 mM TCEP (tris(2-carboxyyethyl)phosphine) in H_2O.
5. 2 mM tetramethylrhodamine (TAMRA) azide (Lumiprobe) in DMF or DMSO.
6. High capacity streptavidin agarose (Thermo Fisher).

2.3 Trypsin Digestion and Peptide Purification

1. 0.2% SDS in PBS.
2. 20 mM Tris–HCl, 500 mM KCl, pH 7.4.
3. 20 mM Tris–HCl, pH 7.4.
4. 6 M urea, 9.5 mM TCEP in PBS.
5. 400 mM iodoacetamide in water.
6. 2 M urea in PBS.
7. 2 M urea in PBS, 1 mM $CaCl_2$.

8. Trypsin (Promega).
9. 10% trifluoroacetic acid (TFA).
10. 90% methanol, 0.1% TFA.
11. 0.1% TFA.
12. 80% acetonitrile, 0.1% TFA.
13. Sep-Pak Vac C18 cartridge (Waters).

2.4 Immuno-precipitation and Fluorescent Labeling

1. NP-40 immunoprecipitation wash buffer: 25 mM Tris–HCl, pH 7.4, 150 mM NaCl, and 0.2% NP-40.
2. 6× Protein loading dye: 60 mM Tris (pH 6.8), 0.12% SDS (w/v), 47% glycerol, 0.6 M DTT, and 0.0006% bromophenol blue (w/v).
3. Coomassie blue stain: 10% acetic acid, 40% methanol, 50% H_2O, and 0.25% Blue R250 Dye (Sigma Aldrich) (w/v).
4. Destaining solution: (v/v) 10% acetic acid, 40% methanol, and 50% H_2O.

3 Methods

3.1 Cell Culture and Alkyne Probe Treatment for Proteomic Profiling

1. Culture sirtuin WT and KD/KO cells in SILAC Light Labeling Medium (medium with L-lysine and L-arginine) and SILAC Heavy Labeling Medium (medium with [$^{13}C_6$, $^{15}N_2$]-L-lysine and [$^{13}C_6$, $^{15}N_4$]-L-arginine) medium, respectively, for six generations (*see* **Notes 1** and **3**). To generate the heavy or light cell lines, culture cells in 6-well dishes, seeding 10×10^4 cells per well. Split the cells at 80% confluency.
2. After passing the cell lines for six generations, culture the cells in 10-cm dishes. Depending on the cell line being used, the experiment may require 5–10 dishes. When cell confluency reaches 80%, treat the cells with fatty acid alkyne probe (Alk12 or Alk14) at a final concentration of 50 μM for 6 h (*see* **Note 4**).
3. Harvest the cells by scraping and centrifuge for 5 min at $500 \times g$ at 4 °C.
4. Wash the cells twice with 15 mL of ice-cold PBS. Centrifuge for 3 min at $500 \times g$ at 4 °C. Remove the supernatant and retain the cell pellet.

3.2 Click Chemistry in Total Cell Lysates

1. Prepare ice-cold methanol, ice-cold chloroform, and ice-cold water before starting the reaction.
2. Lyse the cells with 1% NP-40 lysis buffer by adding 5 mL of lysis buffer supplemented with protease inhibitor cocktail and place on a nutating rocker for 30 min (*see* **Notes 5** and **6**). After 30 min, remove insoluble material by centrifuging at

17, 000 × *g* at 4 °C for 30 min. Quantify the protein concentration of the whole cell lysate (*see* **Note 7**).

3. Mix equal amounts of heavy and light samples. Typically, 5–7.5 mg of each whole cell lysate is needed.
4. To set up the click chemistry reaction, add 10–15 mg of the mixed lysate per tube and make sure the protein concentration is ~2 mg/mL. Add additional 1% NP-40 lysis buffer supplemented with protease inhibitor cocktail to bring the total reaction volume to 4.6 mL. Add 150 μL of 5 mM azide-biotin, followed by 100 μL of 10 mM TBTA, 100 μL of 40 mM TCEP and 100 μL of 40 mM $CuSO_4$. The final concentrations in the reaction should be: 200 μM TBTA, 0.8 mM $CuSO_4$, 0.8 mM TCEP, and 150 μM Azide-biotin (*see* **Notes 8** and **9**).
5. Incubate for 30 min at room temperature.
6. Precipitate the proteins with ice-cold methanol–chloroform–-water (v/v ratio of sample–methanol–chloroform–water is 1:4:1:3) solution. For a 5-mL click chemistry reaction, add 20 mL of methanol, 5 mL of chloroform, and 15 mL of water.
7. Vortex the mixture, and then centrifuge at 4 °C, 17,000 × *g* for 15 min or 4500 × *g* for 30 min.
8. Remove and discard the upper aqueous phase, leaving the protein layer and lower organic phase.
9. Add 20 mL of ice-cold methanol to the sample and mix gently, causing the protein pellet to sink to the bottom of the tube. Centrifuge at 4 °C, 17,000 × *g* for 10 min, or 4500 × *g* for 20 min.
10. Remove the liquid by pipetting, and be careful not to disturb the pellet. Wash the protein pellet by adding 20 mL of ice-cold methanol and inverting the tube. Centrifuge at 4 °C for 10 min at 17,000 × *g* or at 4500 × *g* for 20 min.
11. Remove the methanol by pipetting. Allow the remaining methanol to evaporate by letting the pellet air dry for about 10 min at room temperature. Do not let the protein pellet dry completely. If the protein pellet is allowed to dry completely, it will be difficult to solubilize the precipitated protein.
12. Add SDS Solubilization Buffer and vortex to solubilize the protein pellet (*see* **Note 10**).
13. Centrifuge for 5 min at 14,000 × *g* at room temperature, and collect the supernatant. Dilute the sample to a volume of 10 mL with PBS buffer to ensure that the final SDS concentration is 0.2%.
14. Wash 50 μL of high-capacity streptavidin agarose with 1 mL of PBS (or Brij buffer) three times, and then add all of the streptavidin beads to your sample. Place on a nutating rocker for 90 min at room temperature.

3.3 On-bead Trypsin Digestion

1. Centrifuge the beads from 3.2 for 2 min at 1000 × *g*, room temperature. Remove the supernatant and wash the beads three times with 1 mL of 0.2% SDS in PBS, followed by three washes with 1 mL PBS, three washes with 1 mL of 20 mM Tris–HCl, 500 mM KCl, pH 7.4, and finally three washes with 1 mL of 20 mM Tris–HCl, pH 7.4.
2. Add 400 μL of a PBS solution containing 6 M urea and 9.5 mM TCEP to the beads. Incubate for 20 min at 37 °C with gentle rotation.
3. Add 20 μL of a fresh solution of 400 mM iodoacetamide (dissolved in water) to the suspension of beads. Incubate for 20 min at 37 °C with gentle rotation.
4. Remove the supernatant, and wash the beads with 1 mL of 2 M urea in PBS, and then incubate the beads with 2 μg of trypsin in 200 μL of 2 M urea in PBS with 1 mM $CaCl_2$ at 37 °C overnight with gentle rotation (*see* **Note 11**). Prior to adding the trypsin to the sample, it should be activated for 15 min at 30 °C in the buffer provided by Promega.

3.4 Purify Digested Peptides Using Sep-Pak Vac C18 Cartridge

1. Following trypsin digestion, pellet the beads from 3.3 by centrifugation at 1000 × *g*, room temperature. Transfer the supernatant to a 1.5-mL tube. Wash the beads twice with 300 μL of water. Combine the washes with the supernatant. Dilute the solution to 1 mL by adding water, and adjust the pH to ~2 by adding 15 μL of 10% trifluoroacetic acid (TFA) (*see* **Note 12**).
2. Condition the Sep-Pak Vac C18 cartridge by passing 1 mL of 90% methanol–0.1% TFA through the cartridge three times. Make sure not to let the cartridge dry out.
3. Equilibrate the cartridge by passing 1 mL of 0.1% TFA through the cartridge three times.
4. Load the sample from **step 1** slowly, at a rate of 1 drop/s. To minimize loss, pass the sample through the cartridge three times. Wash the loaded cartridge with 1 mL of 0.1% TFA, three times.
5. Elute the sample by passing 1 mL of 80% acetonitrile–0.1% TFA through the cartridge once. Dry the cartridge with nitrogen to ensure all peptides have been eluted.
6. Lyophilize the eluted sample, and analyze using nano-LC-MS/MS.

3.5 Cell Culture and Alkyne Probe Treatment for Fluorescence Detection

1. Culture Sirtuin WT and KD/KO cells in DMEM Culture Medium (*see* **Note 13**).
2. Add alkyne probe to the cells with a final concentration of 50 μM when the confluency reaches 80% and incubate with cells for 6 h (*see* **Note 3**).

3. Harvest the cells (either by scraping or by treating with trypsin-EDTA) and centrifuge for 5 min at 500 × *g*, 4 °C.
4. Wash the cells using ice-cold PBS. Centrifuge for 3 min at 500 × *g*, 4 °C. Discard the supernatant, saving the cell pellet.

3.6 Fluorescent Labeling of Target Proteins

1. Lyse cells with 1% NP40 lysis buffer supplemented with phosphatase inhibitor cocktail, by adding the appropriate amount of buffer (300 μL per 10-cm dish of HEK-239T cells) and placing on a nutating rocker for 30 min at 4 °C. After 30 min, clear the lysate by centrifugation at 17,000 × *g*, 4 °C for 20 min. After lysing the samples, quantify the protein concentration (*see* **Note 14**).
2. Dispense 500–2000 μg of whole cell lysate for each sample into 1.5 mL microcentrifuge tubes. Bring the volume to 400–1000 μL with 1% NP40 lysis buffer as above (*see* **Note 15**).
3. Carry out standard immunoprecipitation (IP) procedure for the protein of interest with agarose beads conjugated to the appropriate antibody. Wash the agarose beads at least three times with IP washing buffer.
4. Add 10–50 μL of washing buffer or PBS buffer to the pelleted agarose beads (*see* **Note 8**).
5. Add click chemistry reagents into the tubes such that the final concentrations in the reaction are: 200 μM TAMRA-azide, 600 μM TBTA, 2 mM CuSO4, and 2 mM TCEP.
6. Incubate for 30 min at room temperature.
7. Add 6× protein loading dye (final 2×) and heat the sample for 10 min at 95 °C to denature proteins.
8. Centrifuge for 2 min at 2000 × *g*.
9. Transfer the supernatant to another 1.5 mL microcentrifuge tube.
10. Divide the sample in half. To detect noncysteine fatty acylation, add hydroxylamine (pH 7.4, final concentration 400 μM) to the sample and heat for 5 min at 95 °C. To detect all fatty acylation, including cysteine fatty acylation add water (final concentration 400 μM) and heat for 5 min at 95 °C.
11. Load protein samples onto SDS-PAGE gel and resolve by electrophoresis.
12. To reduce background signal, destain the gel by shaking for at least 2 h in destaining solution (*see* **Note 16**).
13. After destaining, change the destaining solution to water. Scan the gel using a fluorescence gel scanner, such as a Typhoon 9400 variable mode imager (GE Healthcare Life Science, Piscataway, NJ). The excitation and emission settings are

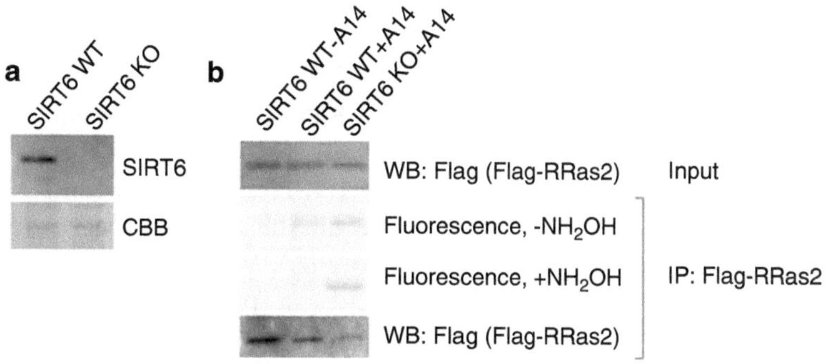

Fig. 2 Typical results for the alkyne-tagged fatty acid labeling of sirtuin targets. (**a**) Western blot to confirm the status of SIRT6 in SIRT6 WT and KO mouse embryonic fibroblast cells (MEFs). (**b**) Fluorescent labeling of RRas-2 in SIRT6 WT and KO MEFs

determined by the fluorescent dye selected. For TAMRA-N_3 the excitation and emission wavelengths are 532/580 respectively.

14. Stain the gel with Coomassie blue staining solution to detect protein loading. Expected results for a SIRT6 target, the small GTPase R-Ras2 are presented in Fig. 2.

4 Notes

1. SILAC labeling medium is deficient in L-lysine and L-arginine, enabling supplementation with the corresponding amino acids with substituted stable isotopic nuclei (^{13}C, ^{15}N) to generate "heavy" medium. "Light" medium is supplemented is supplemented with natural amino acids. DMEM for SILAC is used in this protocol, but a variety of tissue culture media are available commercially.
2. To detect lysine fatty acylation, Alk12 and Alk14 are typically used as these are the best mimics of lysine myristoylation and palmitoylation.
3. To increase confidence in the candidates identified in the proteomics analysis, perform the reverse SILAC experiment, incubating WT cells with SILAC Heavy Labeling Medium and KD/KO cells with SILAC Light Labeling Medium.
4. Alkyne probes at concentrations from 20 to 50 μM show no toxicity to cells; higher concentration might cause cell death. Normally 2–6 h treatment works well for most cells, longer incubation times may lead to metabolism of alkyne probes.
5. Always add a protease inhibitor cocktail into lysis buffer before lysing the cells (Ratio 1:40–1:100).

6. The choice of cell lysis buffer is based on experimental conditions and requirements. A buffer with 4% SDS (4% SDS, 50 mM triethanolamine pH 7.4, and 150 mM NaCl) maximally solubilizes proteins but will also lyse the nuclei, so universal cell nuclease (Thermo) should be included in the buffer. The 1% NP-40 lysis buffer is gentler than 4% SDS but may not solubilize all the hydrophobic proteins (e.g., integral membrane proteins). Vorinostat, or SAHA (suberoylanilide hydroxamic acid), a pan-HDAC inhibitor and nicotinamide, a pan-Sirtuin inhibitor, can be added to the lysis buffer to inhibit zinc-dependent histone deacetylase and sirtuins. This is to prevent deacylation reactions from happening after lysing the cells. Do not add EDTA to the lysis buffer, as it can chelate copper, making it incompatible with click chemistry.
7. If 4% SDS buffer is used as lysis buffer, the protein concentration should be determined using the BCA assay (Thermo). if 1% NP40 lysis buffer is used, protein concentration can be determined by either BCA assay or Bradford assay.
8. Click chemistry works best when the pH of the buffer is 8.
9. The protein concentration for the click chemistry reaction should not exceed 2 mg of lysate per 1 mL of reaction. Exceeding this concentration will hinder the efficiency of the reaction.
10. EDTA is added to chelate any residual copper, allowing the protein to be solubilized more easily.
11. To ensure complete peptide digestion, use proper working concentrations of urea, TCEP, and iodoacetamide to denature, reduce, and alkylate proteins. The concentration of urea in the trypsin digestion solution should be less than 2 M. In general, 2–4 μg of trypsin should be sufficient for the recommended whole cell lysate input. The sample should be incubated with trypsin for at least 8 h to ensure complete digestion. $CaCl_2$ is added to the solution to maximize the activity of trypsin.
12. When using a Sep-Pak Vac C18 cartridge to purify digested peptides, make sure the pH of the digested peptide solution is 2–3 before loading the sample on the C18 cartridge. The sample should be loaded slowly, at a rate of approximately one drop/second. Solutions should be made fresh, and avoid skin contact to prevent the sample from being contaminated by keratin from skin.
13. If the use of sirtuin inhibitors is desired, WT cells can be used and we recommend trying various concentrations and treatment times with the inhibitor. When the alkyne probe is added, it is recommended to add fresh media supplemented with the inhibitor and the alkyne probe.
14. Always add protease inhibitor cocktail into lysis buffer before lysing the cells (Ratio 1:40–1:100). Choose a lysis buffer

compatible with the immunoprecipitation procedure to avoid denaturation of the antibody. Avoid EDTA in the lysis buffer, which is not compatible with click chemistry.

15. The protein amount used in this step depends on the abundance of target proteins; lower-abundance target proteins need more total proteins (same principle as immunoprecipitation).
16. Destaining for a longer time can significantly decrease background. If time allows, destain overnight at 4 °C.

References

1. Jiang H, Zhang X, Chen X et al (2018) Protein lipidation: occurrence, mechanisms, biological functions, and enabling technologies. Chem Rev 118:919–988
2. Aramsangtienchai P, Spiegeleman NA, He B et al (2016) HDAC8 catalyzes the hydrolysis of long chain fatty Acyl Lysine. ACS Chem Biol 11:2685–2692
3. Jiang H, Khan S, Wang Y et al (2013) SIRT6 regulates TNF-alpha secretion through hydrolysis of long-chain fatty acyl lysine. Nature 496:110–113
4. Teng YB, Jing H, Aramsangtienchai P et al (2015) Efficient demyristoylase activity of SIRT2 revealed by kinetic and structural studies. Sci Rep 5:8529. https://doi.org/10.1038/srep08529
5. Tong Z, Wang M, Wang Y et al (2017) SIRT7 is an RNA-activated protein lysine deacylase. ACS Chem Biol 12:300–310
6. Jing H, Zhang X, Wisner SA et al (2017) SIRT2 and lysine fatty acylation regulate the transforming activity of K-Ras4a. elife 6:e32436. https://doi.org/10.7554/eLife.32436
7. Zhang X, Spiegelman NA, Nelson OD et al (2017) SIRT6 regulates Ras-related protein R-Ras2 by lysine defatty-acylation. elife 6: e25158. https://doi.org/10.7554/eLife.25158
8. Zhou Y, Huang C, Yin L et al (2017) N(epsilon)-Fatty acylation of Rho GTPases by a MARTX toxin effector. Science 358:528–531
9. Schey KL, Gutierrez DB, Wang Z et al (2010) Novel fatty acid acylation of lens integral membrane protein aquaporin-0. Biochemistry 49:9858–9865
10. Stevenson FT, Bursten SL, Locksley RM et al (1992) Myristyl acylation of the tumor necrosis factor alpha precursor on specific lysine residues. J Exp Med 176:1053–1062
11. Stevenson FT, Bursten SL, Fanton C et al (1993) The 31-kDa precursor of interleukin 1 alpha is myristoylated on specific lysines within the 16-kDa N-terminal propiece. Proc Natl Acad Sci U S A 90:7245–7249
12. Peng T, Thinon E, Hang HC (2016) Proteomic analysis of fatty-acylated proteins. Curr Opin Chem Biol 30:77–86
13. Tate EW, Kalesh KA, Lanyon-Hogg T et al (2015) Global profiling of protein lipidation using chemical proteomic technologies. Curr Opin Chem Biol 24:48–57
14. Charron G, Zhang MM, Yount JS et al (2009) Robust fluorescent detection of protein fatty-acylation with chemical reporters. J Am Chem Soc 131:4967–4975
15. Hebert N, Beck A, Lennox RB et al (1992) A new reagent for the removal of the 4-methoxybenzyl ether: application to the synthesis of unusual macrocyclic annd bolaform phosphatidylcholines. J Org Chem 57:1777–1783

Part III

zDHHC *S*-Acyltransferases

Chapter 12

siRNA Knockdown of Mammalian zDHHCs and Validation of mRNA Expression by RT-qPCR

Heather McClafferty and Michael J. Shipston

Abstract

The lack of specific pharmacological tools to interrogate the functional role of palmitoyl acyltransferases (zDHHCs) in mammalian cells has significantly hampered the understanding of this important gene family. Gene silencing by RNA interference (RNAi) is a process in eukaryotes that allows specific knockdown of the expression of proteins by targeting their coding mRNA. RNAi can thus be used as a proteomic tool to study the functional role of specific zDHHCs in cells by analyzing the effects of endogenous zDHHC knockdown on their protein targets or pathways. Here we describe the application of short interfering RNA (siRNA), a class of short (20–25 base pairs) double-stranded RNAs, to knockdown endogenous zDHHC enzymes expressed in human embryonic kidney (HEK293) cells and subsequent validation of knockdown efficiency using RT-qPCR to quantify zDHHC mRNA levels.

Key words siRNA, qPCR, zDHHC, BK channel, Primer validation, Knockdown efficiency, MIQE guidelines

1 Introduction

Synthetic small interfering RNA molecules, known as siRNAs, can be introduced into mammalian cells in culture by transfection where they bind to their specific mRNA targets by complementary base pairing [1]. This targets the mRNA for destruction by RNase complexes resulting in knockdown phenotypes that can approach genetic null mutants or gene knockout phenotypes [2–5]. This approach is thus particularly useful where, as for the mammalian zDHHC family of palmitoyl acyltransferases [6–8], (1) there is an almost nonexistent pharmacological toolbox to interrogate specific protein function and (2) interrogation of the functional role of multiple members of a closely related gene family is required.

siRNAs can be designed by the user (for example *see* [9, 10]), or are available to buy commercially for all identified protein coding genes and are sold as predesigned assays, synthesized to order by a variety of manufacturers (*see* **Note 1**). To maximize efficiency of

Maurine E. Linder (ed.), *Protein Lipidation: Methods and Protocols*, Methods in Molecular Biology, vol. 2009, https://doi.org/10.1007/978-1-4939-9532-5_12,

knockdown, two siRNAs designed against different parts of the mRNA sequence are usually transfected into cells together. Effective siRNA concentrations must be determined empirically along with incubation periods post transfection to determine maximum knockdown efficiency for each target and is also dependent on the cell type under investigation.

While knocking down mRNA can be indicative of reduced protein levels it does not always directly correspond to a decrease in protein levels, in particular if turnover of the protein is low over the time course (48–72 h) of typical knockdown assays. Ideally, efficiency of protein knockdown would be validated by analysis at the protein level for example, by antibody detection in either Western blots or by immunocytochemistry. However, in many cases, as for the zDHHC family, high-efficiency and specific antibodies are not necessarily available. In such cases, and where multiple proteins are being knocked down in experiments, an alternative approach is to validate efficiency of knockdown by determining mRNA levels by reverse transcription quantitative polymerase chain reaction (RT-qPCR).

We developed an siRNA screening approach to interrogate which *endogenous* zDHHCs were involved in *S*-acylation of the large conductance calcium- and voltage-activated potassium (BK) channel in mammalian cells [11, 12]. Previously, most studies had exploited use of zDHHC overexpression in cell lines to determine which zDHHC was most likely to control *S*-acylation of target proteins [6]. Use of both overexpression and siRNA mediated knockdown provides powerful validation of zDHHCs involved in controlling *S*-acylation of specific targets. The BK channel is *S*-acylated in two different domains with *S*-acylation having distinct effects on channel function and properties depending on the site *S*-acylated [13–15]. The S0-S1 intracellular loop is *S*-acylated at a cluster of cysteine residues and controls cell surface expression of the channel. In contrast, *S*-acylation of two adjacent cysteine residues in an alternatively spliced domain (STREX) in the intracellular C-terminus controls channel kinetics and regulation by phosphorylation.

We undertook an siRNA knockdown strategy of all 23 human zDHHCs to identify endogenous zDHHCs in HEK293 cells which were involved in the *S*-acylation of recombinantly expressed BK channels. Due to the large number of enzymes involved, and the lack of good antibodies against most zDHHCs, we employed an RT-qPCR approach to monitor the efficiency of zDHHC knockdown by siRNA. In these experiments, we thus transfected HEK293 cells with different variants and site-directed mutants of the BK channels and knocked down endogenous zDHHCs. The effect of individual zDHHC knockdown on BK channel *S*-acylation, properties, and function was monitored biochemically as well as using functional imaging and electrophysiological assays. As such, the protocols described below are designed for cells grown on glass coverslips in 6-well plates while a 24-well plate of cells is prepared in parallel to be used for efficiency checking of the knockdown procedure in *each* experiment. The approach we outline will need to be adapted and validated for other assay formats (e.g., 96-well format) and in particular for different cell types.

Using this approach, we were able to demonstrate that *S*-acylation of the S0-S1 and STREX domains of the BK channel was controlled by distinct zDHHCs [11, 12]. Importantly, our comprehensive siRNA and RT-qPCR validation of all zDHHCs in parallel revealed a level of compensatory feedback in zDHHC mRNA expression for some zDHHCs, in response to the knockdown of other zDHHCs. This revealed an additional level of complexity that must be taken into consideration by investigators when using genetic knockdown, or knockout, approaches to determine the functional roles of zDHHCs.

2 Materials

2.1 HEK293 Cell Culture

1. Human embryonic kidney (HEK293) cells: between passage 12 and 30 (*see* **Note 2**).
2. Tissue culture flasks, 25 cm^3 and 75 cm^3.
3. Tissue culture plates, 24 well and 6 well; 5 mL and 10 mL sterile disposable pipettes.
4. Growth media (GM): Dulbecco's Modified Eagle Medium with L-glutamine and phenol red, 25 mM glucose (Gibco), 10% fetal bovine serum (FBS). Remove 50 mL DMEM from a 500 mL stock and add 50 mL FBS. Mix well by trituration with a stripette.
5. Hank's Balanced Salt Solution (HBSS), no calcium, magnesium or phenol red.
6. Trypsin-EDTA (0.05%), phenol red.
7. Glass coverslips, 14 mm diameter, thickness number 1.

2.2 siRNA and Transfection

1. siRNA design and computational validation [9, 10] can be performed with available online tools or purchased from commercial sources (*see* **Note 1**). Experimental validation of multiple siRNAs is required in all cases.
2. Stock siRNA (Table 1): resuspend lyophilized 1000 pmol siRNA in 100 μL RNase-free water, 10 μM (Qiagen). Shake for 30 mins at 4 °C on a shaking platform. Dilute further to obtain a 2 μM working stock by taking 20 μL and mixing with 80 μL RNase-free water. Aliquot and freeze at −20 °C.

Table 1
siRNA target sequences for human zDHHCs

Gene name	mRNA accession number	siRNA target sequences	Product name
ZDHHC1	NM_013304	CAGCACGCACATGTCATTGAA CCGGCCTGACTCTCTATGCAA	Hs_ZDHHC1_2 Hs_ZDHHC1_3
ZDHHC2	NM_016353	TAGCTACTGCTAGAAGTCTTA TAAGAATTCCTAATAGGTCAA AATGGCCTACCTGATACTCAA CAGGAAGTTCTTAGGCGAGCA	Hs_ZDHHC2_5 Hs_ZDHHC2_6 Hs_ZDHHC2_8 Hs_ZDHHC2_7
ZDHHC3	NM_016598	AACATTGAGCGGAAACCAGAA TCCGTTCTCATGAATGTTTAA	Hs_ZDHHC3_3 Hs_ZDHHC3_5
ZDHHC4	NM_018106	CAGGAGGGTCTCATTGACTCA AAGAACCATGTCCGGCAGCTA	Hs_ZDHHC4_2 Hs_ZDHHC4_6
ZDHHC5	NM_015457	AGGGATTAGAGTGTGCTCCTA ACCACCATTGCCAGACTACAA	Hs_ZDHHC5_5 Hs_ZDHHC5_6
ZDHHC6	NM_022494	AAGGCTAAAGATCGAATTCAG CCGAGAGGTGCTCGGCTTGTA	Hs_ZDHHC6_5 Hs_ZDHHC6_6
ZDHHC7	NM_017740	CCCGTGGTTACTATGAATGTA CAGCAGGGCGTTTACATAGAA	Hs_ZDHHC7_2 Hs_ZDHHC7_3
ZDHHC8	NM_013373	CGCGCCGTGTCTGATGTGTCA CCGGGCTCCGCTGTACAAGAA	Hs_ZDHHC8_5 Hs_ZDHHC8_3
ZDHHC9	NM_001008222 NM_016032	CTCAACCAGACAACCAATGAA CAGGAATAGCAGGCAACGTGA	Hs_ZDHHC9_6 Hs_ZDHHC9_7
ZDHHC11	NM_024786	CCATCAAGGTCCTGCCTGT GGTATGAAGATGTCAAGAA CCAAGAAGATGACCACCTT	SASI_Hs01_00204381 SASI_Hs01_00204382 SASI_Hs01_00204383
ZDHHC12	NM_032799	CAGATACTGCCTGGTGCTGCA CCCACTCTTTGTGGTCTACCT	Hs_ZDHHC12_4 Hs_ZDHHC12_5
ZDHHC13	NM_001001483 NM_019028	CAGCATAGTAGCCTTTCTATA TAGATTGGACATCACAGTACA	Hs_ZDHHC13_3 Hs_ZDHHC13_5
ZDHHC14	NM_024630 NM_153746	ACGCTTGTGGCCAGACTGCAA ATGAGGTCTCGTGTTGAGATA	Hs_ZDHHC14_6 Hs_ZDHHC14_9
ZDHHC15	NM_144969	GTGCTATGTGTGTGTTAAA GATTCTCTTTGGTTACCAT CATCACTGCCCTTGGGTTA	SASI_Hs01_00206035 SASI_Hs01_00206036 SASI_Hs01_00206034
ZDHHC16	NM_032327 NM_198043 NM_198044 NM_198045 NM_198046	TAGCATCGAAAGGCACATCAA AAGGGAGCACAAATAAAGGTA	Hs_ZDHHC16_2 Hs_ZDHHC16_5
ZDHHC17	NM_015336	TAGCGACATCTTATCCTATGA TAGGAACTCCTTTCCTAGTTA	Hs_ZDHHC17_9 Hs_ZDHHC17_10

(continued)

Table 1
(continued)

Gene name	mRNA accession number	siRNA target sequences	Product name
ZDHHC18	NM_032283	AAGCCTGATGCCAGCATGGTA CACGTGCACTGGTCACGAGAA	Hs_ZDHHC18_4 Hs_ZDHHC18_5
ZDHHC19	NM_001039617	CAGCAACTGGTATTTAACA CCGACAAGGCCATCGCCAT GACTGGACATCCATGCCGA	SASI_Hs01_00212348 SASI_Hs01_00212347 SASI_Hs01_00212346
ZDHHC20	NM_153251	CAGCATTGACTTAGAGCTACA TACCTGTTATGAGTTGCTATA	Hs_FLJ25952_2 Hs_ZDHHC20_1
ZDHHC21	NM_178566	AAGCGTAATTTGGACCTCTTT CACCTTCTTATAGTATAGGTA	Hs_ZDHHC21_5 Hs_ZDHHC21_6
ZDHHC22	NM_174976	CCCGCTGATAGCTGCGCAACA TGGGTTCATTTATGCCCTATA	Hs_ZDHHC22_5 Hs_ZDHHC22_6
ZDHHC23	NM_173570	AAGGATCAGGATTGCATGAAA CTGCGAGTACATAGATCGGAA	Hs_ZDHHC23_3 Hs_ZDHHC23_5
ZDHHC24	NM_207340	ACACCTAGACTCAGTAAGGAA CAGGGCTTGGGCCGTGACTTA CTGGTCCGTTGGCCAGATCAA CCGCTGCGTGGGCTTCGGCAA	Hs_LOC254359_3 Hs_LOC254359_1 Hs_LOC254359_2 Hs_ZDHHC24_1

Human zDHHC mRNA accession numbers, siRNA targeted sequences and Qiagen FlexiTube product codes or Sigma MISSION product codes used in these assays. Most knockdowns were achievable using two siRNAs together at 20 nM each; however, others required increasing concentrations and combinations of siRNA (e.g., zDHHC2 was knocked down using four siRNAs at 20 nM each)

3. Negative control siRNA: All Stars Negative Control siRNA (Qiagen).
4. siRNA transfection reagent: HiPerFect (Qiagen).
5. Plasmid transfection reagent: FuGENE-HD (Promega).
6. DMEM (without FBS).
7. 1.5 mL sterile tubes.

2.3 RNA Extraction and cDNA Preparation

1. RNA extraction kit: Reliaprep RNA cell Miniprep System (Promega).
2. Ethanol and isopropanol, added as directed by the kit instructions.
3. Filtered pipette tips.
4. 21G needle and 1 mL syringe.
5. TBE: Make a 10× stock of TBE buffer by dissolving 54 g Tris, 27.5 g Boric Acid and 4.65 g EDTA, pH to 8.3 with HCl, then autoclave.
6. 1% agarose gel: Dilute 100 mL of 10× TBE with 900 mL dH_2O to generate 1× TBE. To 40 mL 1× TBE add 0.4 g

agarose powder creating a 1% w/v solution. Dissolve the agarose by microwaving on full power for approximately 2 min. Adjust the volume with dH_2O to replace evaporated water then add 4 μL of a DNA stain such as Safeview DNA gel stain. Pour the molten agarose into a gel cassette, adding the well comb before the gel sets.

7. UV–VIS spectrophotometer.
8. cDNA first strand kit: SuperScript IV First-Strand Synthesis System (ThermoFisher).
9. PCR tubes/plates and thermocycler.

2.4 Quantitative PCR

1. Users should follow the MIQE guidelines (http://miqe.gene-quantification.info) for correct design and reporting of qPCR experiments [16].
2. Power SYBR™ Green PCR Mastermix (Applied Biosystems).
3. Primers (Table 2): zDHHC primers, briefly centrifuge lyophilized stock forward and reverse primers before resuspending in 1× TE buffer to obtain 100 pmol/μL. Create a combined working assay of primers by diluting the stock primers to 2.5 pmol/μL, adding 5 μL of each primer into 190 μL of dH_2O, vortex to mix. Aliquot and store at −20 °C. Endogenous reference control (ERC) β-actin primer assay (Qiagen): briefly centrifuge tube and resuspend in 1.1 mL TE buffer. Vortex to mix, aliquot and store at −20 °C (*see* **Note 3**).
4. DEPC: Add 1 mL of diethyl pyrocarbonate to 1 L of dH_2O. Shake to mix then vent overnight in a fume hood. Autoclave before use.
5. Centrifuge with rotor for plate spinning.
6. Real-Time PCR machine.
7. 96-well qPCR plates and optical sealing film (Applied Biosystems).

3 Methods

When preparing and working with siRNA take precautions to prevent degradation by environmental ribonucleases and work in a clean environment to avoid cross-contamination. Work in an RNase-free environment, using RNase-free water and tubes. Wear gloves at all stages and use filtered pipette tips throughout. To preserve the integrity of the siRNA, reduce the amount of freeze–thaw cycles they are exposed to by aliquoting stocks into small tubes after resuspension.

Table 2
qPCR primers for human zDHHCs

Gene name	Sense primer	Antisense primer
ZDHHC1	CTCACCCGCTCCAGATTGTG	TGAGGCAGGAGGGGAACAAG
ZDHHC2	TGGAGAGAGAGCCAAGAGGAG	TTATAAGTTGGCATCTGTCACAGTATC
ZDHHC3	CTTAACTTCAGTTCGACAGAGTGC	GAAGAGCAGGCCCTCAAAGC
ZDHHC4	GATTGTCTTCATGCTGGGCTTTG	CTCTGTACCACTCGTTAGTAGTCTG
ZDHHC5	GAAGACTGAAGAAAGATAAGAGACA TTG	GACACTTCAAAAGTTTACTGTGGATG
ZDHHC6	TCTGCTTGTCCTGGGCTGAG	TGTCTTGGTCTGAAACCCAACC
ZDHHC7	AGTCCATGCTCTGATCCTTTGTG	GGAAGATCAACAGGATTACAGTTA TCG
ZDHHC8	TCTGTGACAACTGTGTAGAGGAC	AGCAGGAACAGGAAGAAGTAGC
ZDHHC9	TAGCAATTGGACTTTTGATGATG TTTG	GGGTTTTGCGATTACACGAGAG
ZDHHC11	TCCGTCACCCCAGAAGCCATAC	CGTGAGGCAGGAAGGGAATGAAG
ZDHHC12	TGCTGACCTGGGGAATCACG	CATGAGTGACACAGCGAGGTAG
ZDHHC13	GCATATCTCATCTCAAAGGGACAG	CAACCACATTGAGAGAAGGATTAAAC
ZDHHC14	TCGCTGGCATCCTGTTCTTC	TGCCGTTTGCGATATCTATTTGC
ZDHHC15	TATGAATGAGTCACAGAACCCAC TGCTAG	TAATCTTGGTTGTCATCCTCGTTGTC TTCC
ZDHHC16	ATTGTAGCTATCGCCTACCTGTG	AGTAGTGGAAGACAATCAGGATCAG
ZDHHC17	ATGGTTGTGCAACTAATGAAATATGG	GCTATGAGATAAGCAACAATTGAGG
ZDHHC18	CCACCACCGGCCTCTTCTTC	GAGGATGGCAGCGATGATGG
ZDHHC19	GTCCCTGGTTCCTCCCTAGC	AATGCGAAGAAGAGGCCACTG
ZDHHC20	ATCAACCTTTTCCTATCAAACCAC TTAG	GCCTTCTTCAGCTCCATTCTCC
ZDHHC21	TCTGGGAATTATGTAACAAGTGTAA TTTG	AGAGCCAATGATTATCTTCACCAAC
ZDHHC22	CACCCCTTGGCCTTCCTCAC	ACGAACATTTCAGAACCCAGGAC
ZDHHC23	AAAAGTGAACATCAGCATCATCCC	CCAGGAGGAAATGCCAGGAAG
ZDHHC24	CCCCTTCCGAGCTTCTATTGG	CTAGGCCGCTTCCGTATTGG

3.1 HEK293 Cell Culture, siRNA and Plasmid Transfection

1. Culture HEK293 cells in 25 cm^3 flasks containing 5 mL of GM. Cells are ready for passaging at 90–95% confluency, typically 3–4 days after plating. Remove GM by pipetting using a 5 mL pipette and wash cells quickly with 2 mL HBSS. Add 0.5 mL trypsin–EDTA and incubate the flask for 2–5 mins at

Table 3
siRNA reagent amounts required for transfection

	6-well plate (2.1 mL per well)	24-well plate (0.6 mL per well)
DMEM	100 μL	100 μL
siRNA (2 μM)	10.5 μL each	3 μL each
HiPerFect transfection reagent	12 μL	3 μL

37 °C until cells begin to slide off the surface. Add 1 mL of GM and gently triturate the cells 20–30× using a 1 mL pipette to generate single cells (*see* **Note 4**).

2. Plate HEK293 cells so that they have a confluency of approximately 40–60% after 24 h, 1×10^5 cells in a 24-well plate in 500 μL GM, 3×10^5 cells in a 6-well plate in 2 mL GM, (*see* **Note 5**). Prepare 24-well and 6-well plates (with glass coverslips if required for imaging/electrophysiology experiments) in parallel.
3. Assemble the siRNA transfection complex reagents (Table 3) in 1.5 mL tubes to transfect 10 nM of each siRNA, 20 nM in total (Table 1). (*see* **Note 6**).

   In parallel with siRNA transfections prepare the following control reactions to monitor the procedure (*see* **Note** 7). Include untransfected HEK293 cells in a well to provide cDNA template for primer validations.

   (a) Mock transfection: DMEM + HiPerFect alone.

   (b) Negative control: DMEM + HiPerFect + AllStars Negative Control siRNA.
4. Mix by vortexing and incubate at room temperature for 5–10 min. Add the transfection complex to the cells dropwise and swirl the plate to mix. Incubate the cells at 37 °C for 30 min before carrying out transient DNA transfections of recombinant proteins of interest.
5. (Optional depending on whether target is endogenous or a recombinantly expressed protein in functional assay). Plasmid DNA of the *S*-acylated target protein (in this case BK channel) can be transfected using FuGENE-HD transfection reagent after the introduction of siRNA into the cells. In a 1.5 mL tube, combine 100 μL DMEM with 1 μg plasmid DNA and 3 μL FuGENE-HD. Vortex and incubate for 15 min before adding to the cells dropwise in the 6-well plate, swirling the plate to mix. Return cells to the 37 °C incubator for 48–72 h before use in experiments.

3.2 RNA Extraction and cDNA Preparation

Following the outgrowth period (typically 48–72 h post siRNA transfection for optimal knockdown), extract the RNA from the cells in the 24-well plate using the RNA extraction kit as described in Subheading 2.3, **step 1**, then check the integrity of the RNA before converting into cDNA, using the kit as described in Subheading 2.3, **step 8**. Throughout each step avoid aerosols, keep individual tubes closed between transfers and use new filter-tips for each stage to avoid cross-contamination. Performing these steps in a UV PCR cabinet or workstation will reduce the possibility of RNase degradation and DNA contamination from the environment (*see* **Note 8**).

1. Decant the GM and wash cells in 200 μL of HBSS. Add 100 μL of prepared lysis buffer directly to the well and use a 1 mL pipette tip to dislodge the cells and move the lysate to a 1.5 mL tube. Using a 21G needle and 1 mL syringe, shear the genomic DNA by passing the lysate through the needle up to 10 times. Add 35 μL of isopropanol and vortex to mix, then add the sample to the ReliaPrep Minicolumn supplied in the kit. Centrifuge for 1 min at 12–14,000 × *g*, then wash with RNA Wash Solution, centrifuging 12–14,000 × *g* for 30 s. Empty the collection tube where required. Prepare the DNaseI solution and apply for 15 min at room temperature then wash off directly with Column Wash Solution, centrifuging 15 s at 12–14,000 × *g*. Apply RNA Wash Solution, centrifuge 30 s, 12–14,000 × *g* then transfer the Minicolumn to a fresh collection tube. Give a final wash with RNA Wash Solution and centrifuge for 2 min at 14,000 × *g*. Elute the RNA in 15 μL Elution buffer in a 1.5 mL tube. Keep the purified RNA on ice to preserve integrity. If RNA is to be kept for longer term, precipitate with ethanol and store at −80 °C.
2. Measure the RNA concentration by reading the absorbance at 260 nm on a UV/VIS spectrophotometer, such as a NanoDrop, and check the purity ratio by dividing the absorbance at 260 nm by the absorbance at 280 nm. Pure RNA has a 260:280 ratio of 2.0 (*see* **Note 9**).
3. Visualize 5 μL of RNA on a 1% agarose gel to check the integrity (Fig. 1a). Intact, good quality RNA should appear as two discrete bands comprising 28S and 18S rRNA and should run with an approximate 2:1 intensity ratio (*see* **Note 10**).
4. Reverse transcribe equal amounts of each RNA sample (e.g., 500 ng) into cDNA in a PCR tube. Prime the reaction using a 1:1 mix of 1.25 ng/μL random hexamers and 1.25 μM Oligo d$(T)_{20}$ to enrich for full-length mRNA from the total RNA extraction. Add 10 mM dNTP mix and water up to a final volume of 13 μL. Prepare duplicate reactions for each RNA sample (to act as a negative reverse transcription control,

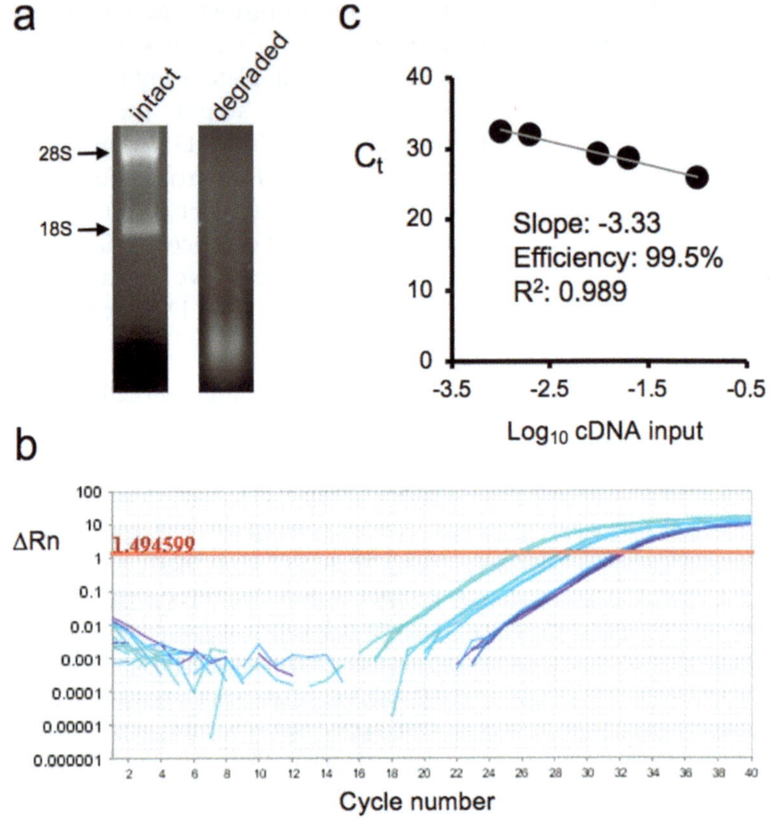

Fig. 1 RNA isolation and qPCR amplification. (**a**) Representative lanes of a stained 1% agarose gel from an intact sample of purified mammalian RNA, where the 28S and 18S RNA bands are clearly visible, and from a degraded RNA sample in which a smear and lower molecular weight staining is observed. (**b**) Representative amplification curves from a qPCR assay. ΔRn is the baseline corrected SYBR™ Green fluorescence signal normalized to the passive reference dye (ROX). The horizontal red line indicates the software applied auto threshold, which in this case is 1.49. C_t values are then calculated as the cycle where the amplification signal crosses the defined threshold. (**c**) Representative qPCR standard curve. C_t values (Data are Means ± SD, with errors within symbol size) are plotted against dilutions of cDNA (Log_{10} cDNA input). The slope of the dilution curve is used to calculate the primer efficiency using the efficiency equation $E = 10^{(-1/\text{slope})} - 1 \times 100$. Efficiency in this example is 99.5% that is within acceptable parameters

No-RT) and a single reaction should be prepared with water instead of RNA for a negative nontemplate control (NTC). Prepare untreated HEK293 cells in parallel to provide template for primer validations. Using a PCR machine, incubate the samples at 65 °C for 5 min then cool to 4 °C.

5. Prepare a master-mix for the total number of RNA samples being reverse transcribed using 5× SuperScript™ IV reaction

buffer (vortexed for 5 s before use), 40 U of the ribonuclease inhibitor, and 100 mM DTT (5 mM final). Pipette thoroughly to mix, then add 6 μL to each sample. Each RT sample and the NTC sample should then receive 200 U of SuperScript™ IV Reverse Transcriptase enzyme, mixing thoroughly. The No-RT samples should receive 1 μL of water in place of enzyme. Incubate the samples at 23 °C for 10 min, 50 °C for 10 min, 80 °C for 10 min. Cool the reaction to 4 °C. Add 2 U of RNase H, mix thoroughly then incubate at 37 °C for 20 min.

3.3 qPCR Primer Validation

The approach below uses a SYBR™ Green based qPCR assay that allows rapid validation of PCR primers based on the intercalation of fluorescent dye (SYBR™ Green) with double-stranded DNA generated during the PCR amplification. For primers against human zDHHCs used in these assays *see* Table 2.

1. Validate all primer assays by PCRing across several dilutions of cDNA. Using cDNA prepared from a sample of HEK293 cells, make the following dilutions:

   1:10—10 μL neat cDNA +90 μL DEPC water

   1:50—2 μL neat cDNA +98 μL DEPC water

   1:100—10 μL of 1:10 cDNA +90 μL DEPC water

   1:500—10 μL of 1:50 + 90 μL DEPC water

   1:1000—10 μL of 1:100 cDNA +90 μL DEPC water.
2. Prepare a mastermix for each template dilution, multiplying the volumes in Table 4 by the number of primer assays to be validated and then by three to PCR in triplicate. Allow a 10% excess of master-mix for pipetting loss.
3. Pipette 18.4 μL of the mastermix into the appropriate wells of a 96-well PCR plate then add 1.6 μL of the primers (*see* **Note 11**).
4. Seal the plate with adhesive film (*see* **Note 12**).
5. Spin the plate briefly at low speed for 1–2 min to collect the samples in the bottom of the wells and to remove any bubbles (*see* **Note 13**).

Table 4
Sample volumes required for one qPCR reaction

Item	Amount for 1 rxn (μL)	Final concentration
2× Power SYBR Green mastermix	10	1×
cDNA template	2	Variable
DEPC water	6.4	Total volume = 20 μL

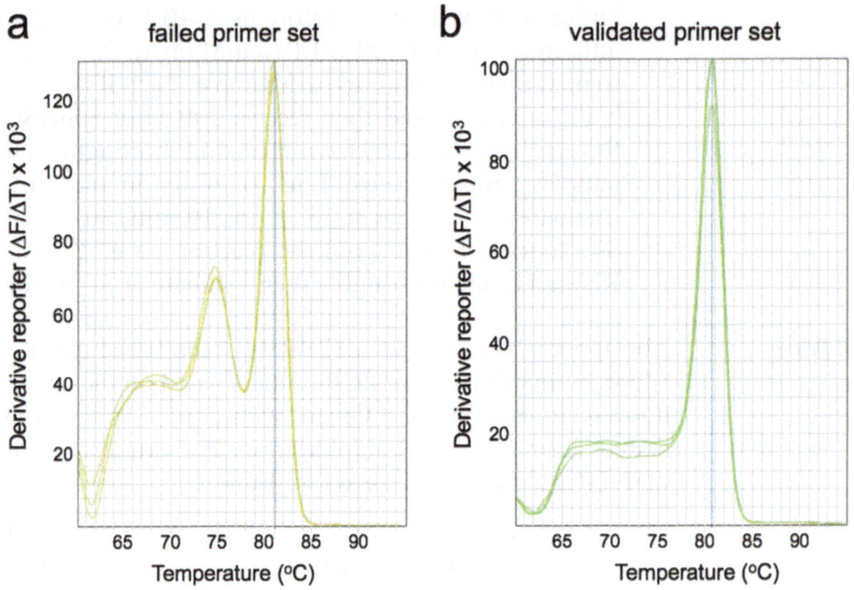

Fig. 2 Dissociation melt curve analysis for primer validation in RT-qPCR. Dissociation melt curve data for zDHHC5 amplified product from (**a**) a qPCR primer set that failed validation and (**b**) a qPCR primer set that passed validation. In both plots the Derivative of the reporter (the change in fluorescence divided by the change in temperature ($\Delta F/\Delta T$) is plotted against temperature. Note that for the qPCR primers that failed validation two clear peaks are seen and these primers cannot be used for qPCR quantification. In the qPCR primers that passed validation, a single peak representing the melting dynamics of a single PCR product is shown, and can be used for qPCR quantification

6. Load the plate into the PCR machine and cycle the PCR as follows: One step at 95 °C for 10 min, followed by 40 cycles of 95 °C for 15 s then 60 °C for 1 min. At the end of the PCR reaction, perform the dissociation (melt) curve by ramping the reaction temperature from 95 °C for 15 s to 60 °C for 1 min then ramp back up to 95 °C for 15 s.
7. Amplification curves generated by qPCR are given a C_t value at the cycle where the amplification crosses an applied threshold (Fig. 1b). Plot the C_t values for each template dilution against the log of the dilution factor to generate a standard curve (Fig. 1c). The R2 value should be >0.98. Using the efficiency equation $E = 10(-1/\text{slope}) - 1 \times 100$, calculate the efficiency of each primer (*see* **Note 14**).
8. Check the plot of the melt curve for each primer (Fig. 2). There should be one single peak corresponding to a single amplification product (*see* **Note 15**).

3.4 SYBR™ Green qPCR

Levels of mRNA remaining are calculated in the cDNA samples by amplifying both the target knockdown gene (zDHHC) and one or more endogenous reference control genes (ERCs), (*see* **Note 16**). The calibrating sample, that is, the All Stars Negative siRNA

Table 5
qPCR reagent amounts required for single reaction

Item	Amount for 1 rxn (μL)	Final concentration
2× Power SYBR Green mastermix	10	1×
Forward and reverse primer 2.5 μM	1.6	0.2 μM each
DEPC water	6.4	Total volume = 20 μL

sample, must be PCR'd with the ERC primers and by all the zDHHC primers being tested.

1. Prepare the templates for PCR by diluting the cDNA and control samples made in Subheading 2.3 in sterile water to give at least a 1 in 10 dilution (e.g., 10 μL cDNA in 90 μL RNase-free water).
2. Using the volumes listed in Table 5, prepare primer master-mixes of SYBR™ Green, primers, and water, allowing a 10% overestimate to allow for pipetting loss (*see* **Note 17**). This should be enough for each cDNA sample to be PCR'd in triplicate plus the NTC and No-RT controls.
3. Vortex and dispense 18 μL of the reaction mix into a 96-well reaction plate. Add 2 μL of the templates individually to the appropriate wells.
4. Seal the plate with sealing film then centrifuge the plate briefly.
5. Load the plate into the PCR machine and cycle as before: 1 step at 95 °C for 10 mins, followed by 40 cycles of 95 °C – 15 s then 60 °C – 1 min. Run the dissociation (melt) curve.
6. Export the C_t data into an Excel spreadsheet for quantitation of knockdown.
7. Confirm the presence of a single melt curve in each triplicate.

3.5 Quantitation of Knockdown Efficiency

1. Calculate the geometric mean of the ERC genes in each sample by multiplying the C_t values together, then using the number of samples as the root, take the *n*th root of the product of the C_t values. For example, the geometric mean of the data in Table 6 would be calculated as follows:

$$\sqrt[6]{20.91 \times 20.95 \times 20.88 \times 13.96 \times 13.87 \times 13.93} = 17.06$$

2. Calculate the delta C_t value (ΔC_t) in each sample by subtracting the respective geometric mean C_t values of the ERCs.

   For example, ΔC_t = (Target zDHHC C_t value) – (geometric mean of ERCs C_t value).
3. Average the ΔC_t values for the triplicates.

Table 6
Example C_t data for two endogenous reference control (ERC) genes

ERC gene	C_t value
β-actin	20.91
β-actin	20.95
β-actin	20.88
Rn18S	13.96
Rn18S	13.87
Rn18S	13.93

Table 7
Example C_t data analysis to calculate percentage siRNA knockdown for zDHHC13

Sample	Primer assay	ΔC_t ($Ct_{sample} - Ct_{ERC}$)	Average ΔC_t	$\Delta\Delta C_t$	$1-2^{-\Delta\Delta Ct} \times 100$	Final average % age knockdown
All stars negative	zDHHC13	7.06 6.79 6.59	6.82			
siRNA-zDHHC13	zDHHC13	11.19 11.82 11.35		4.37 5.00 4.53	95.18 96.88 95.66	95.9%

4. Calculate the $\Delta\Delta C_t$ by subtracting the average ΔC_t value of the calibrating sample (i.e., the All Stars negative control sample) from each of the ΔC_t of the zDHHC siRNA knockdown samples.

   For example, $\Delta\Delta C_t = (\Delta C_t$ zDHHC in siRNA knockdown sample) − (ΔC_t average zDHHC in All Stars negative control sample).

5. Calculate the percentage knockdown using the equation $1-2^{-\Delta\Delta Ct} \times 100$ (*see* **Note 18**).

   In the example data in Table 7, the zDHHC13 mRNA expression was knocked down by 95.9%.

4 Notes

1. Design steps for sequence-specific siRNA and details of publically available free software for siRNA design are available (see for example [9, 10]). Many commercial companies also have design resources online such as http://www.invivogen.com/sirnawizard/ and http://dharmacon.gelifesciences.com/design-center/, or have predesigned siRNA available to buy

(e.g., Qiagens FlexiTube siRNA https://www.qiagen.com/gb/shop/rnai/flexitube-sirna/#orderinginformation and MISSION Predesigned siRNA from Sigma-Aldrich https://www.sigmaaldrich.com/life-science/functional-genomics-and-rnai/sirna/mission-predesigned-sirna.html). Careful consideration of target sequences is required to avoid known/predicted alternatively spliced variants to ensure efficient knockdown.

2. Different variants of human embryonic kidney (HEK293) cells are in common use in laboratories and can display differences in protein expression level and function with time in culture. For example, HEK293-T cells are optimized for high recombinant protein expression. Users should confirm the HEK293 variant in use and establish endogenous zDHHC expression levels and efficiency of knockdown. The following experiments were optimized in parental HEK293 cells between passage number 12 and 30.

3. Users should follow MIQE guidelines (http://miqe.gene-quantification.info) [16] for design, validation, experimental analysis, and reporting throughout. Our zDHHC qPCR primers were custom designed and synthesized by Sigma-Aldrich (Oligotechserv@sial.com) for use in the SYBR™ Green based qPCR assays (*see* Table 2). However, primers can be designed using free online programs such as http://bioinfo.ut.ee/primer3-0.4.0/primer3/ [17]. Design should take into careful consideration possible splice variants and we typically target qPCR primers to targets distant to the siRNA sequence. The primer assays used to measure the ERC genes were predesigned Quantitect Primer Assays from Qiagen. Targets commonly used as reference genes are β-actin, GAPDH, and Rn18S; however, users must validate appropriate ERCs for the cell line and experimental paradigm under consideration to ensure the stably expressed gene(s) remains unchanged in all experimental conditions used. These assays are synthesized to order and supplied as combined forward and reverse primers to be diluted in Tris–EDTA buffer to make 10× stock. These are generally sold as bioinformatically validated at the design stage; however, these must also be validated experimentally.

4. Take care not to introduce too many bubbles at this point.

5. Cells can be counted using a hemocytometer counting chamber for accurate cell number plating.

6. Each gene may require a different amount of siRNA to achieve knockdown of >70%. This will also be dependent on the cell type being used. In HEK293 cells, we were most successful (90–98% mRNA knockdown) when using two independent siRNAs against the same zDHHC at a concentration of 20 nM. In a number of cases we had to validate up to seven

different siRNAs to achieve knockdown efficiencies of greater than 75%. This sometimes required using combinations of three or four siRNAs together at 20 nM each. The final siRNAs selected and validated are given in Table 1. We were able to achieve reliable simultaneous knockdown of two different zDHHCs by multiplexing in the same knockdown although efficiency is generally reduced to the lower end (i.e., 70%). In a limited number of cases, we were able to achieve simultaneous knockdown of three zDHHCs. However, in multiplexing assays extreme care should be taken about the possibility of off-target silencing when using total concentrations of siRNA approaching 80 nM.

7. Mock transfected cells should be monitored for cell death to ensure experiments are within a tolerable range that does not cause cytotoxicity. Negative control siRNA (where the siRNA has no known target) provides an accurate baseline of gene expression in the presence of both transfection reagent and noncoding siRNA molecules.
8. In addition to using UV hoods we carry out these steps in a separate laboratory from where our routine molecular biology work is carried out.
9. Low purity ratios are commonly caused by genomic DNA contamination. If ratio of ~1.8 is detected at this stage, perform a further DNaseI digest and repurify. RNA quality can also be assessed by fluorescent dyes such as QuantiFluor™ RNA system (Promega) or by microfluidics using a 2100 Bioanalyzer.
10. Smearing of RNA on agarose gels is indicative of degradation and/or genomic DNA contamination. It is not recommended to proceed to cDNA synthesis if degradation has occurred as this will affect gene quantitation and reproducibility.
11. Handle qPCR plates with care. Use a PCR tray to keep them raised off the work surface to avoid any autofluorescing environmental contamination being transferred into the block.
12. Firmly press the film onto the plate using a spreader tool to ensure full contact with the raised surface of each well to ensure no evaporation of sample during PCR cycling.
13. Ensure no bubbles remain in the V shape at the bottom of the wells.
14. Using the equation $E = 10^{(-1/\text{slope})} - 1 \times 100$, an ideal primer assay would have an efficiency of 100% representing an exact doubling of product in every cycle. Primers that have efficiencies within 10% of the control, however, can normally be used.
15. Assays must be specific, generating a single product since SYBR™ Green dye will bind and fluoresce any double-stranded DNA. More than one product can be the result of

off target amplification or primer dimer interactions. Single products can be confirmed by running a sample of PCR product on a 2% agarose gel. Nonspecific primers should be discarded and redesigned.

16. Typical ERC genes used are β-actin, GAPDH and Rn18S but these need to be validated for each cell type used and experimental paradigm. The reference control gene is used to normalize expression levels from different RNA extractions allowing a comparison between experiments. The expression of reference genes should remain unchanged by the transfection of siRNA or experimental conditions. Reverse transcribing the same amount of RNA at the cDNA step should give the same C_t value for the reference control gene in each sample. Using two or more genes will increase the accuracy of normalization. All PCRs should be set up in triplicate in 96-well plates. NTCs and No-RT samples are PCR'd singly as a check for genomic DNA or environmental contamination.
17. This is primer mastermix rather than the template mastermix that was prepared at the primer validation stage.
18. For the largest effect of knockdowns in our experiments we only used cells where knockdown was calculated to be greater than 75%. The highest mRNA knockdowns we achieved using this approach was >98%.

Acknowledgments

This work is dedicated to the memory of our colleague Lijun Tian, who developed and optimized the siRNA approach in the laboratory, and died in service before the writing of this chapter. The work was supported by the Wellcome Trust, British Heart Foundation, and Diabetes UK.

References

1. Elbashir SM, Harborth J, Lendeckel W et al (2001) Duplexes of 21-nucleotide RNAs mediate RNA interference in cultured mammalian cells. Nature 411(6836):494–498
2. Ipsaro JJ, Joshua-Tor L (2015) From guide to target: molecular insights into eukaryotic RNA-interference machinery. Nat Struct Mol Biol 22(1):20–28
3. Sioud M (2015) RNA interference: mechanisms, technical challenges, and therapeutic opportunities. Methods Mol Biol 1218:1–15
4. Carthew RW, Sontheimer EJ (2009) Origins and mechanisms of miRNAs and siRNAs. Cell 136(4):642–655
5. Licatalosi DD, Darnell RB (2010) RNA processing and its regulation: global insights into biological networks. Nat Rev Genet 11 (1):75–87
6. Chamberlain LH, Shipston MJ (2015) The physiology of protein S-acylation. Physiol Rev 95(2):341–376
7. Greaves J, Chamberlain LH (2011) DHHC palmitoyl transferases: substrate interactions and (patho)physiology. Trends Biochem Sci 36(5):245–253
8. Fukata Y, Fukata M (2010) Protein palmitoylation in neuronal development and synaptic plasticity. Nat Rev Neurosci 11(3):161–175

9. Tafer H (2014) Bioinformatics of siRNA design. Methods Mol Biol 1097(Chapter 22):477–490
10. Naito Y, Ui-Tei K (2012) siRNA design software for a target gene-specific RNA interference. Front Genet 3:102
11. Tian L, McClafferty H, Knaus H-G, Ruth P, Shipston MJ (2012) Distinct acyl protein transferases and thioesterases control surface expression of calcium-activated potassium channels. J Biol Chem 287(18):14718–14725
12. Tian L, McClafferty H, Jeffries O, Shipston MJ (2010) Multiple palmitoyltransferases are required for palmitoylation-dependent regulation of large conductance calcium- and voltage-activated potassium channels. J Biol Chem 285 (31):23954–23962
13. Tian L, Jeffries O, McClafferty H et al (2008) Palmitoylation gates phosphorylation-dependent regulation of BK potassium channels. Proc Natl Acad Sci U S A 105 (52):21006–21011
14. Jeffries O, Geiger N, Rowe IC et al (2010) Palmitoylation of the S0-S1 linker regulates cell surface expression of voltage- and calcium-activated potassium (BK) channels. J Biol Chem 285(43):33307–33314
15. Shipston MJ, Tian L (2016) Posttranscriptional and posttranslational regulation of BK channels. Int Rev Neurobiol 128:91–126
16. Bustin SA, Benes V, Garson JA et al (2009) The MIQE guidelines: minimum information for publication of quantitative real-time PCR experiments. Clin Chem 55(4):611–622
17. Thornton B, Basu C (2011) Real-time PCR (qPCR) primer design using free online software. Biochem Mol Biol Educ 39(2):145–154

Chapter 13

In Vitro Assays to Monitor the Enzymatic Activities of zDHHC Protein Acyltransferases

David A. Mitchell, Laura C. Pendleton, and Robert J. Deschenes

Abstract

A family of zDHHC protein acyltransferase (PAT) enzymes catalyze the *S*-palmitoylation of target proteins via a two-step mechanism. The first step involves transfer of palmitate from the palmitoyl-CoA donor to the active site cysteine of the zDHHC PAT enzyme, releasing reduced CoA (CoASH). In the second step, the palmitoyl–PAT intermediate thioester reacts with a cysteine side chain within the target substrate to produce the palmitoylated substrate product or, in the absence of a protein substrate, the palmitoyl–PAT intermediate thioester is hydrolyzed and releases palmitate. Formation and resolution of the palmitoyl–PAT intermediate complex (autopalmitoylation) is measured using a coupled enzyme system that monitors the production of CoASH via reduction of NAD^+ by the α-ketoglutarate dehydrogenase complex. This assay can be used to isolate and characterize modulators of autopalmitoylation and is scalable to high-throughput screening (HTS). A second fluorescence-based assay is described that monitors the hydrolysis of the palmitoyl–PAT thioester linked intermediate by thin-layer chromatography using a palmitoyl-CoA analog, BODIPY®-C12:0-CoA, as a substrate. These two assays provide a methodology to quantify the first enzymatic step of the two-step zDHHC PAT reaction.

Key words Palmitate, Palmitoylation, Palmitoyltransferase, zDHHC enzymes, Coupled enzyme assay, Protein acyl transferase (PAT), Coenzyme A (CoA)

1 Introduction

Protein *S*-palmitoylation is a reversible posttranslational modification responsible for the regulated localization of a variety of proteins involved in signal transduction, cell morphology, organelle inheritance, vesicle fusion and protein transport [1, 2]. Although the process of *S*-palmitoylation has been known for over 30 years, the enzymes responsible have only recently been isolated and characterized [3, 4]. First isolated in yeast, *S. cerevisiae*, protein acyltransferases (PATs) are encoded by a relatively large family of zDHHC genes named for their zinc binding motif and conserved Asp-His-His-Cys (DHHC) sequence present in the active site of the enzyme [3, 4]. zDHHC enzymes catalyze *S*-palmitoylation of target proteins via a two-step mechanism [5, 6] (Fig. 1). In the first

Maurine E. Linder (ed.), *Protein Lipidation: Methods and Protocols*, Methods in Molecular Biology, vol. 2009, https://doi.org/10.1007/978-1-4939-9532-5_13,

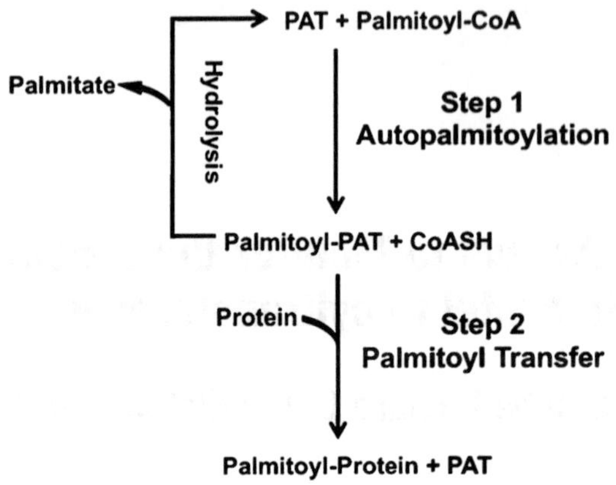

Fig. 1 Schematic diagram of the two-step PAT reaction denoting the reactants (palmitoyl-CoA, protein substrate, H_2O and zDHHC enzyme) and products (palmitate, CoASH and palmitoylated protein substrate). There are two assays which provide quantitative information on the kinetic parameters of the autopalmitoylation reaction. For the TLC-based PAT assay, reactant palmitoyl-CoA and products palmitoyl–enzyme and palmitate are monitored, while for the microtiter plate-based PAT assay, product CoASH is monitored

step, referred to commonly as autopalmitoylation, the palmitoyl-CoA donor transfers palmitate to the zDHHC enzyme to create a palmitoyl–PAT thioester linked intermediate and releases reduced CoA (CoASH). In the second step, the palmitoyl–PAT intermediate transfers palmitate to a protein substrate or undergoes hydrolysis to release palmitate, and in the process, regenerates the zDHHC enzyme to carry out further rounds of the reaction [7, 8].

Measurement of the individual steps of the PAT reaction has been a challenge due to the labile nature of thioester bonds and the low specific activity of ^{3}H-palmitate substrates. However, nonradioactive, fluorescence-based PAT assays allow for interrogation of the individual steps of the reaction [5, 7, 8]. In some cases, these assays have been adapted to high throughput, 96- and 384-well formats, to identify small molecule modulators of zDHHC enzymes [9, 10]. Herein we describe two PAT assays that measure different aspects of the autopalmitoylation reaction. Together, these assays provide a quantitative window into the enzymology of protein palmitoylation.

2 Materials

All reagents are purchased from Sigma-Aldrich, unless noted otherwise.

2.1 Fluorescence-Coupled Protein Acyltransferase Assay

1. 4× Reaction Buffer: 200 mM sodium phosphate Buffer, pH 7.2, 1 mM EDTA, 8 mM α-ketoglutaric acid, 1 mM NAD^+, 0.8 mM thiamine pyrophosphate.

2. 100 mM DTT in dH_2O.
3. Coupled Enzyme: α-KDH (α-ketoglutarate dehydrogenase), mU/mL is determined from the units/mg and the protein concentration in mg/mL of each enzyme prep.
4. zDHHC Enzyme: Purified wild-type (zDHHC) or catalytically inactive mutant zDHHS enzyme.
5. SPB: 50 mM sodium phosphate buffer, pH 7.2.
6. Palmitoyl CoA: Prepare a concentrated stock of 1–2 M in dH_2O and store at −20 °C. Prepare fresh dilutions of 100 μM, 60 μM, and 30 μM in SPB, *see* **Note 1.**
7. 120 μM CoASH in dH_2O—sodium salt hydrate, Sigma-Aldrich C3144.
8. Costar Assay Plate: Flat-bottom, untreated, black polystyrene, 96-well assay plate.

2.2 BODIPY®-C12:0-CoA Synthesis

1. BODIPY®-C12:0 (Invitrogen).
2. Methanol.
3. 1% Triton X-100 in dH_2O.
4. BODIPY®-C12:0-CoA Reaction Buffer: 300 mM MOPS-–NaOH, pH 7.5, 30 mM $MgCl_2$.
5. 100 mM adenosine triphosphate (ATP) in dH_2O.
6. CoASH: trilithium salt (Sigma-Aldrich C3019).
7. 10 mM adenosine-5′-phosphosulfate.
8. 2 U/mL acyl-CoA synthetase.
9. 300 U/mL ATP sulfurylase (New England Biolabs).
10. Perchloric acid.
11. Acetone.
12. Ethyl ether.
13. 10 mM sodium phosphate, pH 6.0.

2.3 Thin-Layer Chromatography

1. Silica G60 TLC plate.
2. zDHHC enzyme.
3. Sodium phosphate, 50 mM, pH 7.2.
4. BODIPY®-C12:0-CoA, synthesized (*see* Subheading 3.2, **step 1**).
5. TLC mobile phase: *n*-butyl alcohol–dH_2O–acetic acid; in a volume ratio of 50:30:20.

3 Methods

3.1 Fluorescence-Coupled Protein Acyltransferase Assay

Coupling the production of CoASH to the reduction of NAD^+ (to NADH) is a time-honored method used in enzyme kinetic studies [11–13]. Herein, it is adapted to measure the first enzymatic step of the zDHHC PAT reaction [5, 8] (Fig. 2a). The assay monitors the production of free CoASH by coupling its production to the production of NADH by α-KDH (α-ketoglutarate

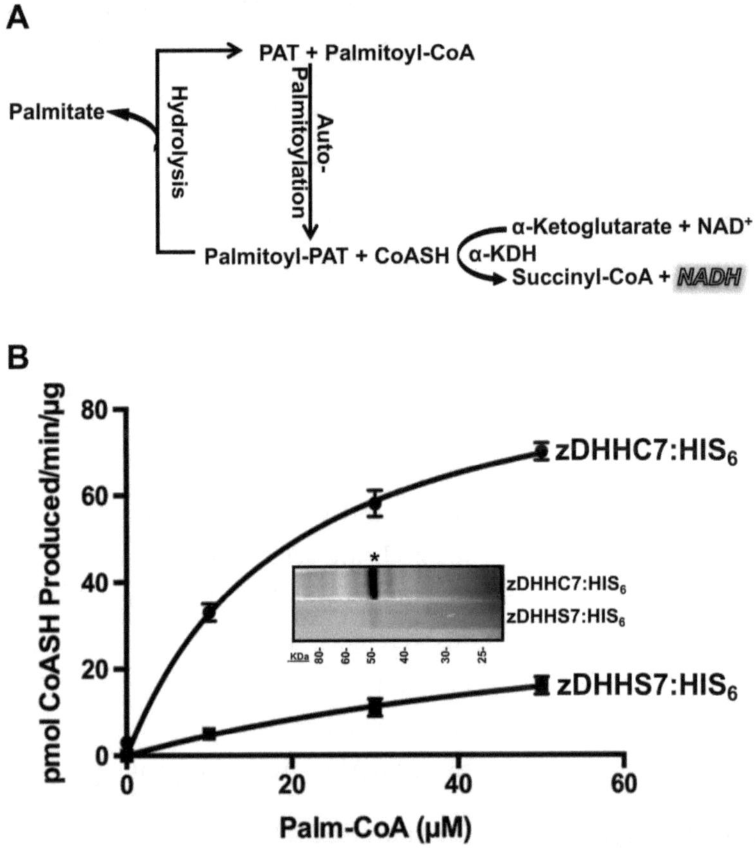

Fig. 2 Real-time fluorescence-coupled PAT Assay. (**a**) Schematic representation of the assay denoting the reactants and products. This assay monitors the production of CoASH [4] from palmitoyl-CoA [2] using an α-ketoglutarate dehydrogenase catalyzed reaction. (**b**) Graphical representation of reaction velocity values for varying concentrations of substrate (palmitoyl-CoA) using wild-type zTgDHHC7 enzyme (closed circles) and catalytically dead zTgDHHS7 enzyme (closed squares) from the Apicomplexa parasite, *Toxoplasma gondii*. Inset picture: Equivalent amounts of zDHHC7 and zDHHS7 were reacted with BODIPY®-C12:0-CoA and the reaction products separated by 10% SDS-PAGE. The band denoted by the * represents the apparent molecular weight of approximately 50 KDa. Controls lacking enzyme were subtracted from both the wild type and the catalytically dead enzyme velocity values

dehydrogenase), which in turn is monitored by a change in fluorescence intensity (340 nm excitation/465 nm emission).

1. Begin by generating the data for a CoASH standard curve. Prepare a reaction cocktail that contains, per well, 25 μL of 4× Reaction Buffer, 0.25 μL 100 mM DTT, 16 mU α-KDH, and dH_2O in a final volume of 50 μL. Make enough cocktail for the number of reactions needed plus 15%. Aliquot 50 μL into each well of a flat-bottom black polystyrene 96-well assay plate. Prepare the plate for triplicate wells of 0, 300, 600, and 3000 pmol of CoASH.
2. Initiate the standard curve reactions with the addition of 50 μL of dH_2O (0 pmol), or 50 μL of 6 μM, 12 μM, or 60 μM of CoASH for final concentrations of 3 μM (300 pmol), 6 μM (600 pmol), or 30 μM (3000 pmol).
3. Monitor the reaction for 30 min in a Biotek Mx fluorometer microplate reader set to 30 °C (340 nm excitation/465 nm emission).
4. Plot the relative fluorescence units (RFU) of the CoASH reactions, determined once equilibrium has been reached, versus the pmol of CoA to generate a standard curve.
5. For each zDHHC enzyme being measured (wild-type zDHHC and zDHHS active site mutant control), prepare a reaction cocktail consisting of 25 μL of 4× Reaction Buffer, 0.25 μL 100 mM DTT, 16 mU α-KDH, 1–3 μg of enzyme, and dH_2O in a final volume of 50 μL. Make enough cocktail for the number of reactions needed plus 15%. Aliquot 50 μL into each well of the assay plate. Prepare the plate for triplicate wells of 0, 10, 30, and 50 μM final concentration of palmitoyl-CoA for each enzyme preparation.
6. Initiate the reactions with the addition of 50 μL of dH_2O, or 50 μL 20 μM, 60 μM, or 100 μM palmitoyl-CoA for final concentrations of 10 μM, 30 μM, or 50 μM. The final volume for all reactions is 100 μL (*see* **Note 2** *regarding palmitoyl-CoA concentrations).*
7. Monitor the reaction for 30 min in the microplate reader set to 30 °C (340 nm excitation/465 nm emission).
8. Analyze the first 10 min of the PAT enzyme reaction to determine the initial rate of CoASH production. Using the standard curve (derived from **steps 1–4**) calculate the pmols of CoASH produced within the first 10 min of the reaction. Typically, the results are expressed as pmol/min/μg zDHHC enzyme as a function of palmitoyl CoA concentration (Fig. 2B).

3.2 Synthesis of BODIPY®-C12:0-CoA

The fluorescent palmitate analog BODIPY®-C12:0 can substitute for palmitate in the PAT reaction, but first it must be conjugated to CoA to serve as a substrate. Currently, there are no commercial sources of BODIPY®-C12:0-CoA. The synthesis of BODIPY®-C12:0-CoA from BODIPY®-C12:0 (Invitrogen) and CoASH (Sigma) was adapted from a procedure originally described to synthesize iodinated palmitoyl-CoA compounds by Berthiaume et al. [14].

1. Dissolve 2 mol of BODIPY®-C12:0 in 500 μL of methanol and 200 μL 1% Triton X-100 in the manufacturer's BODIPY®-C12:0 vial (*see* **Note 3**).
2. Transfer solution to a clean, 5-mL glass vial.
3. Dry the methanol–Triton X-100–BODIPY®-C12:0 solution under a stream of nitrogen at room temperature. Verification of totally evaporated methanol solvent is revealed by the appearance of an orange residue on the bottom of the glass vial.
4. Resuspend the orange residue in 720 μL of BODIPY®-C12:0-CoA Reaction Buffer and sonicate at room temperature for 10 min in a bath sonicator.
5. After sonication, add 200 μL of 100 mM ATP, 8 mg of CoASH, 45 μL of 10 mM adenosine-5′-phosphosulfate, 20 μL of 1 M DTT, 1 unit (500 μL) of acyl-CoA synthetase, and 5 units of ATP sulfurylase in a final volume of 1.6 mL.
6. Cap the vial and incubate the reaction at 35 °C for 14 h with constant, slow stirring.
7. After the incubation period, precipitate the BODIPY®-C12:0-CoA from the reaction by adding perchloric acid to a final concentration of 1.3%.
8. Collect the precipitate using centrifugation, decant the liquid, and wash the pellet once with room temperature acetone to remove the unreacted CoASH.
9. Collect the precipitate using centrifugation, decant the liquid, and wash the pellet twice with ethyl ether, to remove unreacted BODIPY®-C12:0. *This step needs to be performed in a fume hood.*
10. Dry the precipitate under a gentle stream of nitrogen.
11. Resuspend the dried precipitate in 10 mM sodium phosphate, pH 6.0. The amount of compound is determined using absorbance at 260 nm with an extinction coefficient of 15,400 (L mol^{-1} cm^{-1}).

3.3 Analysis of the PAT Autopalmitoylation Reaction by TLC

1. To prepare the TLC plate, use gloved hands to remove a Silica G60 thin-layer chromatography plate from its packaging and lay the plate on the bench.
2. Using a pencil and straight edge, lightly sketch a line approximately 1.5 cm from the bottom of the TLC plate. This is the origin.
3. At equidistant sites along the sketched line, carefully add small tick marks. These will denote the sites where reaction aliquots will be spotted.
4. Mix a known amount of zDHHC enzyme (approximately 1 μg) with 50 mM sodium phosphate, pH 7.2, 0.5 mM DTT in a final volume of 100 μL and place on ice.
5. At room temperature, place a flat-bottom 96-well microplate on a stirring platform and place a small (flea) stir bar in one of the wells. Add the 100 μL of PAT mixture to the well with the stir bar and begin stirring to equilibrate the reaction solution to room temperature (3 min).
6. Remove 2 μL of the stirring reaction for the zero time point and spot on the first tick mark on the TLC plate (Dim the lab lights to minimize the quenching of the fluorescent probe.).
7. To initiate the reaction, BODIPY®-C12:0-CoA is added to the stirring reaction at a final concentration of 40 μM. Immediately begin timing the reaction.
8. At 60 s spot a 2 μL aliquot of the reaction onto the next tick mark on the Silica G60 plate. Spotting the aliquot on the TLC plate terminates the reaction.
9. After the timed spotting is finished, the spotted aliquots are allowed to dry (~5 min) in the dark.
10. The reaction reactants and products are separated on the TLC plate using freshly prepared mobile phase consisting of n-butyl alcohol–water–acetic acid (50:30:20) (v/v/v).
11. When the mobile phase front is approximately 5 cm from the top of the plate, the plate is removed from the TLC tank and placed in the dark on a horizontal drying surface.
12. After the plate is dry, the amounts of BODIPY®-C12:0, BODIPY®-C12:0-CoA, and BODIPY®-C12:0-PAT are determined by fluorescence imaging of the TLC plate using the Typhoon 5 Biomolecular Imager (GE Healthcare) for BODIPY® fluorescence (488 nm excitation/532 nm emission).
13. BODIPY® fluorescence located at the origin corresponds to BODIPY®-C12:0-PAT (R_f 0.0), BODIPY®-C12:0-CoA (R_f 0.5) migrates partially up the plate, and BODIPY®-C12:0 (R_f 0.9) migrates the farthest on the plate (Fig. 3).

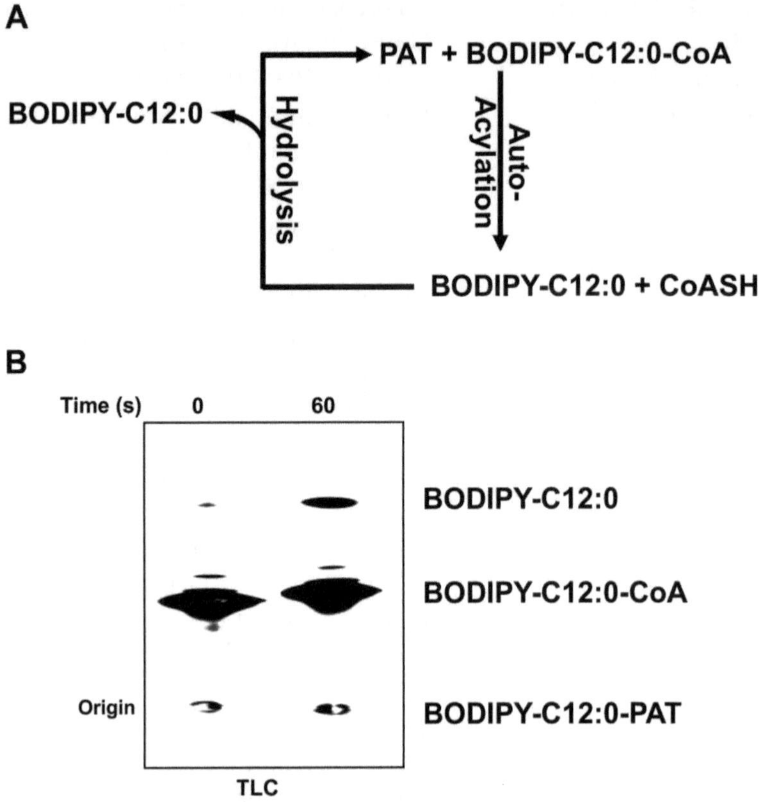

Fig. 3 Identification of the reactants and products of the PAT reaction monitored by thin-layer chromatography [5, 8]. (**a**) Reaction schematic with reactant and product color coding; BODIPY®-C12:0-CoA (Blue), BODIPY®-C12:0-PAT (Orange) and BODIPY®-C12:0 (Green). (**b**) R_f values of the products and reactants of the PAT reaction; BODIPY®-C12:0-CoA (Blue, R_f 0.45), BODIPY®-C12:0-PAT (Orange, R_f 0.00/Origin) and BODIPY®-C12:0 (Green, R_f 0.90)

14. The amount of BODIPY®-C12:0 transferred to PAT was calculated from a standard curve of known amounts of BODIPY®-C12:0.

4 Notes

1. When thawing the solution of palmitoyl-CoA, there is a tendency for the reagent to form micelles. Prepare a stock of 2 M and make fresh dilutions in SPB the day of the experiment.
2. At higher concentrations of palmitoyl-CoA, there will appear to be product inhibition of the enzyme activity. This is most likely due to micelle formation and substrate limitation.
3. BODIPY®-C12:0-CoA has been found to be a better substrate than BODIPY®-C16:0-CoA.

References

1. Linder ME, Deschenes RJ (2007) Palmitoylation: policing protein stability and traffic. Nat Rev Mol Cell Biol 8:74–84
2. Mitchell DA, Vasudevan A, Linder ME, Deschenes RJ (2006) Protein palmitoylation by a family of DHHC protein S-acyltransferases. J Lipid Res 47:1118–1127
3. Lobo S, Greentree W, Linder M, Deschenes R (2002) Identification of a Ras palmitoyltransferase in *Saccharomyces cerevisiae*. J Biol Chem 277:41268–41273
4. Roth AF, Feng Y, Chen L, Davis NG (2002) The yeast DHHC cysteine-rich domain protein Akr1p is a palmitoyl transferase. J Cell Biol 159:23–28
5. Mitchell DA, Mitchell G, Ling Y, Budde C, Deschenes RJ (2010) Mutational analysis of *Saccharomyces cerevisiae* Erf2 reveals a two-step reaction mechanism for protein palmitoylation by DHHC enzymes. J Biol Chem 285:38104–38114
6. Jennings BC, Linder ME (2012) DHHC protein S-acyltransferases use similar ping-pong kinetic mechanisms but display different acyl-CoA specificities. J Biol Chem 287:7236–7245
7. Mitchell DA, Hamel LD, Ishizuka K, Mitchell G, Schaefer LM, Deschenes RJ (2012) The Erf4 subunit of the yeast Ras palmitoyl acyltransferase is required for stability of the Acyl-Erf2 intermediate and palmitoyl transfer to a Ras2 substrate. J Biol Chem 287:34337–34348
8. Mitchell DA, Hamel LD, Reddy KD, Farh L, Rettew LM, Sanchez PR, Deschenes RJ (2014) Mutations in the X-linked intellectual disability gene, zDHHC9, alter autopalmitoylation activity by distinct mechanisms. J Biol Chem 289:18582–18592
9. Hamel LD, Deschenes RJ, Mitchell DA (2014) A fluorescence-based assay to monitor autopalmitoylation of zDHHC proteins applicable to high throughput screening. Anal Biochem 460:1–8
10. Hamel LD, Lenhart BJ, MItchell DA, Santos RG, Giulianotti MA, Deschenes RJ (2016) Identification of protein palmitoylation inhibitors from a scaffold ranking library. Comb Chem High Throughput Screen 19:262–274
11. Massey V (1960) The composition of the ketoglutarate dehydrogenase complex. Biochim Biophys Acta 38:447–460
12. Storer AC, Cornish-Bowden A (1974) The kinetics of coupled enzyme reactions. Applications to the assay of glucokinase, with glucose 6-phosphate dehydrogenase as coupling enzyme. Biochem J 141:205–209
13. Garland PB, Shepherd D, Yates DW (1965) Steady-state concentrations of coenzyme A, acetyl-coenzyme A and long-chain fatty acyl-coenzyme A in rat-liver mitochondria oxidizing palmitate. Biochem J 97:587–594
14. Berthiaume L, Peseckis SM, Resh MD (1995) Synthesis and use of iodo-fatty acid analogs. Methods Enzymol 250:454–466

Chapter 14

Purification of Recombinant DHHC Proteins Using an Insect Cell Expression System

Martin Ian P. Malgapo and Maurine E. Linder

Abstract

DHHC enzymes are a family of integral membrane proteins that catalyze the posttranslational addition of palmitate, a 16-carbon fatty acid, onto a cysteine residue of a protein. While the library of identified palmitoylated proteins has grown tremendously over the years, biochemical and mechanistic studies on DHHC proteins are challenged by the innate difficulty of purifying the enzyme in large amounts. Here we describe our protocol for preparing recombinant DHHC proteins tagged with a hexahistidine sequence and a FLAG epitope that aid in the purification. This procedure has been tested successfully in purifying several members of the enzyme family; DHHC3 and its catalytically inactive cysteine mutant, DHHS3 are used as examples. The recombinant protein is extracted from whole cell lysates using the detergent dodecylmaltoside (DDM) and is subjected to a two-column purification. Homogeneity and monodispersity of the purified protein are checked by size exclusion chromatography (SEC). A preparation from a 400-mL infection of Sf9 insect cell culture typically yields 0.5 mg of DHHC3 and 1.0 mg of catalytically inactive DHHS3. Both forms appear monodisperse up to a concentration of 1 mg/mL by SEC.

Key words DHHC protein, Palmitoylation, Protein acyltransferase, Immobilized metal affinity chromatography, Affinity chromatography, Size exclusion chromatography

1 Introduction

Palmitoylation refers to the posttranslational attachment of a fatty acid onto a cysteine residue of a protein. Although fatty acids of varying degrees of unsaturation and chain lengths can be added to cysteines, the 16-carbon fatty acid, palmitate, is the most commonly observed and hence most well studied. Palmitoylation can occur in both soluble and integral membrane proteins. In soluble proteins, the addition of palmitate increases the hydrophobic character of the protein at the site of palmitoylation, allowing for specific membrane targeting. The consequences of palmitoylating an integral membrane protein include alterations in the conformation of transmembrane domains, association with specific membrane domains such as lipid rafts, controlled interactions with

Maurine E. Linder (ed.), *Protein Lipidation: Methods and Protocols*, Methods in Molecular Biology, vol. 2009,
https://doi.org/10.1007/978-1-4939-9532-5_14,

other proteins, and controlled interplay with other posttranslational modifications [1]. Current estimates suggest that more than 10% of the human proteome undergoes palmitoylation [2].

Although protein palmitoylation had been known for several decades, the identity of the family of enzymes that catalyze palmitoylation in cells remained unknown until 2002, with the discovery of the two founding members of the DHHC enzyme family in yeast. The Erf2/Erf4 (Effect on Ras function) protein complex was found to palmitoylate the Ras2 protein [3], and ankyrin repeat-containing protein 1 was found to palmitoylate yeast casein kinase 2 [4]. DHHC enzymes act as protein acyltransferases and are highly conserved throughout eukaryotes with families ranging in size from 5 members in *Schizosaccharomyces pombe* [5] to 23 in humans [6].

DHHC enzymes are polytopic integral membrane proteins characterized by their highly conserved DHHC (Asp-His-His-Cys) motif within its catalytic 51-amino acid cysteine-rich domain (CRD) [7]. The typical membrane topology is four transmembrane domains with the N- and C-termini exposed to the cytoplasm. A few DHHC enzymes have six transmembrane helices [8]. The N- and C-terminal regions vary in sequence and size, potentially allowing specific protein–protein interactions. It is now appreciated that DHHC proteins generally function via a kinetic ping-pong mechanism [9, 10]. The enzyme first undergoes autopalmitoylation on the cysteine of the DHHC motif using palmitoyl-coenzyme A (CoA) as a palmitoyl donor, the palmitate is then transferred to a cysteine residue on the substrate. Mutating the cysteine in the catalytic DHHC motif into a serine or alanine typically inactivates the enzyme, although there are exceptions [11].

The innate difficulty of expressing and purifying sufficient quantities of DHHC proteins has served as a bottleneck in the biochemical, mechanistic, and structural characterization of these enzymes. Standard bacterial expression systems lack the enzymes required for posttranslational modification, and hence are not suited for purifying these eukaryotic membrane enzymes. Moreover, preservation of their biological and functional activities during the isolation process can be compromised. One key consideration is the use of an appropriate detergent such as dodecylmaltoside (DDM) during solubilization to prevent the protein from aggregating upon extraction from the membrane. Solubilized with amphiphilic detergents and purified as protein–detergent complexes, the detergent micelle covers the hydrophobic surface of the membrane protein as would the hydrophobic lipids in the native membrane bilayer.

In this chapter we describe a protocol for purifying recombinant DHHC proteins. We have used this protocol to routinely prepare DHHC2, DHHC3, DHHC20, and their catalytically inactive DHHS mutants. The DHHC constructs are cloned to include

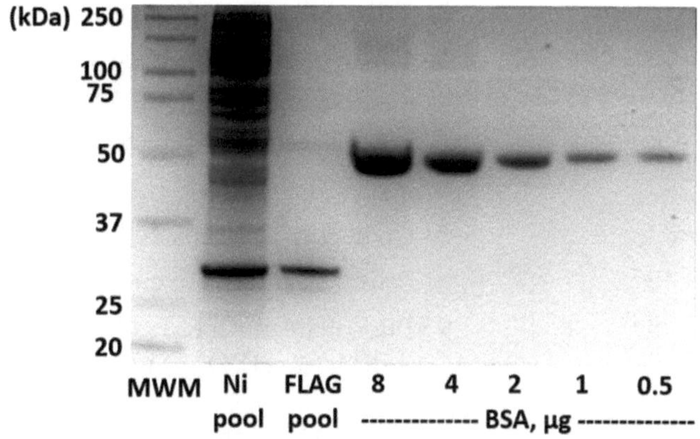

Fig. 1 Coomassie Blue-stained gel of DHHC3 Ni and FLAG pools with a BSA standard curve for protein yield quantification

a hexahistidine sequence and a FLAG tag at the C-terminus and are expressed and purified using a two-step column purification to give a preparation of high yield and good protein purity. Immobilized metal affinity chromatography (IMAC) is used as the first step. Histidine residues in the hexahistidine tag of the DHHC protein bind to the vacant positions in the coordination sphere of the nickel ions immobilized on the nitrilotriacetic acid agarose resin. After washing to remove nonspecifically bound proteins, the enzyme is eluted using 200–500 mM imidazole as a binding competitor under nondenaturing conditions. While Ni IMAC purification is effective in removing most contaminants in the whole cell lysate, following it with FLAG affinity purification significantly increases the purity of the enzyme (Fig. 1). The FLAG (N-DYKDDDDK-C epitope) tag of DHHC binds to the FLAG M2 monoclonal antibody covalently cross-linked to agarose. FLAG-tagged proteins bind to the resin while contaminants pass through. After washing to remove nonspecifically bound proteins, FLAG-DHHC3 is eluted using 100 μg/mL FLAG peptide as a binding competitor under nondenaturing conditions. A final step of size exclusion chromatography (SEC) enables removal of FLAG peptide and buffer exchange, as well as serving to check the monodispersity of the preparation.

2 Materials

Prepare all solutions and buffers using analytical grade reagents and ultrapure water (18.2 MΩ•cm at 25 °C).

2.1 Baculovirus Stocks and Insect Cell Culture

1. Bacmid encoding His- and FLAG-tagged DHHC protein (*see* **Note 1**).
2. Bacvector Sf9 cells (Novagen).

3. TriEx Sf9 cells (Novagen).
4. Unsupplemented Grace's Insect Medium.
5. Growth Medium: Grace's Medium, 10% heat-inactivated fetal bovine serum (HI-FBS) (*see* **Note 2**).
6. ESF 921 medium (Expression Systems).
7. Cellfectin® II transfection reagent.
8. 6-well sterile tissue culture plates.
9. Phosphate-buffered saline (PBS): 137 mM NaCl, 2.7 mM KCl, 8 mM Na_2HPO_4, and 2 mM KH_2PO_4.

2.2 Ni-NTA Purification

1. Nickel-nitrilotriacetic acid (Ni-NTA) agarose (Qiagen).
2. 1.5 × 15 cm glass chromatography column (cross-sectional area 1.77 cm^2).
3. DDM Extraction Buffer: 50 mM Tris–HCl, pH 7.4, 200 mM NaCl, 10% glycerol, 1 mM (tris(2-carboxyethyl)phosphine) (TCEP), pH 7.4, 1% dodecylmaltoside (DDM), 1 μg/mL aprotinin, 1 μg/mL leupeptin, 1 μg/mL pepstatin A, 0.25 mM PMSF, 1 mM benzamidine (*see* **Note 3**).
4. Ni Equilibration Buffer: 50 mM Tris–HCl, pH 7.4, 200 mM NaCl, 10% glycerol, 1 mM TCEP, pH 7.4, 0.05% DDM, 1 μg/mL aprotinin, 1 μg/mL leupeptin, 1 μg/mL pepstatin A, 0.25 mM PMSF, 1 mM benzamidine.
5. Ni Wash Buffer: 50 mM Tris–HCl, pH 7.4, 200 mM NaCl, 10% glycerol, 0.5 mM TCEP, pH 7.4, 0.05% DDM, 1 μg/mL aprotinin, 1 μg/mL leupeptin, 1 μg/mL pepstatin A, 0.25 mM PMSF, 1 mM benzamidine, 15 mM imidazole, pH 7.4 (*see* **Note 4**).
6. Ni Elution Buffer A: 50 mM Tris–HCl, pH 7.4, 100 mM NaCl, 10% glycerol, 0.25 mM TCEP, pH 7.4, 0.05% DDM, 200 mM imidazole.
7. Ni Elution Buffer B: 50 mM Tris–HCl, pH 7.4, 100 mM NaCl, 10% glycerol, 0.25 mM TCEP, pH 7.4, 0.05% DDM, 500 mM imidazole.
8. 5× Protein Sample Buffer: 250 mM Tris–HCl, pH 6.8, 50% glycerol, 5% SDS, 0.1% bromophenol blue.

2.3 FLAG Purification

1. Anti-FLAG M2 Affinity Gel (Sigma).
2. 1.5 × 15 cm glass chromatography column (cross-sectional area 1.77 cm^2).
3. Tris-buffered Saline (TBS) Buffer: 50 mM Tris–HCl, pH 7.4, 150 mM NaCl.
4. 30 mg/mL 3× FLAG peptide (Sigma) in TBS.

5. FLAG Equilibration Buffer: 50 mM Tris–HCl, pH 7.4, 150 mM NaCl, 10% glycerol, 1 mM EDTA, 1 mM TCEP, pH 7.4, 0.05% DDM.
6. FLAG Elution Buffer: 50 mM Tris–HCl, pH 7.4, 150 mM NaCl, 10% glycerol, 1 mM EDTA, 1 mM TCEP, pH 7.4, 0.05% DDM, 100 μg/mL FLAG peptide.
7. FLAG Column Regeneration Buffer: 100 mM glycine, pH 3.5.
8. FLAG Column Storage Buffer: TBS, 50% glycerol, 0.02% sodium azide.

2.4 Size Exclusion Chromatography

1. Superdex® 200 10/300 GL analytical SEC column.
2. Running Buffer: 50 mM Tris–HCl, pH 7.4, 150 mM NaCl, 1 mM TCEP, pH 7.4, 0.05% DDM.

3 Methods

3.1 Baculovirus Stock and Insect Cell Culture

1. Create a suspension culture of Bacvector Sf9 cells at a concentration of 4×10^5 cells/mL with unsupplemented Grace's Insect Medium (*see* **Note 5**).
2. Seed each 2-cm well with 8×10^5 cells.
3. Allow the cells to attach onto the bottom of the well for 15 min.
4. Dilute 8 μL of Cellfectin in 100 μL of unsupplemented Grace's Insect Medium in a 1.5 mL-eppendorf tube.
5. Dilute 1 μg of bacmid DNA in 100 μL unsupplemented Grace's Insect Medium in a separate 1.5 mL-eppendorf tube.
6. Combine the two solutions and incubate for 15–30 min at room temperature.
7. Pipette the combined DNA–Cellfectin–Grace's mixture dropwise onto the cells (*see* **Note 6**).
8. Incubate for 3–5 h.
9. Vacuum aspirate the transfection mixture and replace with 2 mL of Growth Medium.
10. Incubate the cells at 27 °C for 4–6 days or until signs of viral infection become apparent (*see* **Note 7**).
11. Centrifuge the cells at $500 \times g$ for 10 min and collect the supernatant. This is designated P1 virus.
12. Add HI-FBS to the P1 virus to a concentration of 10% and store at 4 °C in the dark (*see* **Note 8**).
13. Create a 50-mL suspension culture of Triex Sf9 cells in ESF 921 medium at a logarithmic growth phase (*see* **Note 5**).
14. Dilute the P1 virus 1:1000 into the Triex Sf9 cell culture.

15. Incubate the cells at 27 °C with shaking at 125 rpm for 4–6 days or until signs of viral infection become apparent (*see* **Note 7**).
16. Centrifuge the cells at 500 × *g* for 10 min and collect the supernatant. This is designated P2 virus.
17. Add HI-FBS to the P2 virus to a concentration of 10% and store at 4 °C in the dark (*see* **Note 9**).
18. Use the P2 virus to infect Sf9 cells in ESF 921 medium for a large-scale preparation.
19. Infect a 400-mL culture of Triex Sf9 cells in logarithmic growth phase at a density of 1.5–2.0 × 10^6 cells/mL using a ratio of 1:50 (P2 virus–cell culture volume).
20. Incubate the culture with shaking (125 rpm) in a 27 °C incubator for 48 h for optimal expression while limiting proteolysis. Before harvesting, check the cells for signs of infection (i.e., enlarged nuclei, enlarged cells) and count using a hemocytometer. The cell counts postinfection typically range from 2.5 to 3.5 × 10^6 cells/mL.
21. Collect the cells by centrifugation at 1000 × *g* for 10 min at 4 °C. Wash twice with 40 mL PBS.
22. Flash freeze the cell pellets with liquid nitrogen and store at −80 °C.

3.2 Ni IMAC Purification

Carry out all procedures at 4 °C in the cold room unless otherwise specified.

1. Pour the Ni agarose resin into a column. Let the resin settle at the bottom of the column and wash with 10 column volumes (CV) of water to remove the ethanol from the beads (*see* **Notes 10–12**).
2. Wash the resin with 10 CV of Ni Equilibration Buffer (*see* **Notes 13** and **14**).
3. Thaw the frozen cell pellet in a 25 °C water bath, flicking frequently to ensure homogeneous thawing. Once thawed, keep the tube on ice.
4. Resuspend pellet in the DDM Extraction Buffer by pipetting up and down. For DHHC3, the tested ratio of cell pellet to volume of extraction buffer is 3.5 × 10^6 cells/100 μL (*see* **Notes 15** and **16**).
5. Transfer the suspension into a conical tube and rotate end over end for 120 min to lyse the cells.
6. Centrifuge the lysate at 100,000 × *g* for 35 min (*see* **Note 17**).
7. Collect the supernatant (S100).

8. Resuspend a fraction of the pellet (P100) with protein sample buffer containing 0.5 mM TCEP for later **analysis** (*see* **Note 18**).
9. Apply S100 to the column and allow to pass through dropwise at a flow rate of less than 3 mL/min. Collect flow-through (FT) (*see* **Note 19**).
10. Apply FT to column and allow to pass through in a dropwise fashion; collect and reapply for a total of three times. Save the final FT.
11. Wash the column with 10 CV of DDM Extraction Buffer (contains no imidazole) to elute unbound proteins. Save XT wash.
12. Wash the column with 10 CV of Ni Wash Buffer (contains 15 mM imidazole) to elute proteins weakly bound to resin. Save buffer wash (*see* **Notes 20** and **21**).
13. Collect a total of six 2-mL elutions, the first two using Ni Elution Buffer A (containing 200 mM imidazole) and the remaining four using Ni Elution Buffer B (containing 500 mM imidazole) (*see* **Note 22**).
14. If continuing with the FLAG purification on the same day, hold Ni elutions at 4 °C. Otherwise, flash freeze the elutions and store at −80 °C (*see* **Note 23**).
15. Save 40-μL aliquots of each elution. Run SDS-PAGE on half of the gel samples of the six Ni elutions with bovine serum albumin (BSA) protein standards. Stain with Coomassie. Estimate DHHC3 protein concentrations in each of the elutions by comparing it with a linear protein band concentration curve generated with known concentrations of BSA standards (*see* **Note 24**).
16. Run a western blot with aliquots of the P100, XT, Ni FT, and the other half of the elutions to determine the efficiency of extraction, efficiency of binding to the Ni column, and elution profile.
17. Identify peak fractions for FLAG purification based on purity and DHHC3 protein concentration.
18. The Ni-NTA agarose resin can be regenerated by successively washing with 5 CV of the following: water, 0.1 M EDTA pH 8.0, water, 0.1 M $NiSO_4$, 20% ethanol. Store the resin at 4 °C.

3.3 FLAG Purification

Carry out all procedures at 4 °C in the cold room unless otherwise specified.

1. Rinse an empty glass column twice with 5 mL of FLAG Equilibration Buffer. Drain the column and leave about 1 mL residual buffer in the column to aid in packing the gel.
2. Pipette 500 μL FLAG resin (50% suspension) into the column (*see* **Note 25**).
3. Allow the gel bed to drain. The resin buffer will flow slowly and the flow rate will increase as the glycerol is removed.
4. Prime the gel by applying three sequential CV of FLAG Column Regeneration Buffer (*see* **Note 26**).
5. Wash the resin three times with 10 CV of FLAG Equilibration Buffer.
6. If Ni elution samples were frozen, quickly thaw in a 25 °C water bath and place on ice.
7. Dilute the pooled fractions: 1 part Ni pool–4 parts FLAG Equilibration Buffer.
8. Apply pooled fractions to the column resin under gravity flow and collect the FT. Reapply the FT for a total of three times as multiple passes improve binding.
9. Wash the column with 10 CV of FLAG Equilibration Buffer. This should remove contaminating proteins nonspecifically bound to the FLAG M2 antibody.
10. Collect a total of six 500-μL elutions using FLAG Elution Buffer. Save 40 μL aliquots for gel analysis.
11. Elute any remaining protein from the FLAG resin by washing the column three times with 1 CV of FLAG Column Regeneration Buffer and save the fractions (*see* **Note 27**).
12. Immediately add 500 μL of FLAG Equilibration Buffer to each fraction to achieve neutral pH.
13. Rinse the column with FLAG Equilibration Buffer until the pH of effluent is neutral.
14. Prepare the column for storage by washing with 10 CV of FLAG Column Storage Buffer.
15. Add 5 mL of FLAG Column Storage Buffer. Do not drain. Label and store the column at 4 °C for future use (*see* **Note 28**).
16. Run SDS-PAGE on half of the gel samples of the six nickel elutions with bovine serum albumin (BSA) protein standards. Stain with Coomassie. Estimate DHHC3 protein concentrations in each of the elutions by comparing it with a linear protein band concentration curve generated with known concentrations of BSA standards (*see* **Note 29**).

3.4 Size Exclusion Chromatography

1. Pool FLAG elutions and concentrate to 1 mg/mL.
2. Apply 25 μg of protein diluted to 100 μL with running buffer to a size exclusion column at a flow rate of 0.5 mL/min.
3. Check for monodispersity based on the intrinsic tryptophan fluorescence at A280.

4 Notes

1. Bacmids are constructed using the Bac-to-Bac Insect Cell Expression System from Invitrogen.
2. Fetal bovine serum is heat inactivated by incubating at 56 °C for 30 min.
3. Protease Inhibitors are made up as 1000× stocks as follows: 1 mg/mL aprotinin in H_2O, 1 mg/mL leupeptin in H_2O, 1 mg/mL pepstatin A in absolute ethanol, 250 mM PMSF in absolute ethanol, 1 M benzamidine in H_2O.
4. Adding 1 M imidazole to the wash and elution buffers will change the overall pH of the solution. Adjust the final pH of the solution to 7.4 with 12 M HCl.
5. Before setting up the transfection or infection, ensure that the BacVector/Triex Sf9 cells are in the log phase of cell growth (target concentration = 1.5–2.5 × 10^6 cells/mL).
6. It is important to add the solution dropwise to not detach the cells from the bottom of the well.
7. Signs of viral infection include clearings in the well, enlarged cells, and enlarged cell nuclei. Having a negative control (without the transfection mixture) in a separate well is recommended. In the control well, the cells will keep growing and minimal clearing of the monolayer will be seen.
8. We have determined that the P1 virus remains potent for at least 3 months if stored in the dark at 4 °C.
9. We have determined that the P2 virus remains potent for 2 weeks if stored in the dark at 4 °C.
10. For efficient column packing, make sure that both the column and the resin are equilibrated at 4 °C before starting. For a 400 mL cell culture, we typically use 2 mL of Ni-NTA agarose packed in a 1.5 × 15 cm, cross-sectional area 1.77 cm^2 glass chromatography column. The theoretical binding capacity of the Qiagen Ni-NTA agarose beads is 50 mg/mL.
11. Ni-NTA agarose comes as a 50% suspension in 20% ethanol. Resuspend the resin by inverting the bottle several times.
12. Do not let the agarose resin run dry at any point in the purification.

13. Ni purification buffers can be prepared a day in advance in the bench and kept at 4 °C until the next day. However, the reducing agent (TCEP), detergent (DDM), and protease inhibitors must be added just before use on the same day of purification.
14. Increasing the pH of the equilibration buffer to pH 8 will increase protein binding to the NTA beads but will also increase nonspecific binding of other proteins.
15. For a 400 mL prep harvested at 3.5×10^6 cells/mL, we extract with about 40 mL of extraction buffer.
16. Pipetting up and down ensures that there are no clumps of unbroken cells left during the extraction.
17. Thick-wall polycarbonate tubes can be partially filled and centrifuged. Thin-walled ultracentrifuge tubes must be filled completely, or they will collapse during centrifugation. For ultracentrifugation, the tubes must be balanced such that the masses of the tubes with the sample must be within ±0.1 g.
18. If P100 is hard to solubilize, use a syringe needle (25G) to let the SDS get into the pellet.
19. Ensure that S100 clear and not cloudy to avoid clogging the column.
20. It is recommended to include 15 mM imidazole in the Ni Wash Buffer to remove proteins nonspecifically bound to the resin.
21. Although 10 CV of wash buffer is usually sufficient, the total protein concentration in the washes can be monitored by measuring the absorbance at 280 nm (A_{280}). Ensure that the final wash fraction has a baseline A_{280} value before proceeding with the elution.
22. Most of the protein typically elutes in Ni elutions 2 and 3.
23. It is recommended to keep the Ni column at 4 °C until the results of the gel analysis are available.
24. We use the VersaDoc™ 5000 imaging system to quantify protein bands in a gel using volume density determination followed by linear regression analysis.
25. The ANTI-FLAG M2 affinity gel resin is stored in 50% glycerol. The glycerol must be removed just prior to use and the resin equilibrated with the equilibration buffer. Because the gel is very viscous, it is very important to thoroughly resuspend the resin. Make sure the bottle is a uniform suspension of the gel beads before pipetting. Vortex if necessary.
26. Avoid disturbing the gel bed while applying the FLAG Column Regeneration Buffer. Let each aliquot drain completely before adding the next. Do not leave the column in glycine HCl for longer than 20 min.

27. This step serves a dual purpose. The glycine in the FLAG Column Regeneration Buffer elutes any remaining protein and regenerates the FLAG resin for future use.
28. Do not allow the resin to remain in TBS buffer for >24 h without an antimicrobial agent, 0.02% NaN_3.
29. For a 400-mL preparation, the typical yields are 0.5 mg for DHHC3 and 1 mg of DHHS3.

Acknowledgements

This work was supported by the National Institutes of Health (R01 GM121540).

References

1. Sanja B, Mathieu B, Gisou GF (2013) What does S-palmitoylation do to membrane proteins? FEBS J 280(12):2766–2774
2. Blanc M et al (2015) SwissPalm: Protein Palmitoylation database [version 1; referees: 3 approved]. F1000Res 4:261
3. Lobo S et al (2002) Identification of a Ras Palmitoyltransferase in *Saccharomyces cerevisiae*. J Biol Chem 277(43):41268–41273
4. Roth AF et al (2002) The yeast DHHC cysteine-rich domain protein Akr1p is a palmitoyl transferase. J Cell Biol 159(1):23–28
5. Zhang MM et al (2013) Quantitative control of protein S-palmitoylation regulates meiotic entry in fission yeast. PLoS Biol 11(7): e1001597
6. Mitchell DA et al (2006) Thematic review series: lipid posttranslational modifications. Protein palmitoylation by a family of DHHC protein S-acyltransferases. J Lipid Res 47 (6):1118–1127
7. Gottlieb CD, Linder ME (2017) Structure and function of DHHC protein S–acyltransferases. Biochem Soc Trans 45(4):923–928
8. Politis EG, Roth AF, Davis NG (2005) Transmembrane topology of the protein palmitoyl transferase Akr1. J Biol Chem 280 (11):10156–10163
9. Mitchell DA et al (2010) Mutational analysis of *Saccharomyces cerevisiae* Erf2 reveals a two-step reaction mechanism for protein palmitoylation by DHHC enzymes. J Biol Chem 285 (49):38104–38114
10. Jennings BC, Linder ME (2012) DHHC protein S-acyltransferases use similar ping-pong kinetic mechanisms but display different acyl-CoA specificities. J Biol Chem 287 (10):7236–7245
11. González Montoro A, Chumpen Ramirez S, Valdez Taubas J (2015) The canonical DHHC motif is not absolutely required for the activity of the yeast S-acyltransferases Swf1 and Pfa4. J Biol Chem 290(37):22448–22459

Chapter 15

Bioinformatic Identification of Functionally and Structurally Relevant Residues and Motifs in Protein *S*-Acyltransferases

Rodrigo Quiroga and Javier Valdez Taubas

Abstract

DHHC palmitoyltransferases (DHHC-PATs) are very peculiar in that, outside the DHHC domain, they are very divergent even across orthologs from closely related species. This represents a challenge for the bioinformatic analyses of these proteins. Sequence-based analyses and predictions require a valid sequence alignment, which for this family of proteins requires extensive manual curation and this is difficult to attain for the nonspecialist. Here we present a simple method for the in silico analysis of the sequence of a particular PAT, that would allow for the identification of important structural features and functional residues in a PAT or PAT family.

Key words *S*-acylation, Palmitoyltransferases, PATs, 3D structure, PaCCT motif, DHHC motif, Hidden Markov Model, Homology modeling, Sequence alignment

1 Introduction

S-acylation is carried out by *S*-acyltransferases or palmitoyltransferases (PATs). These are multispanning membrane proteins containing 4–6 transmembrane domains (reviewed in [1, 2]). They share a conserved 50-amino acid domain called the DHHC domain [3]. In a previous work, we predicted this domain to contain two structural zinc fingers in a C3H configuration [4] by alignment and homology-based structure modeling. Additionally, it has been shown experimentally that the DHHC domain binds two zinc atoms [4–6]. Other prevalent motifs in PATs are the TTxE and DPG, named after their conserved residues [7]. An interesting feature of PATs is the presence of the PaCCT (Palmitoyltransferase Conserved C-Terminus) motif, which is moderately conserved in most PATS [8]. This motif is not easy to spot and a dedicated Hidden Markov Model (HMM) or a very careful manual curation

---

Electronic supplementary material: The online version of this chapter (https://doi.org/10.1007/978-1-4939-9532-5_15) contains supplementary material, which is available to authorized users.

Maurine E. Linder (ed.), *Protein Lipidation: Methods and Protocols*, Methods in Molecular Biology, vol. 2009, https://doi.org/10.1007/978-1-4939-9532-5_15,

of multiple alignments, are required for its identification [8]. This motif was shown to be essential for the function and stability of at least three PATS, while the importance of individual residues in the motif was analyzed by mutagenesis, and an aromatic residue in position 3, as well as the conserved Asp residue in position 11, proved essential [6, 8].

Other motifs are present only in some PATS, such as ankyrin repeats, or SH3 domains [2]. The recent crystal structure of human DHHC20 has confirmed the predicted structural zinc fingers of the DHHC domain and provided information on the catalytic pocket, the transmembrane domains, and the structural and functional relevance of the PaCCT and TTxE motifs [6]. Many conserved residues, such as the hydrophobic residues present in the PaCCT motif, interact profusely with transmembrane domains. Downstream of the PaCCT motif, an additional short alpha helix followed by a hydrophobic loop in the cytosolic C-terminus of DHHC 20 was identified in the structure [6]. We now find this motif to be present in many PATs (*see* below), and we will hereby refer to it as MACCT (Membrane Associated Conserved in the C-Terminus) motif.

Lastly, the work by Rana et al. [6] also postulated the structural basis for the preference of a given PAT for fatty acid chains of a certain length, owing mainly to the effect of conserved bulky amino acids present in certain positions of transmembrane domains [6].

The bioinformatic analysis of PATs is challenging due to several reasons: (1) There is little conservation outside the DHHC domain and the TTxE motif. (2) The PaCCT motif is present in most PATs but requires either a dedicated HMM or a carefully curated alignment to be detected, since it is only moderately conserved; the same is true for the MACCT motif. (3) The presence of PAT-specific domains. (4) The presence of repetitive sequences such as the ankyrin repeats. (5) A variable number and composition of transmembrane domains (TMDs) (*see* **Note 1**). (6) Loop sequences are highly variable in both length and composition.

These difficulties preclude the proper alignment of multiple PAT homologs by simply feeding PATsequences to an alignment software. To make any significant inferences about residue and motif conservation and function, a carefully curated alignment is required. Moreover, automated 3D-structure prediction based on pairwise alignment with the available DHHC20 structure fails to produce satisfactory results (our unpublished results).

Here we provide the necessary tools and methods for a researcher that is interested in a particular PAT, but not an expert in bioinformatics or in the operation of command-line software, to predict which residues are relevant to the structure and function of a PAT. To this end, we provide a highly curated PAT family alignment and the methods to incorporate a PAT of interest into that alignment. This allows identification of the DHHC domain and within this domain, which cysteines are predicted to be involved in zinc coordination. The method also allows for the identification of the TTxE motif, the DPG motif, the PaCCT motif (a sequence logo of the motif in the PAT family alignment is shown in Fig. 1a), the number of transmembrane domains, as well as the MACCT motif. The MACCT motif is composed of a short hydrophilic helix and a hydrophobic loop which is found inserted in the lipid bilayer in the C-terminus of the crystal structure of DHHC20 [6]. We find this motif to be present in many PATS of the alignment, and while

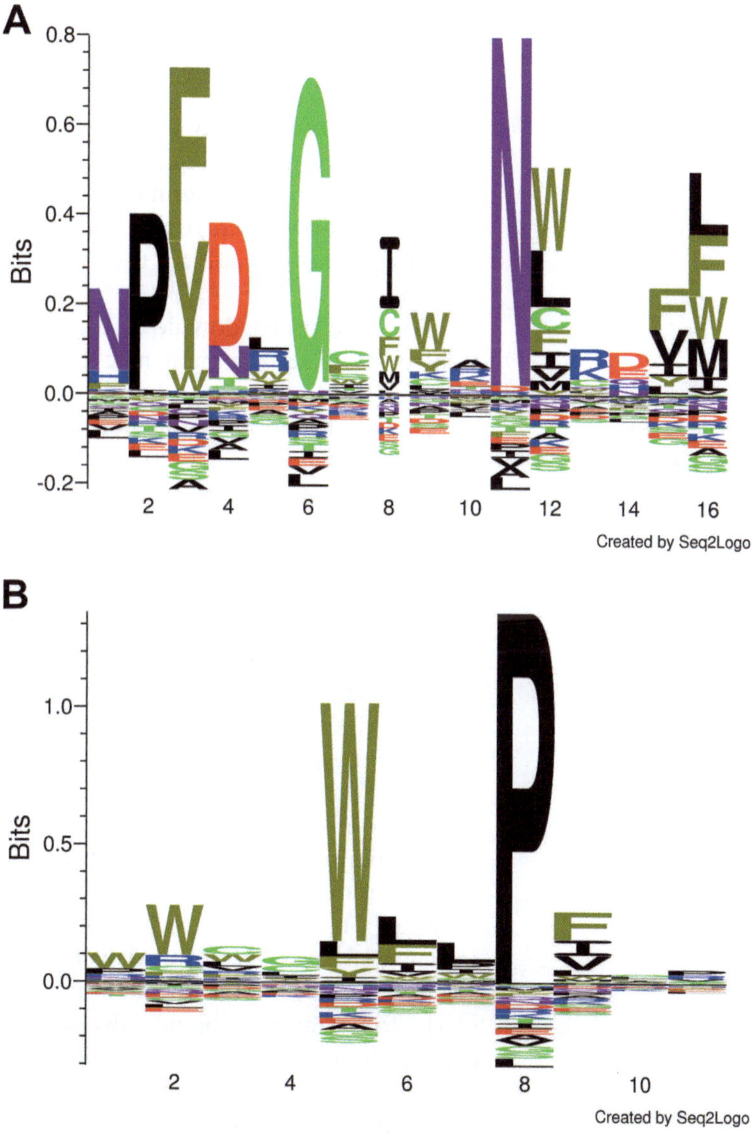

Fig. 1 Sequence logos depicting conservation (positive on the *Y*-axis) and depletion (negative values on the *Y*-axis) of residues at each position of the moderately conserved PaCCT motif (**a**) and MACCT motif (**b**)

the sequence conservation is quite low, there is a high prevalence of hydrophobic residues for positions 5, 6, 7, and 9. A sequence LOGO of this motif is depicted in Fig. 1b. Lastly, this alignment will be useful in the generation of 3D models for any PAT of interest by homology modeling.

2 Materials

2.1 Software

The proposed methodology for aligning any PAT to the PAT family alignment requires the installation of the following software, which is freely available for any operating system:

AliView—Alignment visualization and editing. Available at https://github.com/AliView

(AliView requires Java).

Mafft—Multiple sequence alignment suite. Available at https://mafft.cbrc.jp/alignment/software/

For Windows users, download the all-in-one version of Mafft and unzip the archive. Place the uncompressed "mafft-win" folder in "C:\".

For Mac users, download the all-in-one package, unzip the archive, and place the mafft-mac folder in the Applications folder.

2.2 PAT Family Alignment

We provide a multiple sequence alignment of the PAT family in FASTA format (Data S1). This alignment includes PAT protein sequences from *Homo sapiens, Danio rerio, Caenorhabditis elegans, Saccharomyces cerevisiae, Ashbya gossypii, Schizosaccharomyces pombe, Debaromyces hanseni, Candida albicans, Kluyveromyces lactis*, and *Yarrowia lipolytica*. The multiple alignment will provide the user with a number of "dummy" sequences which consist of consensus sequence motifs and domains with which to compare the PAT of interest. Below these motifs, domains and residues of interest, the user will find the multiple alignment of the PAT family, where we conveniently placed DHHC20 at the top for direct comparison with a given PAT sequence. Manual curation was carried out using a combination of alignment software and other criteria such as transmembrane domain detection, HMM searching and structural considerations. MUSCLE [9], MAFFT [10], and Clustal Omega [11] were used for alignment, while AliView [12] and Jalview [13] were used for visualization and editing. HMMER3 [14] was used for HMM generation and searching, and Phobius [15] for transmembrane domain prediction. All regions corresponding to structural domains in the DHHC20 crystal structure were taken into account for gap placement. Conserved regions in this alignment include, and therefore allow identification of up to six transmembrane domains, DPG motif, DHHC domain (Zn binding and catalytic residues), TTxE motif, PaCCT motif, and the MACCT motif. Sequence logos were created with

Seq2logo using the *P*-weighted Kullback–Leibler method with heuristic clustering [16].

3 Methods

3.1 Software Setup for WINDOWS

1. Configure Aliview to run Mafft (you may need to pause your antivirus software, or add the mafft-win folder to antivirus exceptions if running on Windows). Open Aliview, go to the menu and click "Align > Change default aligner program > when pasting new sequences (profile-alignment)".
2. Select the second option "Mafft –add".
3. On the right side, below the box that reads "cmd.exe", modify "C:\Program Files\mafft-win\mafft.bat" to "C:\mafft-win \mafft.bat".
4. After the path and before "--add" enter the following text "--maxiterate 1000". Your screen should look like Fig. 2a.
5. Click "OK."

3.2 Software Setup for Mac

1. Open Aliview, go to the menu and click "Align > Change default aligner program > when pasting new sequences (profile-alignment)".
2. Select the second option "Mafft –add" and browse for the "Mafft.bat" file in the Mafft folder in Applications.
3. To the command parameters box, add the text " --maxiterate 1000" as shown in Fig. 2b.
4. Click OK.

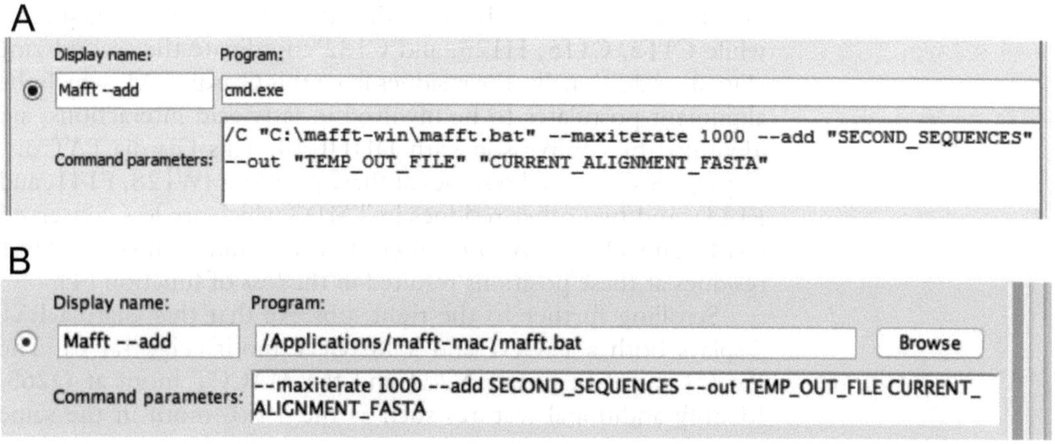

Fig. 2 Screenshots depicting the necessary setup of Mafft aligner within AliView. Instructions are given for both Windows (**a**) and Macintosh (**b**) operating systems

3.3 Sequence Alignment

1. In the AliView menu, click "File > Open File" to open the PAT-family-ali.fasta file which should be downloaded from Data S1.
2. To introduce a PAT sequence of interest to the PAT family alignment, add the new sequence in FASTA format from a file using the "Edit>Add sequences from file" from the menu. Alternatively, copy and paste the sequence into the alignment window. The sequence of interest will appear at the top of the alignment.
3. To align the PAT sequence to the PAT family alignment, left click the name of your sequence, then right-click it, and select "Realign selected sequence(s)".
4. If the sequence of interest is already in the alignment (such as any human or S. cerevisiae PAT), the sequence of interest can be brought to the top of the alignment for ease of analysis by left-clicking the sequence name and then clicking "Selection>Move selected sequence to top".

3.4 Analysis of Conserved Domains and Motifs in PATs

As an example, search the UniProt site (www.uniprot.org) for a *Giardia lamblia* PAT (accession number V6U0K4) and use this protein sequence to carry out the entire alignment procedure and subsequent analysis (*see* **Note 2**).

By scrolling the alignment left to right, compare the sequence of the Giardia PAT with interesting residue positions and conserved domains and motifs. For the Giardia PAT, observe that no residues align with TMDs -2 and -1, while TMDs 1 and 2 are present. The DPG motif is also present. Clicking the D in the Giardia "DPG" sequence, AliView informs us that it corresponds to residue 83 (see the bottom of AliView window, it is called "Pos (ungapped)"). Scrolling to the right, observe that this Giardia PAT has both zinc pockets conserved. Clicking on the residues, identify C98, C101, H111, and C118, which are predicted to coordinate the first zinc, while C112, C115, H125, and C132 coordinate the second zinc (Fig. 3). Additionally, the residues from this Giardia PAT, which the alignment postulates to be involved in fatty acid interactions, are identified by comparison with DHHC20. This Giardia PAT displays three conserved residues at these positions (W128, F141, and F144), and two other residues in TMD3 which are less conserved (A148 and Q152). Also of interest is that mutation of the Swf1 residues at these positions resulted in the loss of function [4].

Scrolling further to the right, observe that this Giardia PAT displays both a PaCCT and a MACCT motif (Fig. 4a, b). The PaCCT motif begins at N224 and the MACCT motif at Q265. Identify additional features such as the TTxE motif in the same manner.

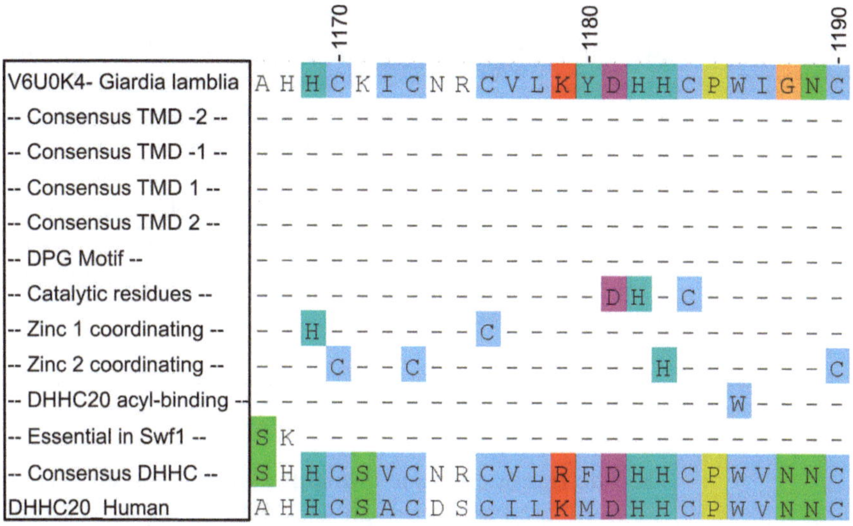

Fig. 3 Multiple sequence alignment depicting the DHHC domain. This view allows comparison of the Giardia PAT V6U0K4, consensus and DHHC20 DHHC domains sequences, as well as identification of zinc coordinating residues, catalytic residues, and putative acyl-binding residues

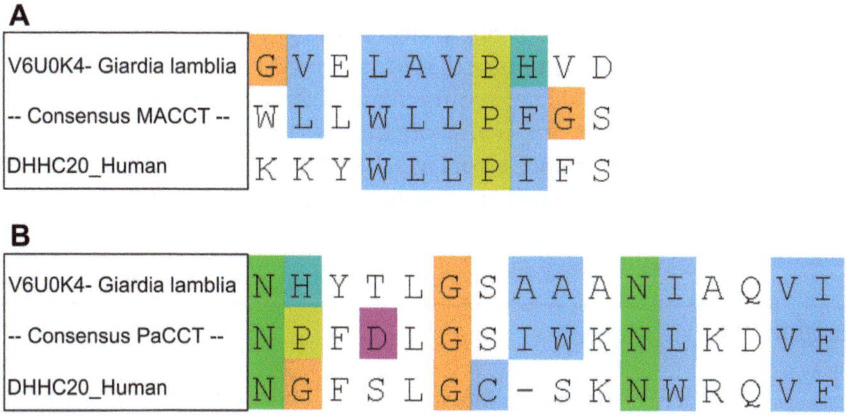

Fig. 4 Multiple sequence alignment depicting the PaCCT (**a**) and MACCT motifs (**b**) of Giardia V6U0K4, consensus, and DHHC20 sequences

3.5 Extracting Pairwise Alignments with DHHC20

In many cases when using conventional alignment programs, direct pairwise alignment of a PAT of interest to DHHC20 or other PATs fails to produce satisfactory results. In the example of this Giardia PAT, a direct pairwise alignment fails to align both the PaCCT and the MACCT motif. To produce pairwise a comparison with DHHC20 guided by the family PAT alignment herein, select both sequences in AliView within the PAT family alignment by clicking one sequence name, and then holding down the CTRL key and clicking the second sequence name. The pairwise alignment is produced by clicking "File > Save selection as Fasta" in the menu. This pairwise alignment extracted from the PAT family alignment is apt for using it as a seed to generate a homology-based structural model. Homology-based modeling programs such as MODELLER [17], or web-based servers that accept input in the form of an alignment, such as ProtMod [18], may be used for this purpose. Depending on the software and/or server, format conversions might be required. The quality of this 3D model will exceed that of an automated procedure that uses direct pairwise alignment between DHHC20 and the PAT to be modeled. However, the generation of refined 3D models for a PAT is beyond the scope of this chapter.

4 Notes

1. The transmembrane domains conserved in all PATs have been numbered as 1 through 4 by Mitchell et al. [7] and the same notation has been used by other authors [4, 6]. However, many PATS have one or two additional TMDs in the N-terminal region. We named these TMD-2 and -1.
2. The procedure indicated above will work well with the majority of PATs. To highlight this fact, we have tested PATs from a very divergent organism such as *Giardia lamblia*. However, always keep in mind that for some PATs the alignment may be suboptimal, and require manual curation. Finally, it should be noted that for many organisms, there is always the possibility of sequence errors which might arise from sequencing, gene models, alternative splicing, and/or in the annotation of the protein.

Acknowledgments

We thank Dr. Agustín Carbajal and Consuelo Coronel for testing the method described in this work and critical discussion of the manuscript.
Funding: This work was funded by grants from the Argentinian Ministry of Science and technology (MINCyT) (PICT 2013 0288 and PICT 2015 1316) and by Córdoba National University. RQ and JVT are career researchers at the National Research Council of Argentina (CONICET).

References

1. Chamberlain LH, Shipston MJ (2015) The physiology of protein S-acylation. Physiol Rev 95(2):341–376
2. Gottlieb CD, Linder ME (2017) Structure and function of DHHC protein S-acyltransferases. Biochem Soc Trans 45(4):923–928
3. Putilina T, Wong P, Gentleman S (1999) The DHHC domain: a new highly conserved cysteine-rich motif. Mol Cell Biochem 195 (1-2):219–226
4. Gonzalez Montoro A, Quiroga R, Valdez Taubas J (2013) Zinc co-ordination by the DHHC

cysteine-rich domain of the palmitoyltransferase Swf1. Biochem J 454(3):427–435
5. Gottlieb CD, Zhang S, Linder ME (2015) The cysteine-rich domain of the DHHC3 palmitoyltransferase is palmitoylated and contains tightly bound zinc. J Biol Chem 290 (49):29259–29269
6. Rana MS, Kumar P, Lee CJ et al (2018) Fatty acyl recognition and transfer by an integral membrane S-acyltransferase. Science 359 (6372)
7. Mitchell DA, Vasudevan A, Linder ME, Deschenes RJ (2006) Protein palmitoylation by a family of DHHC protein S-acyltransferases. J Lipid Res 47(6):1118–1127
8. Gonzalez Montoro A, Quiroga R, Maccioni HJ, Valdez Taubas J (2009) A novel motif at the C-terminus of palmitoyltransferases is essential for Swf1 and Pfa3 function in vivo. Biochem J 419(2):301–308
9. Edgar RC (2004) MUSCLE: multiple sequence alignment with high accuracy and high throughput. Nucleic Acids Res 32 (5):1792–1797
10. Katoh K, Standley DM (2014) MAFFT: iterative refinement and additional methods. Methods Mol Biol 1079:131–146
11. Sievers F, Higgins DG (2018) Clustal Omega for making accurate alignments of many protein sequences. Protein Sci 27(1):135–145
12. Larsson A (2014) AliView: a fast and lightweight alignment viewer and editor for large datasets. Bioinformatics 30(22):3276–3278
13. Waterhouse AM, Procter JB, Martin DM, Clamp M, Barton GJ (2009) Jalview Version 2—a multiple sequence alignment editor and analysis workbench. Bioinformatics 25 (9):1189–1191
14. Eddy SR (2011) Accelerated Profile HMM Searches. PLoS Comput Biol 7(10):e1002195
15. Kall L, Krogh A, Sonnhammer EL (2007) Advantages of combined transmembrane topology and signal peptide prediction—the Phobius web server. Nucleic Acids Res 35 (Web Server issue):W429–W432
16. Thomsen MC, Nielsen M (2012) Seq2Logo: a method for construction and visualization of amino acid binding motifs and sequence profiles including sequence weighting, pseudo counts and two-sided representation of amino acid enrichment and depletion. Nucleic Acids Res 40(Web Server issue):W281–W287
17. Webb B, Sali A (2014) Comparative protein structure modeling using MODELLER. Curr Protoc Bioinformatics 47:5.6.1–5.6.37
18. Jaroszewski L, Li Z, Cai XH, Weber C, Godzik A (2011) FFAS server: novel features and applications. Nucleic Acids Res 39(Web Server issue):W38–W44

Part IV

Online Resource

Chapter 16

SwissPalm 2: Protein *S*-Palmitoylation Database

Mathieu Blanc, Fabrice P. A. David, and F. Gisou van der Goot

Abstract

Protein *S*-palmitoylation is increasingly recognized as an important posttranslational modification, present in all eukaryotic organisms, involved in the regulation of many biological processes. The SwissPalm database centralizes the large and increasing number of published palmitoyl-proteome datasets, provides tools to compare them, and includes curated data from the literature on the identification and analysis of palmitoylated proteins. SwissPalm 2 provides an updated version, with 38 palmitoyl-proteomes at the time of release, from 17 different species, and new features such as the inclusion of orthologs.

Key words Palmitoylation, Proteomics, Database, Confidence, DHHC, *S*-acylation, Palmitoyl-proteomes, Posttranslational modification

1 Introduction

1.1 Definition of *S*-Palmitoylation

S-acylation, generally referred to as *S*-palmitoylation, is defined as the attachment of an acyl chain moiety (C14 to C18) to a cysteine by a thioester bond [1]. First described for viral proteins in 1979 [2], *S*-palmitoylation has since been shown to occur in all eukaryotic organisms [3]. Importantly, palmitoylation is, to date, the only known lipid modification that is enzymatically reverted. It is mediated by DHHC-domain containing palmitoyltransferases (PATs), which catalyze the attachment of the lipid moiety to the cysteine via a thioester bond [4]. These enzymes are polytopic transmembrane proteins that harbor their catalytic site on the cytoplasmic face of the membrane they reside in [5, 6]. No other *S*-acylating enzymes have so far been identified, and thus *S*-palmitoylation appears restricted to the cytosol. The reversal of *S*-acylation is mediated by thioesterases of the serine hydrolase family [7, 8]. Jointly palmitoyltransferases and thioesterases drive dynamic *S*-palmitoylation, thereby regulating protein function. Thus, *S*-

Mathieu Blanc and Fabrice P. A. David contributed equally to this work.

Maurine E. Linder (ed.), *Protein Lipidation: Methods and Protocols*, Methods in Molecular Biology, vol. 2009, https://doi.org/10.1007/978-1-4939-9532-5_16,

palmitoylation is a regulatory posttranslational modification (PTM) alike phosphorylation or ubiquitination.

In human, there are 23 DHHC palmitoyltransferases and so far five identified acyl protein thioesterases [6]. Orthologs of palmitoyltransferases and thioesterases have been identified in all eukaryotic species. The enzyme-substrate specificity remains unclear as well as whether palmitoyltransferase orthologs recognize the same substrates, rendering site prediction difficult [6, 9]. Various prediction software have been developed based on sequence similarity [10] but their reliability remains limited.

1.2 Functional and Biological Consequences of S-Palmitoylation

Due to the hydrophobic nature of the acyl moieties, protein *S*-palmitoylation induces drastic changes in proteins biochemical properties and consequently their functions. For cytosolic proteins, *S*-palmitoylation drives protein attachment to membrane compartments, inducing changes in protein localization [11]. As a consequence, *S*-palmitoylation of soluble proteins drives membrane targeting and subsequent specific protein–protein interactions [12]. *S*-palmitoylation of membrane proteins may induce conformational changes, affect protein–protein interactions and regulate the association of proteins with membrane domains [1]. For both protein types, *S*-palmitoylation may regulate protein turnover rates and cross talk with other posttranslational modifications such as phosphorylation and ubiquitination [13, 14]. Over the years, *S*-palmitoylation has been associated with an increasing number of biological processes such as signaling, transport, cell cycle, immune response, lipid metabolism, and host–pathogen interaction [15, 16].

Not surprisingly, *S*-palmitoylation is increasingly associated with an number of human diseases, including neurological disorders (Huntington's disease, schizophrenia, Alzheimer's disease), cancer, and infectious diseases [16–18]. In addition, palmitoyltransferase-deficient mice present marked phenotypic and metabolic alterations [19, 20]. Thus, *S*-palmitoylation has raised attention for its potential in therapeutic interventions.

1.3 Techniques Used to Study S-Palmitoylation

There are two major classes of techniques to detect protein *S*-palmitoylation: metabolic labeling of palmitate and chemical modification of *S*-palmitoylated cysteines [21]. The labeled palmitate is incorporated into the cells as the natural palmitate and is utilized for direct detection (radiography or fluorescent probe) or for further capture (for example with biotin). Palmitate labeling permits the analysis of *S*-palmitoylation dynamics but is dependent on the labeling time and restricted to cell lines.

The second method is based on the chemical modification of the *S*-palmitoylated cysteines. This is achieved by a multistep procedure that involves blocking the free cysteine by alkylation, followed by selectively breakage of the thioester bond with hydroxylamine and the final capture of newly freed cysteines. Several methods based on this principle have been developed such as Acyl biotin exchange (ABE) or Acyl resin assisted capture (ARAC) [22, 23]. These approaches allow the capture of the whole population of palmitoylated proteins and can be applied to cells, tissues and organs.

1.4 Emergence of Palmitoyl-Proteomes

Detection of protein *S*-palmitoylation by radioactive palmitate labeling has been considered as the gold standard for decades. However, the technical and safety constraints associated with this technique render the study of *S*-palmitoylation tedious and time-consuming . It is moreover low-throughput and not always the most suitable method. The emergence of new techniques (ABE, Acyl RAC, metabolic labeling with clickable palmitate) allowed for large-scale capture of palmitoylated protein and their identification by mass spectrometry. The first palmitoyl-proteome was reported by Roth et al. in 2006 identifying 48 palmitoylated proteins in *Saccharomyces cerevisiae* using the ABE method [24]. Since then more than 35 palmitoyl-proteomes from 10 species have been published (for the complete list: https://swisspalm.org/articles). Thousands of proteins were identified revealing the extent of cellular palmitoylation and opening new areas of research.

However, the hydroxylamine-switch capture techniques used for mass spectrometry lead to the isolation of all proteins containing thioester bonds thus leading to potential false positives in terms of palmitoylation. False positives may also originate from incomplete alkylation of free cysteines or unspecific binding to chemical beads. The clearest evidence that these methods lead to false positives is the presence amongst the palmitoyl-proteome hits of proteins that are devoid of any cysteines. When labeling with clickable palmitate, *O*- and *N*-palmitoylated protein may also be identified. Thus, proteins identified in palmitoyl-proteomes require validation using independent approaches.

1.5 Development of SwissPalm as a S-Palmitoylation Community Resource

The information contained in the palmitoyl-proteome studies and the literature constitute valuable resources that give the opportunity to better apprehend the biological importance of *S*-palmitoylation and discover new research avenues.

However, these resources have not been exploited fully due to the lack of confidence in the results and the difficulty to access the information often hidden in complex supplementary tables. To offer the community an efficient tool to explore and analyze palmitoyl-proteome datasets, in 2015 we launched SwissPalm, the first *S*-palmitoylation database, [25]. SwissPalm curates and cross-references evidence of protein *S*-palmitoylation from the literature and palmitoyl-proteome studies. The first release contained approximately 5000 putative *S*-palmitoylated proteins from 19 palmitoyl-proteomes and around 500 specific sites from 7 species.

Table 1
Comparison between SwissPalm and SwissPalm 2

	SwissPalm initial release	SwissPalm 2 release
Number of palmitoyl proteomes	19	38
Number of curated articles	303	632
Number of proteins from curated articles	365	664
Number of unique proteins	5199	9797
Number of curated sites	535	1062

SwissPalm also offers curated and filtered information based on validated studies, increasing confidence and providing information on palmitoylation sites. SwissPalm reveals that more that 10% of the proteome of a given organism, in particular human, may undergo *S*-palmitoylation. Since its launch, SwissPalm has attracted hundreds of visits per month, revealing its use for the community, and motivating the current update.

1.6 SwissPalm 2

For this update, SwissPalm 2 (https://swisspalm.org), we have doubled the number of large-scale palmitoyl-proteome studies, now 38, of curated articles, and of validated sites (*see* Table 1). SwissPalm 2 also utilizes the evolutionary relationship between proteins and the comparison with datasets of orthologs, to increase the confidence of palmitoyl-proteome hits. In addition, a new user-friendly interface has been created offering new features and tools to help the analysis of protein palmitoylation datasets.

2 Materials

2.1 SwissPalm Architecture

SwissPalm is a Ruby on Rails 5 application that both supports and uses a PostgreSQL relational database for the storage and retrieval of data records. The application implements a double web interface, one for curators to add new annotations and one for the public usage. Periodic updates of the database are performed using a series of Rake tasks, which are scripts using the Rails application's environment.

2.2 Data Structure for S-Palmitoylation Annotation

Four types of objects have been created to characterize a palmitoylation event (protein or site) (Fig. 1).

1. An experiment is defined by a specific organism, subcellular fraction, cell type, and a list of techniques described in a given article.

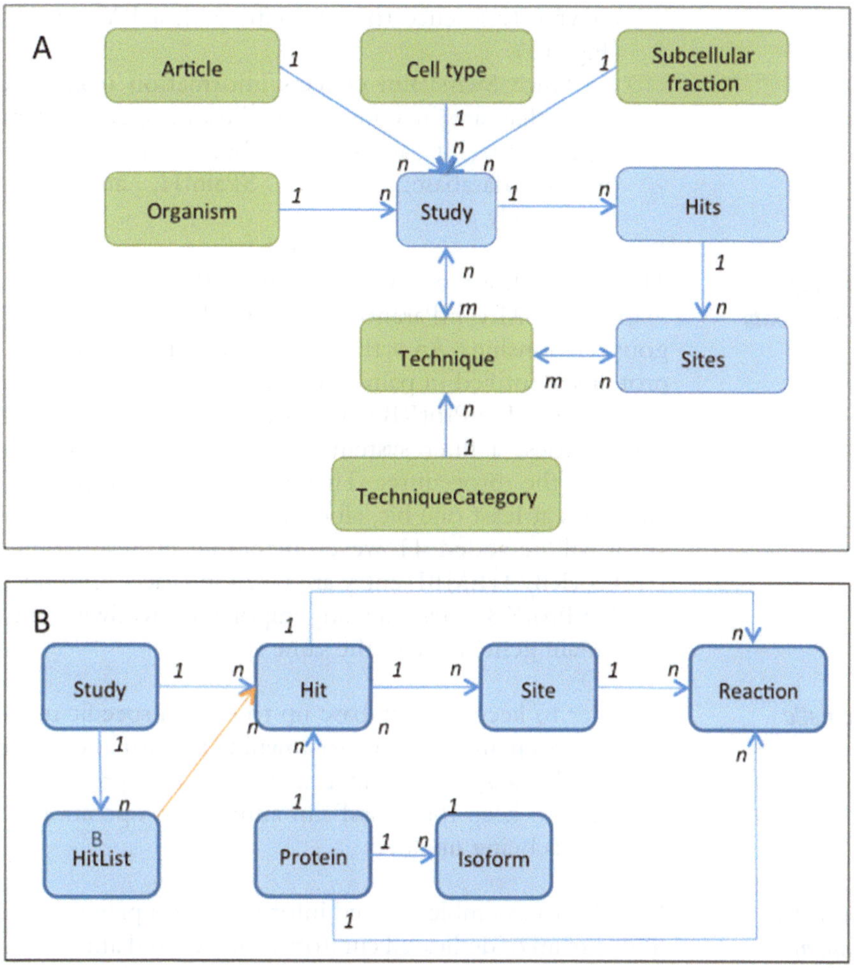

Fig. 1 Diagrams representing elements of SwissPalm data structure. Each box corresponds to an object in the application and a table in the database. Arrows show the relationships between the objects and tables. (**a**) Definition of a study and associated hits in the context of manual curation in SwissPalm. (**b**): Core data structure needed for the annotation of palmitoylation events. Data source is both literature and MS large-scale study results. At the level of the database, the orange links correspond to optional references of HitList and Isoform in the Hit table

2. A hit is indicated if a protein was identified to be palmitoylated. Its attributes are a given experiment and a reference to the palmitoylated protein (this reference can be isoform specific—see orange arrows in Fig. 1). One or several hits can be attached to an experiment.
3. A site that is associated to a given hit. A site identifies the position of a palmitoylated cysteine in a specific protein sequence that can be isoform specific.
4. A reaction corresponds to the palmitoylation event itself as it is built with a reference to a hit or a site, plus the definition of an

enzyme (reference to a protein) with a PAT or APT activity (Fig. 1b).

Since SwissPalm collects information from several databases that are in constant evolution, specific identifiers for experiments, hits, and sites have been created to ensure stability of the database: SPalmE, SPalmH, and SPalmS IDs, respectively.

2.3 Orthology and Computed Data

The new release of SwissPalm uses three different sources of mapping (OMA, inParanoid, and MetaPhors) to identify orthologous relationships. An orthologous relationship is defined solely for proteins identified in palmitoyl-proteomes or in validation experiments. Since UniProtKB/TrEMBL database contains many redundant entries, a filter system was implemented to optimize and simplify the procedure. Therefore, if a UniProtKB/TrEMBL entry has at least one hit, all of its orthologs in UniProtKB/SwissProt will be added. However, orthologs corresponding to a new UniProtKB/TrEMBL entry are taken into account only if none of the UniProtKB entries already imported into SwissPalm have the same main gene name in the same species.

2.4 Automatic Update

In order to keep the database up to date, a specific update procedure has been implemented to download information from external sources. Details are explained in Fig. 2. This procedure has been designed to keep the SwissPalm annotation operational while the database is being updated.

2.5 SwissPalm Source Content

2.5.1 Manually Curated Dataset from the Literature

In order to assemble curated information on protein *S*-palmitoylation, a team of dedicated curators collected and annotated the peer-reviewed literature. A protein is *S*-palmitoylated if at least one of the following techniques was used in the study to show direct evidence of *S*-palmitoylation: metabolic labeling (click chemistry, radioactive labeling), direct identification by mass spectrometry or chemical modification of cysteine (ABE, ACYL RAC) and if loss of *S*-palmitoylation signal upon hydroxylamine treatment is demonstrated. For a cysteine to be identified as an *S*-palmitoylation site, cysteine mutagenesis experiments leading to loss or decrease of *S*-palmitoylation signal are required. To date, 671 articles containing direct evidence for protein *S*-palmitoylation have been curated resulting in the annotation of 663 proteins and 787 unique sites (https://swisspalm.org/hits).

Additional information such as species, cell type, or tissues was also collected when available, and users can download protein datasets corresponding to these specific criteria.

2.5.2 Palmitoyl-Proteomic Datasets

Datasets from 38 published palmitoyl-proteomic articles were collected (https://swisspalm.org/palmitoylproteomes). Information provided by specific articles such as: putative *S*-palmitoylation

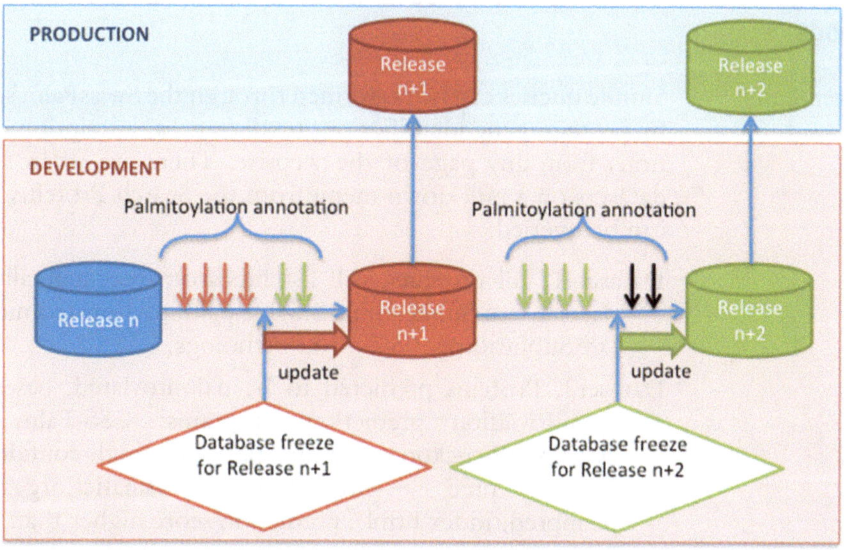

Fig. 2 Description of the update procedure. Curators edit the development version of SwissPalm. Small colored arrows represent annotations that are produced. The color of the arrow (red or green) indicates in which release the annotation will be incorporated (release $n + 1$ (red) or $n + 2$ (green), respectively). For release $n + 1$, the update procedure starts with the freeze of the database, which consists of adding the pending annotations (Studies, Proteins, Hits, Sites, and Reactions) in the current database. After this initial step, the update procedure which is composed of many steps is executed as a pipeline, until the release is ready. The new release database is then deployed on a production server for serving the public version of SwissPalm

sites identified by mass spectrometry, putative PAT candidates, or changes in palmitoylation levels in response to a specific stimulus was also included into SwissPalm 2. Consequently, 9940 proteins, from 17 species, presenting evidence of *S*-palmitoylation, and 3669 putative sites have been incorporated.

2.5.3 Downloadable Datasets

For each release, SwissPalm provides a list of downloadable files that contain all the *S*-palmitoylation related information (experiments, proteins, hits, sites and enzymatic reactions) present in the database (https://swisspalm.org/releases, "Downloads" tab). These files are available in tab-separated and JSON format for a manual and programmatic use, respectively. Complete palmitoyl-proteome comparison results are also available as a zip archive containing tab-separated files for each reference species.

In addition to these files, at the end of each release, a series of advanced statistics and related datasets are automatically generated for a series of protein features (e.g., number of transmembrane domain) or experiment features (e.g., cell type). These datasets are directly available for download from the "Statistics & Datasets" tab in the "Release & Downloads" section.

3 Methods

3.1 Protein Search

Simple queries can be performed through the SwissPalm search bar, by protein, gene identifier or "text" (e.g., a description or a function) from any page of the website. There are seven searchable datasets on a pull-down menu from the Search Proteins page that can be queried.

Dataset 1, All proteins, will list hits from the manually curated dataset (*see* Subheading 2.5.1), palmitoyl-proteomic datasets (*see* Subheading 2.5.2), and orthologs.

Dataset2, Proteins predicted to be palmitoylated, based on the palmitoylation prediction programs: CSS-Palm (http://csspalm.biocuckoo.org/index.php; only high confidence) and Palm pred (http://proteininformatics.org/mkumar/palmpred/index.html); prediction score higher than 0.4).

Dataset 3 includes proteins that have been validated as palmitoyl-proteins or are found in at least one palmitoyl-proteome.

Dataset 4 is the manually curated dataset.

Dataset 5 includes proteins from the manually curated dataset or proteins that have been identified in palmitoylproteome studies using two independent methods.

Dataset 6 includes only proteins found in palmitoylproteome studies using two independent methods. Dataset 7 is dataset 6 grouped by gene.

For complex queries, a dedicated protein search has been created (https://swisspalm.org/proteins). A list of proteins or gene identifiers can be uploaded, specific species can be selected or the query can be restricted to a predefined dataset. Using the Advanced tab of the search bar, a set of helpers (**Reviewed** in UniProt, sequence **Motif**, **GO** term, **Subcellular** compartment) is available to increase the query specificity.

The query result appears as a list of proteins, with a direct link to the SwissPalm page for each protein. The table in addition displays *S*-palmitoylation-relevant information for each protein (*see* Fig. 3).

A cart now allows the selection of proteins from the query result. The cart can be used as an input for the protein search and the comparison of palmitoyl-proteomes (*see* Subheading 3.4, comparison of palmitoyl proteomes) or to view and download the list of selected proteins.

3.2 Statistics

The new functionality "Get statistic" provides graphical views from various analyses of a protein input list. As an example, a list of 2287 proteins was queried in Swiss Palm to identify those proteins with evidence of palmitoylation (using dataset 3). A list of 500 proteins

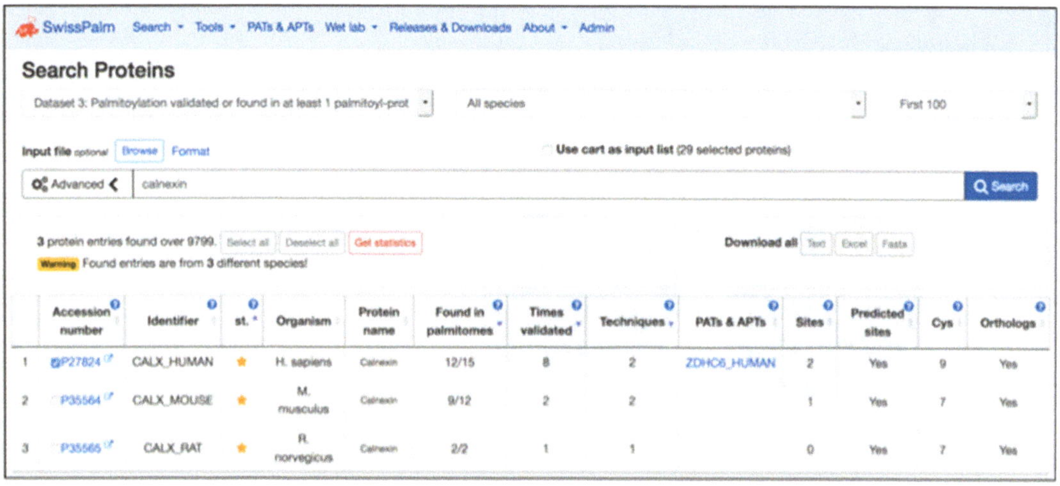

Fig. 3 Screenshot of the search protein page: The database was queried for calnexin protein in all species. The query was restricted to the predefined dataset (Dataset 3): palmitoylation validated or found in at least one palmitoyl-proteome. The resulting table contains all the calnexin orthologs that were identified. Additional information related to *S*-palmitoylation is displayed in the different columns

is generated. Selecting the Get Statistics tab brings up the "General statistic section," where a pie chart displays the percentage of proteins from the query that are identified in SwissPalm with evidence of palmitoylation (SwissPalm annotated proteins). It also displays the overall statics for the chosen species, providing a view of the enrichment/de-enrichment in SwissPalm annotated proteins in the input list (*see* Fig. 4). Other information, such as the % of transmembrane vs. cytosolic proteins, the number of SwissPalm annotated proteins by occurrence in palmitoyl-proteomes, the percentage of protein found in palmitoyl-proteomes per techniques used or information on membrane topology are also available. All protein lists, from the various statistical analyses can be added independently to the cart for further specific analysis or simply to display and download a specific subset of proteins.

3.3 Protein Information Page

The protein information page displays more detailed information related to the *S*-palmitoylation of a specific protein. Included in the protein information page are summary boxes, information on the protein, its topology, the existence of isoforms, alignment of orthologous sequences, palmitoylation site predictions, and a list of articles related to the *S*-palmitoylation of the protein. The information page for human calnexin can be found at https://swisspalm.org/proteins/P27824. SwissPalm 2 offers a new visual representation of the topology and the curated or potential sites. Information on the involvement of identified PATs and APTs is also included. With a click, the list of palmitoyl proteomic studies where the protein appeared can be accessed.

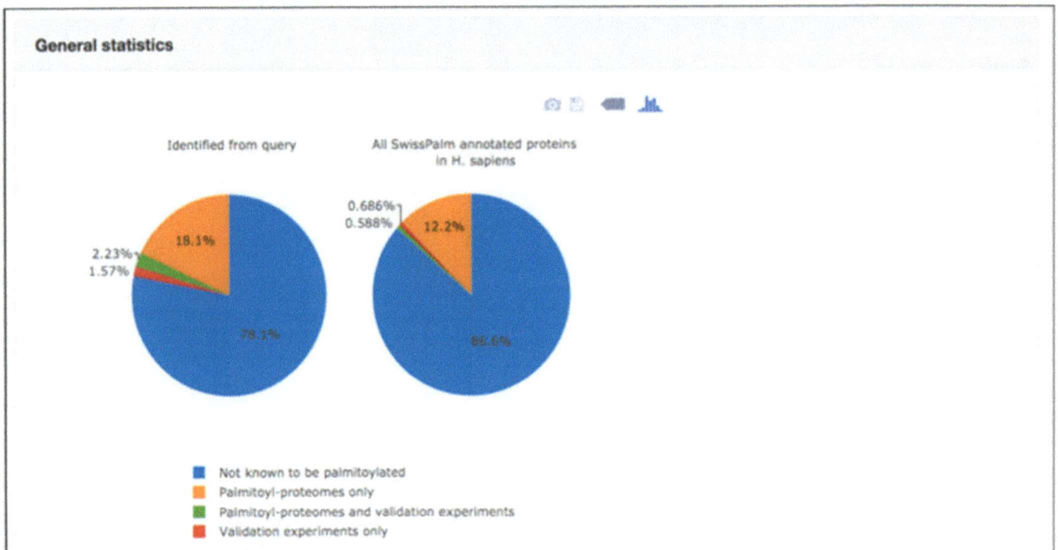

Categories	Protein #	%	To cart
Total number of proteins identified	2287	100%	
Total number of SwissPalm annotated proteins	500	21.86%	+
Number of proteins in palmitoyl-proteomes	464	20.29%	+
Number of proteins in validation experiments	87	3.8%	+
Number of SwissPalm annotated proteins with known associated PAT enzyme	29	1.27%	+
Number of SwissPalm annotated proteins with known associated APT enzyme	4	0.17%	+

Fig. 4 Screenshot of the general statistics section: SwissPalm database was queried for a list of 2287 human proteins. The left pie chart "Identified from query" shows the percentage of proteins from the input list that are identified in palmitoyl-proteomes only, from validation experiments only, from both palmitoyl-proteome and validation experiment or not known. The right pie chart "All SwissPalm annotated proteins in *H. sapiens*" shows the same percentage applied to the whole human proteome. All subsets from the proteins corresponding to the input file can be specifically added to the cart (blue + button) for further analysis

3.4 Palmitoyl-Proteome Comparisons

A novel tool allows the comparison of a list of proteins against any of the palmitoyl-proteomes published. The orthologous relationship is used to identify proteins present in palmitoyl-proteomes from different species. In this release, a list of protein or gene identifiers can be uploaded in the palmitoyl proteomes comparison tool, and several parameters can be defined (experiments to be compared or minimum number of studies in which the proteins was found).

As a result, the proteins from the palmitoyl-proteomes matching the query list will be displayed (green for the same species, light orange if it is an ortholog). The comparison is made at the gene level, and a UniProtKB/TrEMBL entry was merged with the corresponding UniProtKB/SwissProt entry. Hits from validation experiments are also reported when the minimum number of palmitoyl-proteome studies is set to 0.

3.5 PATs and APTs

An increasing number of articles describe substrate–enzyme relationships. A dedicated page that displays all the known substrates of each PAT and APT/thioesterase from the curated literature has been created: https://swisspalm.org/enzymes.

4 Conclusion

SwissPalm 2 is an open-access database that provides comprehensive information and tools for the study of protein *S*-palmitoylation. The database contains up-to-date information from manually curated articles and large-scale studies. This release contains more than 9000 proteins from 38 palmitoyl-proteome screens, covering 17 species, 664 validated proteins, and more than 1000 sites.

In addition to centralizing published information related to *S*-palmitoylation, SwissPalm uses this information to help users estimate the likelihood of a protein of interest to be a bona fide target of *S*-palmitoylation. Confidence can be estimated based on the SwissPalm provided information, that is, the number of occurrences of a hit across all palmitoyl-proteomes of the same species, across palmitoyl-proteomes of other species, the number of independent types of techniques used to identify the hit, whether *S*-palmitoylation of this protein was validated in any given system. A tool is provided that allows for the direct comparison of a list of proteins to a chosen list of published palmitoyl-proteomes, across species. SwissPalm 2, with its updated information and new analysis tools, should accelerate the study and understanding of *S*-palmitoylation.

Acknowledgments

This work benefited from funding from the European Research Council under the European Union's Seventh Framework Programme (FP/2007-2013)/ERC Grant Agreement no. 340260—PalmERa. This work was also supported by grants from the Swiss National Science Foundation (to G.v.d.G), the Swiss National Centre of Competence in Research (NCCR) Chemical Biology, and the Swiss SystemsX.ch initiative evaluated by the Swiss National Science Foundation (LipidX) (to G.v.d.G). M.B. was a recipient from an EMBO Long Term Fellowship.

References

1. Blaskovic S, Blanc M, van der Goot FG (2013) What does S-palmitoylation do to membrane proteins? FEBS J 280:2766–2774
2. Schmidt MF, Schlesinger MJ (1979) Fatty acid binding to vesicular stomatitis virus glycoprotein: a new type of post-translational modification of the viral glycoprotein. Cell 17:813–819
3. Chamberlain LH, Shipston MJ (2015) The physiology of protein S-acylation. Physiol Rev 95:341–376
4. Ohno Y, Kihara A, Sano T et al (2006) Intracellular localization and tissue-specific distribution of human and yeast DHHC cysteine-rich domain-containing proteins. Biochim Biophys Acta 1761:474–483
5. Gottlieb CD, Linder ME (2017) Structure and function of DHHC protein S-acyltransferases. Biochem Soc Trans 45:923–928
6. Tabaczar S, Czogalla A, Podkalicka J et al (2017) Protein palmitoylation: palmitoyltransferases and their specificity. Exp Biol Med 242:1150–1157
7. Yokoi N, Fukata Y, Sekiya A et al (2016) Identification of PSD-95 depalmitoylating enzymes. J Neurosci 36:6431–6444
8. Davda D, Martin BR (2014) Acyl protein thioesterase inhibitors as probes of dynamic S-palmitoylation. Medchemcomm 5: 268–276
9. Lemonidis K, Salaun C, Kouskou M et al (2017) Substrate selectivity in the zDHHC family of S-acyltransferases. Biochem Soc Trans 45:751–758
10. Ren J, Wen L, Gao X et al (2008) CSS-Palm 2.0: an updated software for palmitoylation sites prediction. Protein Eng Des Sel 21:639–644
11. Salaun C, Greaves J, Chamberlain LH (2010) The intracellular dynamic of protein palmitoylation. J Cell Biol 191:1229–1238
12. Levental I, Lingwood D, Grzybek M et al (2010) Palmitoylation regulates raft affinity for the majority of integral raft proteins. Proc Natl Acad Sci U S A 107:22050–22054
13. Gauthier-Kemper A, Igaev M, Sündermann F et al (2014) Interplay between phosphorylation and palmitoylation mediates plasma membrane targeting and sorting of GAP43. Mol Biol Cell 25(21):3284–3299
14. Perrody E, Abrami L, Feldman M et al (2016) Ubiquitin-dependent folding of the Wnt signaling coreceptor LRP6. elife 5
15. Yount JS, Zhang MM, Hang HC (2013) Emerging roles for protein S-palmitoylation in immunity from chemical proteomics. Curr Opin Chem Biol 17:27–33
16. Blanc M, Blaskovic S, van der Goot FG (2013) Palmitoylation, pathogens and their host. Biochem Soc Trans 41:84–88
17. Hornemann T (2015) Palmitoylation and depalmitoylation defects. J Inherit Metab Dis 38:179–186
18. Resh MD (2017) Palmitoylation of proteins in cancer. Biochem Soc Trans 45:409–416
19. Napoli E, Song G, Liu S et al (2017) Zdhhc13-dependent Drp1 S-palmitoylation impacts brain bioenergetics, anxiety, coordination and motor skills. Sci Rep 7:12796
20. Du K, Murakami S, Sun Y et al (2017) DHHC7 palmitoylates glucose transporter 4 (Glut4) and regulates Glut4 membrane translocation. J Biol Chem 292:2979–2991
21. Martin BR (2013) Chemical approaches for profiling dynamic palmitoylation. Biochem Soc Trans 41:43–49
22. Forrester MT, Hess DT, Thompson JW et al (2011) Site-specific analysis of protein S-acylation by resin-assisted capture. J Lipid Res 52:393–398
23. Drisdel RC, Green WN (2004) Labeling and quantifying sites of protein palmitoylation. BioTechniques 36:276–285
24. Roth AF, Wan J, Bailey AO et al (2006) Global analysis of protein palmitoylation in yeast. Cell 125:1003–1013
25. Blanc M, David F, Abrami L et al (2015) SwissPalm: protein palmitoylation database. F1000Res 4:261

Part V

Fatty Acylation in the Lumen of the Secretory Pathway

Chapter 17

Probing Interaction of Lipid-Modified Wnt Protein and Its Receptors by ELISA

Aaron H. Nile and Rami N. Hannoush

Abstract

Wnts are lipid-modified proteins that regulate stem cell signaling via Frizzled receptors on the cell surface. Determination of binding interactions between lipid-modified Wnt proteins and their Frizzled receptors has been challenging due to the lack of availability of facile detection methods and technical hurdles associated with generating the relevant reagents. Here we report an enzyme-linked immunosorbent assay to measure the binding of a biotinylated form of lipid-modified Wnt3a to the extracellular cysteine-rich domain of Frizzled receptor. The method described herein is robust and rapid, uses minimum volumes of reagents, and can be conducted in a high-throughput format. Because of these attributes, the method could find utility in drug discovery applications such as characterizing the effect of pharmacological inhibitors on Wnt signaling without the need for sophisticated biophysical instrumentation.

Key words Frizzled, Wnt, Interaction, Fatty acylation, Lipid, ELISA

1 Introduction

Wnt proteins are essential signaling molecules that regulate developmental processes and adult tissue homeostasis [1, 2]. There are 19 secreted, lipid-modified Wnt glycoproteins in the human genome. Wnt proteins signal through at least three independent pathways including planar cell polarity (PCP) pathway, Ca^{2+}, and β-catenin [3]. Signaling is initiated at the plasma membrane upon binding of lipid-modified Wnt ligand to the extracellular cysteine-rich domain (CRD) domain of Frizzled (FZD) receptor [2, 4, 5] and recruitment of other co-receptors. There are ten Frizzled receptors encoded in the human genome and they are divided into four subfamilies (FZD1/2/7, FZD3/6, FZD4/9/10, and FZD5/8) based on their amino acid sequence conservation [6].

Biochemical methods for characterizing the interactions between lipid-modified Wnts and their Frizzled receptors have been limited. Technical challenges with purifying recombinant Wnt proteins in good yields [5, 7, 8], along with the hydrophobic

Maurine E. Linder (ed.), *Protein Lipidation: Methods and Protocols*, Methods in Molecular Biology, vol. 2009, https://doi.org/10.1007/978-1-4939-9532-5_17,

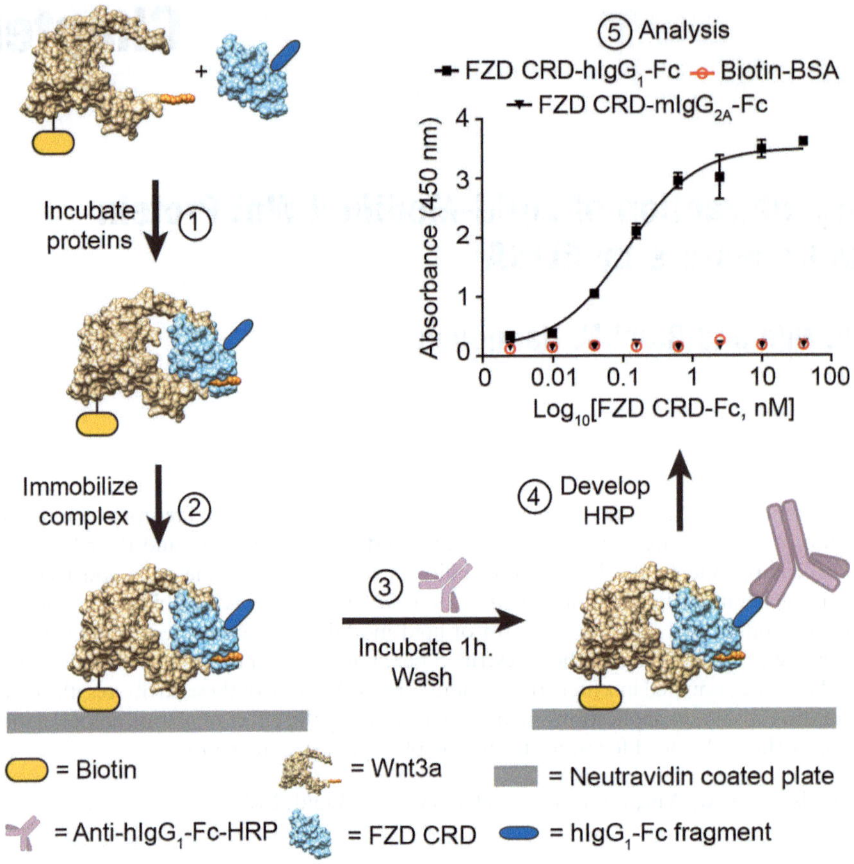

Fig. 1 Diagram of method for probing interactions between lipid-modified Wnt3a and Frizzled CRD by ELISA. Outline of experimental design: (**1**) Biotin-Wnt3a and FZD CRD-hIgG$_1$-Fc are mixed together; (**2**) the formed complex between Wnt3a and FZD CRD complex is then captured on a NeutrAvidin-coated 384-well plate, followed by a washing step to get rid of excess unbound reagents and then (**3**) incubation with anti-hIgG$_1$-HRP antibodies and another wash step. (**4**) Immobilized FZD CRD-hIgG$_1$-Fc is detected with TMB reagent and the signal is developed by measuring absorbance at 450 nm. (**5**) The extent of Wnt binding signal is plotted as a function of FZD CRD-hIgG$_1$-Fc concentration

nature of their lipid modification, have hindered the generation of modified versions of Wnt proteins that could be utilized in the development of such methods. These challenges have also limited our understanding of the molecular interaction networks between various Wnts and FZDs [9, 10]. Here we report an enzyme-linked immunosorbent assay (ELISA) to monitor Wnt–FZD CRD interactions in a robust, rapid, and high-throughput format (*see* Fig. 1). The method utilizes a biotinylated form of Wnt3a protein and is suitable for characterizing Wnt binding to different Frizzled partners. It could also find applications in drug screening and validation of pharmacological Wnt inhibitors.

2 Materials

Make all solutions using analytical grade reagents and ultrapure water (which has a resistivity of 18 MΩ.cm at 25 °C). Prepare and store all reagents at 4 °C (unless indicated otherwise). Follow the material safety data sheet (MSDS) to use and properly dispose of reagents. Personal protective equipment (PPE) should be worn to minimize exposure to potentially hazardous materials.

1. LoBind Microcentrifuge Tubes (Eppendorf).
2. Phosphate-buffered saline (PBS): 137 mM NaCl, 2.7 mM KCl, 10 mM Na_2HPO_4, 1.8 mM KH_2PO_4. Prepare PBS by dissolving 8 g NaCl, 0.2 g KCl, 1.13 g Na_2HPO_4, and 0.2 g KH_2PO_4 in 800 mL of H_2O. Adjust the pH to 7.2 ± 0.1 with 6 N HCl. Add H_2O to a final volume of 1 L. Filter the solution through a 0.22 μm filter and store at 4 °C.
3. PBS-T: PBS with 0.05% (vol/vol) Tween 20.
4. Dilution buffer: PBS + 0.5% bovine serum albumin (BSA), fraction V + 15 ppm ProClin 300 (Sigma). Dissolve 5 g BSA, 500 μL ProClin 300 in 1 L PBS and filter through a 0.22 μm filter.
5. Goat anti-human IgG_1-Fc antibody HRP (ThermoFisher Scientific). Prepare 1:10,000 dilution stock of anti-human IgG_1-Fc antibody in dilution buffer (final concentration, 1 μg/mL).
6. Blocking Buffer: PBS, 2% BSA, and 15 ppm ProClin 300.
7. Immobilization buffer: NeutrAvidin solution at 2 μg/mL in PBS. Mix 50 μL NeutrAvidin stock (2 mg/mL in 50% glycerol; stored at −20 °C) in 50 mL PBS. Store solution at 4 °C.
8. Developing reagent: TMB 2-Component Microwell Peroxidase Substrate Kit (SeraCare).
9. 1 M phosphoric acid. Mix 67.5 mL of 83% H_3PO_4 with 932.5 mL of H_2O.
10. Biotinylated horseradish peroxidase (Invitrogen).
11. Frizzled CRD-Fc proteins (R&D Systems). Dilute protein reagents in PBS according to manufacturer's recommendations (*see* **Note 1**). Human (h) FZD7 CRD-hIgG_1-Fc (cat. no. 6178-FZ; MW = 86.6 kDa), hFZD8 CRD-hIgG_1-Fc (cat. no. 6129-FZ; MW = 86.6 kDa), mouse (m) FZD9 CRD-mIgG_{2A}-Fc (cat. no. 2440-FZ; MW = 90 kDa). For short-term storage, keep proteins at 4 °C.
12. 384-Well Maxisorp plate (Thermo Scientific).
13. 96-Well plate.
14. Biotin-mWnt3a solution (R&D Systems custom synthesis; MW = 37 kDa) (*see* **Note 2**).

15. Biotinylated BSA at 1 mM in PBS.
16. Topseal A PLUS clear adhesive seal for microplates (PerkinElmer).
17. Plate washer with 384-well plate adapter (MDS Analytical Technologies, AquaMax 4000).
18. Centrifuge with microplate adapter.
19. Plate reader for measuring absorbance at 450 nm.

3 Method

Carry out all procedures on ice, unless otherwise specified.

3.1 Determination of Binding of Lipid-Modified Wnt to FZD CRD-Fc

This protocol will provide enough material for part of a 384-well plate (*see* Fig. 2 and **Note 3**). Volumes can be adjusted accordingly depending on desired scale of experiments.

3.1.1 Preparing the Capture Plate

1. Add 30 μL/well immobilization buffer to a 384-well Maxisorp plate.
2. Centrifuge the plate at 130 g for 0.5 min. Seal plate with Topseal A and incubate overnight at 4 °C.

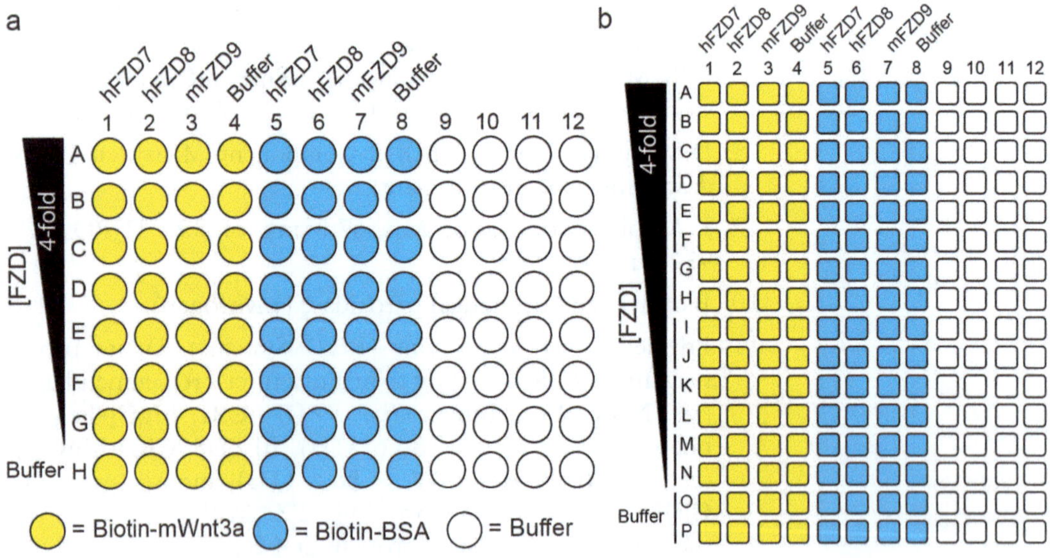

Fig. 2 Recommended plate layout to probe biotin-mWnt3a and hFZD CRD-hIgG$_1$-Fc interactions. (**a**) Layout of a 96-well plate depicting the indicated combinations of biotin-mWnt3a (0.675 nM), biotin-BSA (0.675 nM) and hFZD CRD-hIgG$_1$-Fc (fourfold serial dilutions). (**b**) Liquid from the 96-well plate is transferred to a NeutrAvidin-coated 384-well capture plate for protein immobilization in technical duplicates

3. Centrifuge the plate at 130 g for 0.5 min. Remove Topseal A slowly (*see* **Note 4**).
4. Aspirate fluid from the 384-well plate then wash seven times with PBS-T (70 μL/wash) using a plate washer.
5. Centrifuge the plate at 130 × *g* for 0.5 min, then aspirate residual fluid.
6. Add 100 μL/well of blocking buffer to the 384-well plate. Centrifuge the plate at 130 × *g* for 0.5 min. Seal with Topseal A film, then incubate plate at room temperature for at least 2 h (*see* **Note 5**).

3.1.2 Preparing Wnt and FZD Binding Assay

1. Add 25 μL/well of either biotin-mWnt3a or biotin-BSA (1.35 nM in dilution buffer; 2× stock) to a 96-well plate (*see* Fig. 2a for suggested layout).
2. Perform fourfold serial dilution of FZD CRD-Fc proteins in dilution buffer starting at 100 nM in a separate 96-well plate (100 μL/well).
3. Add 25 μL/well of the diluted FZD CRD-Fc to the 96-well plate containing -mWnt3a and biotin-BSA (*see* Fig. 2a for suggested layout).
4. Centrifuge the 96-well plate at 130 × *g* for 0.5 min. Mix by pipetting three times and then seal with Topseal A film. Incubate plate for at least 1 h at 4 °C.
5. Centrifuge the 384-well plate at 130 × *g* for 0.5 min. Aspirate fluid from the 384-well plate and wash seven times (70 μL/wash) with PBS-T using a plate washer.
6. Centrifuge the 96-well plate at 130 × *g* for 0.5 min. Transfer 20 μL/well of the Wnt-FZD solution to two wells (technical duplicates) in the 384-well capture plate (*see* Fig. 2b for suggested layout).
7. Centrifuge the 384-well plate at 130 × *g* for 0.5 min. Seal the plate with Topseal A and incubate at 4 °C for at least 1 h with gentle shaking. Aspirate fluid and then wash seven times (70 μL/wash) with PBS-T using a plate washer.
8. Aspirate fluid. Add 20 μL/well of anti-hIgG$_1$-Fc-HRP (1 μg/mL; not cross-reactive with IgG$_{2A}$) and incubate at RT for 1 h.
9. *See* **Note 6** for optional step.

3.1.3 Signal Development

1. Aspirate fluid from the 384-well plate and wash seven times (70 μL/wash) with PBS-T.
2. Aspirate fluid. Add TMB reagent (30 μL/well) following manufacturer's recommendation and incubate for ca. 15 min at RT. As the blue color develops, tap gently to mix.

3. Add 1 M phosphoric acid (30 μL/well) and incubate for ca. 10 min at RT (*see* **Note 7**).
4. Measure absorbance at 450 nm using a plate reader.
5. Values were collected and plotted using Prism Graphpad software. Absorbance was plotted against the $\log_{10}$ transformed FZD CRD-Fc molar concentration (Fig. 3a, b).

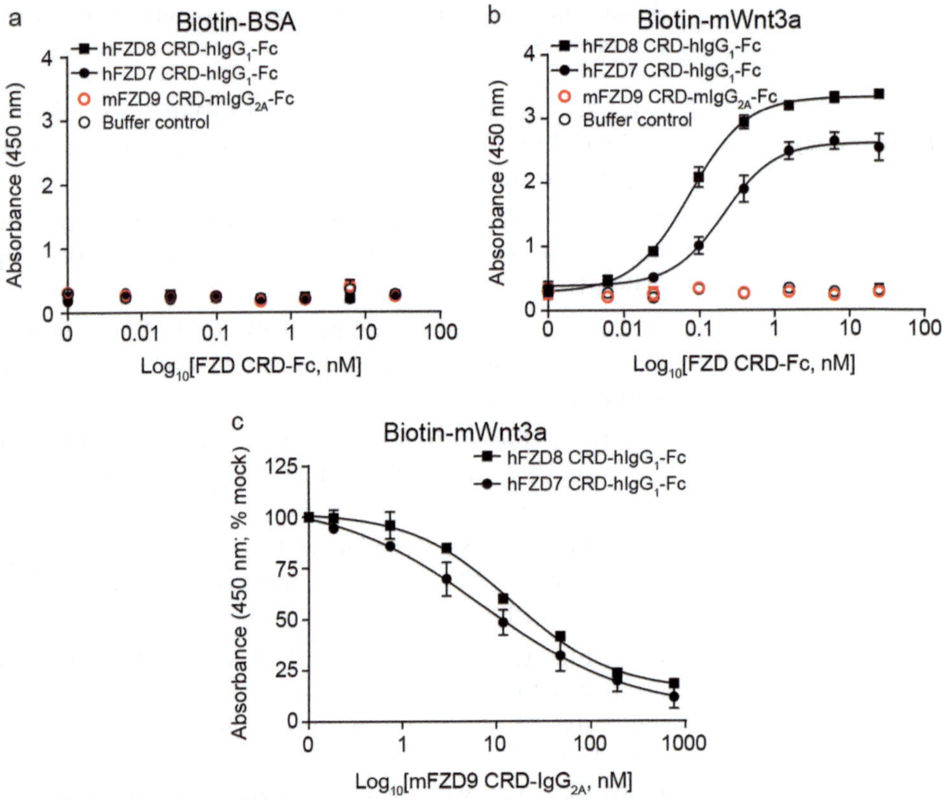

Fig. 3 Biotin-mWnt3a interacts with hFZD7/8 CRD-Fc as measured by ELISA. hFZD7 CRD-hIgG_1-Fc, hFZD8 CRD-hIgG_1-Fc or mFZD9 CRD-mIgG_2A-Fc were incubated in the presence of (**a**) biotin-BSA (0.675 nM) or (**b**) biotin-mWnt3a (0.675 nM), and then developed by ELISA. Plotted values represent the mean ± s.e.m. of three independent experiments, each done in technical duplicates. EC_{50}: hFZD7 CRD-hIgG_1-Fc, 0.23 ± 0.09 nM; hFZD8 CRD-hIgG_1-Fc, 0.075 ± 0.016 nM. EC_{50} values represent the mean ± 95% confidence interval. mFZD9 CRD-mIgG_{2A}-Fc is not detected since the mouse species is not recognized by the anti-hIgG_1-Fc-HRP antibody. No signal is detected in biotin-BSA treated samples (**a**) which serve as a negative control in the assay. (**c**) hFZD7 CRD-hIgG_1-Fc (1 nM) or hFZD8 CRD-hIgG_1-Fc (0.5 nM) and biotin-mWnt3a (0.675 nM) were incubated in the presence of increasing concentration of mFZD9 CRD-mIgG_{2A}-Fc (fourfold serial dilution). IC_{50}: hFZD7 CRD-hIgG_1-Fc, 8.5 ± 4.3 nM; hFZD8 CRD-hIgG_1-Fc, 15.9 ± 5.7 nM. IC_{50} values represent the mean ± 95% confidence interval. Plotted values represent the mean ± s.e.m. of three experiments in technical duplicate

3.2 Competition ELISA to Assess the Effect of Inhibitors of Wnt–FZD CRD Interaction

To establish proof-of-concept application of the ELISA method toward assessing the effect of inhibitors on Wnt–FZD interaction, we chose to test the commercially available mouse (m) FZD9 CRD-IgG_{2A}-Fc fusion protein as a competing ligand. Since it comprises a mouse IgG_{2A}-Fc, it does not react with the anti-hIgG_1-Fc antibody used in the assay (*see* **Note 8**).

1. Select a concentration of FZD CRD-hIgG_1-Fc that is at ~60–70% of signal saturation (Fig. 3b).
2. Prepare a 384-well capture plate by following **steps 1–6** of the above protocol.
3. In a 96-well plate, add 16.7 μL of biotin-mWnt3a (3× stock; 2 nM) to column 1 and column 2. Centrifuge plate for 0.5 min at 130 × *g*.
4. In the same 96-well plate, generate a fourfold serial dilution of mFZD9 CRD-mIgG_{2A}-Fc in column 12. Mix 25 μL of 4.5 μM stock with 75 μL dilution buffer. Aliquot 25 μL from the resulting solution into 75 μL of dilution buffer. Repeat for a total of seven dilutions. The last well serves as a control and should contain buffer only (*see* **Note 9**).
5. Add 16.7 μL of mFZD9 CRD-mIgG_{2A}-Fc serial dilution from column 12 to column 1 and column 2.
6. Centrifuge plate at 130 × *g* for 0.5 min. Mix reagents by pipetting up and down three times. Seal with Topseal A film and then incubate plate for at least 10 min at 4 °C.
7. After incubation, add 16.7 μL of hFZD7 CRD-hIgG_1-Fc (3× stock of your determined concentration from **step 1**) to column 1 and 16.7 μL of hFZD8 CRD-hIgG1-Fc (3× stock of your determined concentration from **step 1**) to column 2.
8. Centrifuge 96-well plate at 130 × *g* for 0.5 min. Mix reagents by pipetting up and down three times. Seal with Topseal A film and then incubate for 1 h at 4 °C with gentle shaking.
9. Transfer solution to 384-well NeutrAvidin coated plate and follow **steps 7–9** and **1–5** from the above protocol Subheadings 3.1.2 and 3.1.3, respectively.
10. Plot values as $\log_{10}$-function of mFZD9 CRD-IgG_{2A}-Fc molar concentration to the normalized untreated control wells. Loss of absorbance at 450 nm is a readout for the loss of hFZD7/8 CRD-IgG_1-Fc binding to biotin-mWnt3a (Fig. 3c).

4 Notes

1. Multiple human (h) and mouse (m) FZD CRD-Fc fusion constructs are commercially available or could be generated by recombinant expression methods [4, 6, 11]. hFZD CRD proteins from the FZD7 and FZD8 subfamilies, incorporating

C-terminal fusions to $hIgG_1$ Fc fragment, were chosen for method development. The FZD3/6 CRD-Fc subclass is not commercially available, and our attempts at generating these proteins from insect cells were unsuccessful.

2. Biotinylated mWnt3a (Lot no. HTR113) was generated by R&D systems (www.rndsystems.com). In brief, purified recombinant mWnt3a (cat. no. 5036-WN) was dialyzed into PBS containing 0.5% CHAPS, pH 7.4 and biotinylated according to the manufacturer's recommendation (Pierce, cat no. 21338), yielding an average biotinylation efficiency of 1.4 mol of biotin per mol of protein as detected by MS.
3. Individual preparations of mWnt3a and FZD CRDs may have slightly different activity or biotinylation efficiency. A trial run is advisable when employing new proteins or lots to determine assay working concentrations. Additional Wnt and Frizzled CRDs have been tested using this assay, indicating that this approach is broadly applicable to measuring interactions between a range of Wnt and FZD CRD proteins.
4. Removing Topseal A film should be done gently to avoid liquid splashing and material loss.
5. If convenient, blocking procedure could be done by incubating overnight at 4 °C. Plates have been stored for months at 4 °C with no noticeable deterioration in signal quality.
6. This is an optional step that serves as a control for determining NeutrAvidin coating capacity and signal saturation. Perform fourfold serial dilution of biotin-HRP in dilution buffer in the 96-well plate (starting concentration 220 ng/mL; for example, dilute 25 μL biotin-HRP (220 ng/mL) in 75 μL dilution buffer and repeat for a total of seven dilutions). Transfer 20 μL/well of biotin-HRP serial dilution to two wells (technical duplicate) in the 384-well plate. Centrifuge the plate at 130 × *g* for 0.5 min, seal with Topseal A and incubate at RT for 1 h. Visual inspection of the biotin-HRP control may be used to determine signal saturation. It is recommended to keep at least one column separating the biotin-HRP containing wells to reduce cross-well contamination due to its strong signal.
7. Acidification will halt blue color development and will turn the TMB substrate yellow. Stopped reactions should be read within 30 min. Incubation time should be adjusted to avoid signal saturation in the experimental wells.
8. Alternative ligands, such as small molecules, peptides, and antibodies, may be employed in the assay for determining their mechanism of inhibition.
9. If competition is not observed using this concentration range, an increased concentration may be used depending on the ligand used.

Acknowledgments

We would like to thank Kevin Callahan and Mitch Brabec from R&D systems for technical support and Simon Hansen and Xinxin Gao for review of the manuscript.

References

1. Clevers H, Nusse R (2012) Wnt/β-catenin signaling and disease. Cell 149:1192–1205
2. Willert K, Nusse R (2012) Wnt proteins. Cold Spring Harb Perspect Biol 4:a007864. https://doi.org/10.1101/cshperspect.a007864
3. Wang J, Sinha T, Wynshaw-Boris A (2012) Wnt signaling in mammalian development: lessons from mouse genetics. Cold Spring Harb Perspect Biol 4(5):a007963. https://doi.org/10.1101/cshperspect.a007963
4. Janda CY, Waghray D, Levin AM, Thomas C, Garcia KC (2012) Structural basis of Wnt recognition by frizzled. Science 337:59–64
5. Nile AH, Hannoush RN (2016) Fatty acylation of Wnt proteins. Nat Chem Biol 12:60–69
6. Nile AH, Mukund S, Stanger K, Wang W, Hannoush RN (2017) Unsaturated fatty acyl recognition by frizzled receptors mediates dimerization upon Wnt ligand binding. Proc Natl Acad Sci U S A 114:4147–4152
7. Willert K, Brown JD, Danenberg E, Duncan AW, Weissman IL, Reya T, Yates JR, Nusse R (2003) Wnt proteins are lipid-modified and can act as stem cell growth factors. Nature 423:448–452
8. Bourhis E, Tam C, Franke Y, Bazan JF, Ernst J, Hwang J, Costa M, Cochran AG, Hannoush RN (2010) Reconstitution of a frizzled8. Wnt3a.LRP6 signaling complex reveals multiple Wnt and Dkk1 binding sites on LRP6. J Biol Chem 285:9172–9179
9. Dijksterhuis JP, Baljinnyam B, Stanger K, Sercan HO, Ji Y, Andres O, Rubin JS, Hannoush RN, Schulte G (2015) Systematic mapping of WNT-FZD protein interactions reveals functional selectivity by distinct WNT-FZD pairs. J Biol Chem 290:6789–6798
10. Voloshanenko O, Gmach P, Winter J, Kranz D, Boutros M (2017) Mapping of Wnt-frizzled interactions by multiplex CRISPR targeting of receptor gene families. FASEB J 31:4832–4844
11. Dann CE, Hsieh J-C, Rattner A, Sharma NJ, Leahy DJ (2001) Insights into Wnt binding and signalling from the structures of two frizzled cysteine-rich domains. Nature 412:86–90

Chapter 18

Biochemical Assays for Ghrelin Acylation and Inhibition of Ghrelin *O*-Acyltransferase

Michelle A. Sieburg, Elizabeth R. Cleverdon, and James L. Hougland

Abstract

Ghrelin *O*-acyltransferase (GOAT) is an enzyme responsible for octanoylating and activating ghrelin, a peptide hormone that plays a key role in energy regulation and hunger signaling. Due to its nature as an integral membrane protein, GOAT has yet to be purified in active form which has complicated biochemical and structural studies of GOAT-catalyzed ghrelin acylation. In this chapter, we describe protocols for efficient expression and enrichment of GOAT in insect cell-derived microsomal fraction, HPLC-based assays for GOAT acylation activity employing fluorescently labeled peptides, and assessment of inhibitor potency against GOAT.

Key words Ghrelin *O*-acyltransferase (GOAT), Ghrelin, Protein acylation, Protein octanoylation, Enzyme inhibition, HPLC, Membrane-bound *O*-acyltransferase (MBOAT)

1 Introduction

Ghrelin is a 28-amino acid peptide hormone that plays a role in multiple physiological processes including, but not limited to, energy regulation and metabolism [1]. During its maturation process prior to secretion, ghrelin undergoes a serine octanoylation modification near its N-terminus that is required for this hormone to be recognized by the GHS-R1a receptor [2]. The enzyme responsible for this modification, ghrelin *O*-acyltransferase (GOAT), was identified as a protein-modifying member of the membrane-bound *O*-acyltransferase (MBOAT) family [3–5]. Ghrelin is the only predicted substrate for GOAT within the human proteome [5, 6], which increases the attractiveness of GOAT as a target for regulating ghrelin signaling.

GOAT is an integral membrane protein found mainly in the endoplasmic reticulum, although recent studies suggest GOAT may also be present on the plasma membrane in some cell lines

Maurine E. Linder (ed.), *Protein Lipidation: Methods and Protocols*, Methods in Molecular Biology, vol. 2009,
https://doi.org/10.1007/978-1-4939-9532-5_18, © Springer Science+Business Media, LLC, part of Springer Nature 2019

[7–9]. GOAT has a complex membrane topology, with 11 transmembrane helices and one reentrant loop based on epitope tagging and selective permeabilization studies [9]. Despite extensive detergent studies, GOAT has proven intractable to solubilization and purification in active form [9, 10]. This limitation has restricted the ability to investigate the substrate selectivity and catalytic mechanism of GOAT.

Early studies of GOAT demonstrated the enzyme accepts synthetic peptides as ghrelin-mimetic substrates for octanoylation [11–13]. Using short peptides mimicking the N-terminal sequence of ghrelin, our group and others have defined the peptide substrate selectivity of GOAT [6, 10–14]. Our group has utilized a fluorescently tagged ghrelin mimetic peptide ($GSSFLC_{AcDan}$) to explore GOAT structure and function using mutation analysis and a novel fluorescence-based assay for GOAT activity (Fig. 1) [6, 13, 15]. In this chapter, we describe protocols to express and enrich active human ghrelin *O*-acyltransferase (hGOAT) in the microsomal fraction of Sf9 insect cells and evaluate its ability to octanoylate ghrelin in the presence of inhibitors.

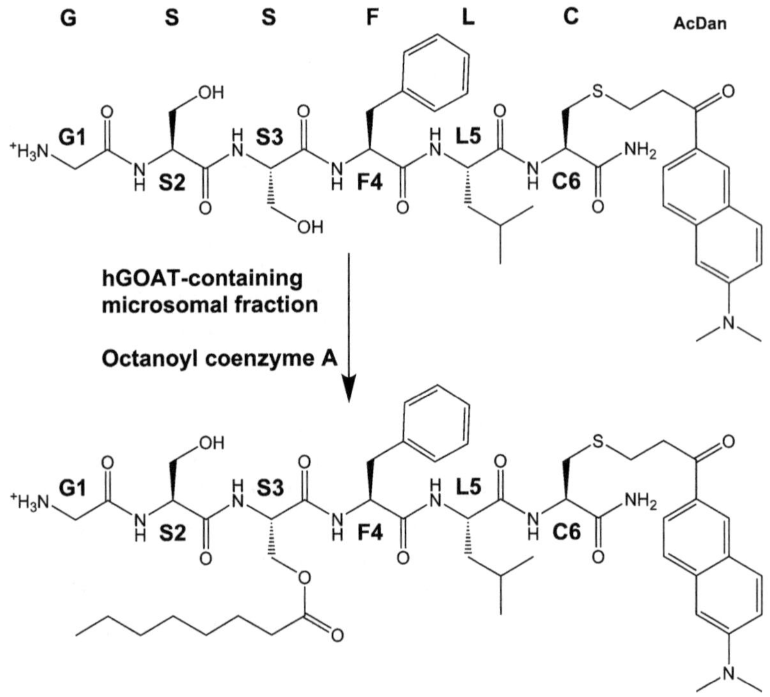

Fig. 1 hGOAT-catalyzed octanoylation of fluorescently labeled $GSSFLC_{AcDan}$ ghrelin mimetic peptide substrate

2 Materials

2.1 Instruments and Other General Materials

1. Refrigerated benchtop shaker (e.g., Excella E24 Incubator Shaker and holders).
2. Refrigerated benchtop centrifuge (e.g., Sorvall Legend XTR with bucket rotor and inserts for 250 mL centrifuge bottles and conical tubes).
3. 250 mL polypropylene (PP) centrifuge bottles (e.g., Nalgene Centrifuge Bottles).
4. Ultracentrifuge capable of holding an SW-41 Ti rotor or similar for 100,000 × *g* sample ultracentrifugation.
5. Clear 9/16″ × 3 ½″ ultracentrifuge tubes rated for 100,000 × *g*.
6. Analytical balance (e.g., Mettler Toledo™ New Classic MS Analytical Balance).
7. Low adhesion 0.65 mL microcentrifuge tubes.
8. 18 G 1.5″ blunt needles.
9. 1 mL Luer lock syringes.
10. Heat block capable of holding 50 °C.
11. Mini protein gel electrophoresis system (e.g., Bio Rad Mini PROTEAN Tetra Cell).
12. Gel electrophoresis transfer system (e.g., Bio Rad Trans-Blot Turbo Blotting System).
13. PVDF membrane (e.g., Trans-Blot Turbo Mini-size PVDF membrane).
14. Transfer blotting paper (e.g., Trans-Blot Turbo Mini-size transfer stacks).
15. Power supply capable of 300 V/400 mA/75 W (e.g., Bio Rad PowerPac Basic).
16. Variable speed rocking shaker.
17. Gel imaging system (e.g., Bio Rad Gel Doc XR+ System).
18. Gel imaging software (e.g., Bio Rad Image Lab Software).
19. 1.5 mL low-adhesion black microcentrifuge tubes.
20. Analytical HPLC (e.g., Agilent Infinity HPLC system series 1260 or equivalent, equipped with a fluorescence detector and autosampler [optional]).
21. Semi-Prep HPLC (e.g., Agilent Infinity HPLC system series 1260 or equivalent, equipped with a UVVis absorbance detector).
22. HPLC analysis software (e.g., Agilent ChemStation Software version C.01.06 or equivalent).

23. Analytical reversed phase HPLC column (e.g., ZORBAX Eclipse C18 (4.6 × 150 mm) analytical column or equivalent equipped with a guard column).
24. Semi-prep reversed phase HPLC column (e.g., ZORBAX Eclipse C18 (9.4 × 250 mm) semiprep column or equivalent).
25. NanoDrop 2000c Spectrophotometer.
26. 1.5 mL semi-micro disposable PMMA cuvettes.
27. Vacuum concentrator (e.g., Eppendorf Vacufuge Plus).
28. Autosampler vials for HPLC (e.g., 250 μL capacity conical bottom polypropylene vial with snap cap with a PTFE/silicone septa).
29. Vortex (e.g., Vortex Genie-2).
30. MALDI-TOF mass spectrometer (e.g., Bruker Autoflex III).
31. MTP 384 massive target T plate (e.g., Bruker Daltonik GmbH).

2.2 Cells and Medium

1. *Spodoptera frugiperda* (Sf9) cells (ThermoFisher Scientific).
2. Sf900-III medium (ThermoFisher Scientific).

2.3 Glassware/Plasticware for hGOAT Expression and Microsomal Fraction Enrichment

1. 250 mL sterile polypropylene Erlenmeyer flask with vented lid, unbaffled.
2. Two-liter sterile glass Erlenmeyer flasks (e.g., Pyrex 2 L flask, unbaffled).
3. Dounce homogenizer (40 mL) (e.g., Wheaton glass pestles and tube).

2.4 Chemicals and Miscellaneous Reagents

1. Lysis Buffer: 150 mM NaCl, 50 mM Tris–HCl (pH 7.0), 1 mM sodium ethylenediamine tetraacetate (NaEDTA), 1 mM dithiothreitol (DTT), Complete protease inhibitor tab (Roche), 10 μg/mL pepstatin A, 100 μM bis(4-nitrophenyl) phosphate.
2. 1M 4-(2-hydroxyethyl)-1-piperazineethanesulfonic acid (HEPES)–NaOH pH 7.0.
3. Resuspension Buffer: 50 mM 4-(2-hydroxyethyl)-1-piperazineethanesulfonic acid (HEPES)–NaOH pH 7.0.
4. 5 mM Octanoyl coenzyme A, free acid in 10 mM Tris–HCl pH 7.0; Advent Bio (*see* **Note 1**).
5. Ghrelin(1–5) peptide substrate, free N-terminus and amidated C-terminus: [H]GSSFLC[NH_2].
6. 500 μM 5,5′-dithiobis(2-nitrobenzoic acid) (DTNB) in 100 mM Na_2HPO_4, 1 mM EDTA, pH 7.5.

7. MALDI matrix solution: 53 mM cyano-4-hydroxycinnamic acid (CHCA) and 50 mM ammonium phosphate monobasic dissolved in 1:1 acetonitrile–aqueous 0.1% trifluoroacetic acid.
8. Bradford dye reagent (e.g., Quick Start Bradford 1× Dye Reagent (Bio-Rad)).
9. 3× Sample buffer: 150 mM Tris pH 6.8, 300 mM DTT, 6% (w/v) SDS, 0.3% (w/v) bromophenol blue dye, 30% (v/v) glycerol.
10. Tris–glycine running buffer: 25 mM Tris base, 150 mM glycine, 0.1% (w/v) SDS, pH 8.1–8.8.
11. Tris–glycine transfer buffer: 25 mM Tris base, 190 mM glycine, 20% (v/v) ethanol.
12. TBST buffer: 20 mM Tris–HCl pH 7.5, 150 mM NaCl, 0.1% (v/v) Tween 20.
13. Blocking buffer: 5% nonfat dry milk in TBST.
14. HRP-conjugated Flag (DYKDDDDK) Tag Antibody (Invitrogen; Product #PA1-984B-HRP).
15. Amino-terminal FLAG-BAP fusion protein (FLAG positive control, Sigma).
16. Prestained molecular weight standards.
17. Chemiluminescent substrate (e.g., Thermo Scientific SuperSignal West Pico).
18. Plastic wrap.
19. Aluminum foil.
20. 50 μM methyl arachidonyl fluorophosphonate (MAFP) in DMSO; Cayman Chemical.
21. HPLC grade acetonitrile.
22. Trifluoroacetic acid.
23. Stop solution: 20% (v/v) acetic acid in isopropanol.
24. TCA solution: 20% trichloroacetic acid (w/v) in ultrapure water.
25. Acrylodan (6-acryloyl-2-dimethylaminonaphthalene).

3 Methods

3.1 Human Ghrelin O-Acyltransferase Expression

1. Generate a high-titer P2 hGOAT baculoviral stock using the Bac-to-Bac Baculoviral Expression System and the pFastBac Dual plasmid containing hGOAT C-terminally tagged with a 3×HA/FLAG/His_6 triple affinity tag under control of the polyhedrin promoter and the TagRFP red fluorescent protein under the p10 promoter (*see* **Note 2**).

2. Propagate Sf9 cells in suspension in a 250 mL PP Erlenmeyer flask with a vented lid at a density of 0.5–5 × 10^6 cells/mL in log-phase growth in Sf900-III medium at 28 °C with shaking (140 rpm) in a refrigerated shaker incubator.
3. Seed Sf9 cells at a density of 1.5 × 10^6 cells/mL in 500 mL Sf900™-III medium in a sterile 2 L glass Erlenmeyer flask with an aluminum foil cap.
4. Inoculate expression culture with P2 hGOAT baculoviral particles at a multiplicity of infection (MOI) of 10 (*see* **Note 3**).
5. Express hGOAT for 42 h at 28 °C with shaking (140 rpm) in a refrigerated shaker incubator (*see* **Note 2**).
6. Harvest 500 mL expression culture using two 250 mL PP bottles via centrifugation at 500 × *g* for 5 min at room temperature.
7. Aspirate the supernatant from the pink-colored pellet and freeze pellets in bottles at −80 °C overnight (or until microsomal fraction enrichment). Pellets can be stored at −80 °C for at least 1 month.

3.2 Microsomal Fraction Enrichment

1. Thaw cell pellets on ice (*see* **Note 4**).
2. Resuspend both pellets together in 25 mL lysis buffer and homogenize the suspension on ice in a prechilled 40 mL Dounce homogenizer using 30 strokes with the loose pestle and 30 strokes with the tight pestle.
3. Transfer cell lysate suspension to a cold 50 mL conical centrifuge tube and clarify the suspension via centrifugation at 3000 × *g* for 10 min at 4 °C.
4. Record the mass of four empty ultracentrifuge tubes. Transfer the supernatant to ultracentrifuge tubes, taking care not to disturb the pellet of cell debris.
5. Pellet the microsomal fraction (MF) by ultracentrifugation at 100,000 × *g* for 1 h at 4 °C.
6. Aspirate the supernatant and determine the mass of the MF pellets using the tared mass for each ultracentrifuge tube. Calculate total mass for the four MF pellets.
7. Combine the MF pellets and add Resuspension Buffer, 4 μL/mg of MF pellet mass (*see* **Note 5**).
8. Resuspend MF pellets well by pipetting using a 1000 μL micropipettor (avoid vigorous pipetting leading to frothing/bubbles), then homogenize the MF resuspension by passing 10 times through an 18 G blunt end needle attached to an appropriate sized syringe.

9. Maintain the MF suspension at 4 °C and prepare 100 μL aliquots of MF suspension in siliconized low-adhesion 0.65 mL microcentrifuge tubes.
10. Passively freeze MF suspension aliquots at −80 °C and store at −80 °C until use (*see* **Note 6**).

3.3 GOAT Expression Verification by Western Blot

1. Prepare a 10% Tris–glycine SDS-PAGE gel (7 × 10 cm; 1 mm thick) with 10 wells (*see* **Note 7**).
2. Prepare MF samples for gel analysis: 20 μg MF in 15 μL 50 mM HEPES–NaOH pH 7.0 and 7.5 μL 3× Sample Loading Buffer. Analyze hGOAT MF with empty SF9 MF and MF expressing only RFP under the p10 promoter as negative controls.
3. Prepare a FLAG positive control sample for gel analysis: 0.1 μg Amino Terminal FLAG-BAP Fusion Protein in 15 μL 50 mM HEPES–NaOH pH 7.0 and 7.5 μL 3× Sample Loading Buffer.
4. Heat samples at 50 °C for 5 min, followed by centrifugation at 1000 × *g* for 10 s (*see* **Note 8**).
5. Apply samples and prestained molecular weight markers to the gel. Run gel at 150 V until the 50 kDa prestained marker band runs near the middle of the gel, for about 2 h.
6. Transfer the protein to a PVDF membrane for 30 min at 25 V using the Trans-Blot Turbo transfer system.
7. Check transfer efficiency—indicated by the presence of the prestained ladder proteins on the membrane, before transferring the membrane to a small blotting container.
8. Rinse membrane briefly in ~10 mL TBST and decant wash solution.
9. Block the membrane in ~15 mL Blocking buffer for 3–4 h while rocking at room temperature.
10. Decant the blocking solution and incubate overnight at 4 °C in 1:1000 HRP-conjugated Flag Antibody diluted in 5% nonfat dry milk in TBST with rocking.
11. Decant antibody solution and save for further use (*see* **Note 9**).
12. Wash the membrane 6 times in ~10 mL TBST for 5 min each time.
13. Prepare chemiluminescent substrates according to the manufacturer's protocol.
14. Decant wash solution and develop membrane in chemiluminescent substrates according to the manufacturer's protocol.
15. Expose the membrane for 30 s in a gel imaging system.
16. Capture the image using the Image Lab software and assess hGOAT expression (*see* **Note 10**).

3.4 Ghrelin Peptide Concentration

1. The ghrelin (1–5) mimetic peptide GSSFLC comes from the manufacturer as a lyophilized solid. Upon receipt, dissolve the peptide (5–9 mg) in 200 μL of 1:1 acetonitrile: ultrapure water to make peptide stock.
2. Determine the concentration of the ghrelin peptide stock by DTNB assay by diluting 5 μL of the GSSFLC peptide stock into 1 mL 500 μM DTNB (*see* **Note 11**).
3. Measure the concentration of 2-nitro-5-mercaptobenzoic acid produced by absorbance at 412 nm ($\varepsilon_{412\ \mathrm{nm}} = 14{,}150\ \mathrm{M}^{-1}\ \mathrm{cm}^{-1}$) [16] on the NanoDrop using a cuvette, with GSSFLC peptide concentration in the DTNB reaction equal to the concentration of 2-nitro-5-mercaptobenzoic acid.
4. Calculate GSSFLC peptide stock concentration by multiplying by 200 to account for the dilution into the DTNB reaction (5 μL into 1 mL reaction volume).

3.5 Acrylodan Labeling of Ghrelin Peptide

1. Acrylodan is provided by the manufacturer as a powder. Dissolve ~2 mg acrylodan in 600 μL 100% acetonitrile to a concentration of 10–20 mM (*see* **Note 12**).
2. Prepare a 500 μL peptide labeling reaction in an opaque black 1.5 mL microcentrifuge tube: 50% (v/v) acetonitrile, 300 μM GSSFLC peptide, 500 μM acrylodan, 50 mM HEPES–NaOH pH 7.0.
3. Incubate at room temperature 16–18 h with shaking at medium speed on a vortexer.
4. Purify the acrylodan-labeled $\mathrm{GSSFLC_{AcDan}}$ peptide by HPLC using a semipreparative C18 reversed phase column (Zorbax Eclipse XDB column, 9.4 × 250 mm) and gradient elution by acetonitrile (Table 1), with peptide detection using the acrylodan absorbance at 360 nm.

Table 1
Reversed phase HPLC method for semipreparative purification of acrylodan-labeled ghrelin peptide

Time [min]	0.05% TFA in water [%]	Acetonitrile [%]	Flow [mL/min]
0.00	98.0	2.00	4.2
35.00	0.00	100.0	4.2
44.00	0.0	100.0	4.2
45.00	0.0	100.0	0.1

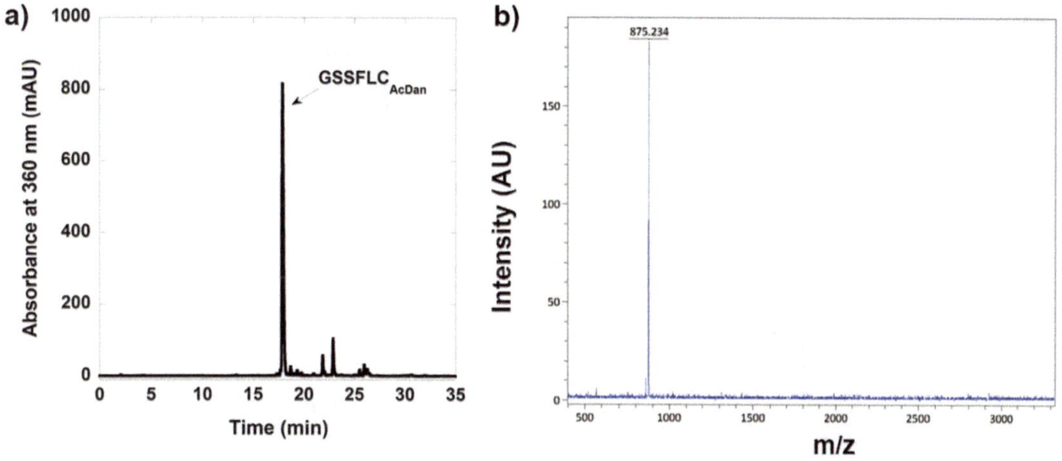

Fig. 2 Labeling and purification of the ghrelin-mimetic $GSSFLC_{AcDan}$ peptide substrate for hGOAT. (**a**) Semipreparative reversed phase HPLC purification of acrylodan-labeled $GSSFLC_{AcDan}$ ghrelin peptide substrate. The fluorescently labeled peptide elutes with a retention time of 17.5 min. (**b**) Verification of $GSSFLC_{AcDan}$ labeling by MALDI-MS. Expected mass 837.40 Da [M+H], observed mass 875.234 Da [M+K]

5. Collect the peak for $GSSFLC_{AcDan}$ (retention time of 17.5 min under conditions provided) into clear 1.5 mL microcentrifuge tubes (Fig. 2a).
6. Cover tubes with foil and evaporate to dryness using a vacuum concentrator at room temperature for 10 h.
7. Once completely dried, dissolve $GSSFLC_{AcDan}$ peptide in 30 μL of 1:1 acetonitrile–ultrapure water (*see* **Note 13**).
8. Determine the concentration of the $GSSFLC_{AcDan}$ stock by UV absorbance ($\varepsilon_{360\ nm} = 13{,}300\ M^{-1}\ cm^{-1}$) in 1:1 acetonitrile–ultrapure water; measure absorbance for 1:5 and 1:10 dilutions of peptide stock and calculate the average.
9. Prepare a $GSSFLC_{AcDan}$ peptide sample for MALDI mass spectrometry analysis by mixing 2 μL of $GSSFLC_{AcDan}$ stock with an equal volume of MALDI matrix solution and spotting 2 μL of the 1:1 peptide–matrix solution on a MTP 384 massive target T plate.
10. Verify acrylodan labeling of the $GSSFLC_{AcDan}$ peptide by positive mode MALDI–TOF mass spectrometry (Fig. 2b); peptides are often observed as sodium ($GSSFLC_{AcDan}$•Na^+, M + 22) or potassium ($GSSFLC_{AcDan}$•K^+, M + 38) adducts.

3.6 hGOAT Activity Assay and MF Activity Titration

1. Warm 1× Bradford Dye reagent to room temperature.
2. Thaw 100 μL aliquot of MF suspension on ice.
3. Pass the MF suspension 10 times through an 18 G needle and 1 mL syringe.

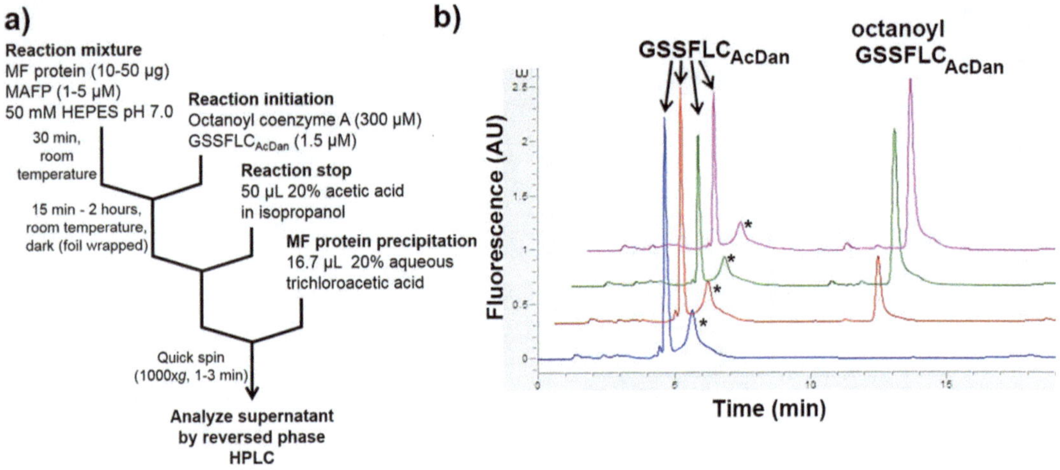

Fig. 3 Titration of hGOAT octanoylation activity. (**a**) Flow chart for hGOAT activity assays. (**b**) HPLC analysis of hGOAT octanoylation reactions. The octanoylated GSSFLC$_{AcDan}$ peptide exhibits a retention time of 12.3 min. Blue trace, hGOAT (25 μg MF protein) with no octanoyl coenzyme A; Red trace, hGOAT (25 μg MF protein) with 300 μM octanoyl coenzyme A; Green trace, hGOAT (50 μg MF protein) with 300 μM octanoyl coenzyme A; Purple trace, hGOAT (75 μg MF protein) with 300 μM octanoyl coenzyme A; traces are horizontally offset by 0.3 min for visual clarity, and the asterisk (*) denotes a peak arising from MF constituents

4. Determine the MF protein concentration of a 1:5 dilution of MF suspension in 50 mM HEPES–NaOH pH 7.0 using the standard protocol for Bio Rad 1× Bradford Dye Reagent.
5. Prepare four 50 μL reactions in 0.65 mL siliconized microcentrifuge tubes (consult flow chart for steps and incubation times of each component of reaction; Fig. 3a). Note that octanoyl CoA and the GSSFLC$_{AcDan}$ peptide are added to start the reactions in **step 6** (*see* **Note 14**).

   Reaction 1: 25 μg MF protein, 50 mM HEPES–NaOH pH 7.0, 1 μM MAFP, 1.5 μM acrylodan-labeled ghrelin mimetic peptide (GSSFLC$_{AcDan}$).

   Reaction 2: 25 μg MF protein, 50 mM HEPES–NaOH pH 7.0, 1 μM MAFP, 300 μM octanoyl coenzyme A, 1.5 μM GSSFLC$_{AcDan.}$

   Reaction 3: 50 μg MF protein, 50 mM HEPES–NaOH pH 7.0, 1 μM MAFP, 300 μM octanoyl coenzyme A, 1.5 μM GSSFLC$_{AcDan.}$

   Reaction 4: 75 μg MF protein, 50 mM HEPES–NaOH pH 7.0, 1 μM MAFP, 300 μM octanoyl coenzyme A, 1.5 μM GSSFLC$_{AcDan.}$

6. Following a 30 min preincubation of MF protein, HEPES buffer, and MAFP, initiate reactions by adding the octanoyl coenzyme A and the GSSFLC$_{AcDan}$ and incubate for 15 min to 2 h under foil at room temperature.

Table 2
Reversed phase HPLC method for analysis of $GSSFLC_{AcDan}$ octanoylation by hGOAT

Time [min]	0.05% TFA in water [%]	Acetonitrile [%]	Flow [mL/min]
0.00	70.0	30.0	1.0
14.00	37.0	63.0	1.0
15.00	0.0	100.0	1.0
25.00	0.0	100.0	1.0
25.10	70.0	30.0	1.0
30.20	70.0	30.0	0.1

7. Stop reactions by addition of 50 μL of stop solution, precipitate MF proteins by addition of 16.7 μL TCA solution, and clarify reaction by centrifugation at 1000 × *g* for ~1 min.
8. Transfer 100 μL of each reaction supernatant to an HPLC autosampler vial.
9. Analyze reactions by HPLC on an analytical C18 reversed phase column using a gradient elution method (Table 2), with detection of the $GSSFLC_{AcDan}$ substrate and octanoylated $GSSFLC_{AcDan}$ product by fluorescence (λ_{ex} = 360 nm, λ_{em} = 485 nm). The unacylated $GSSFLC_{AcDan}$ substrate elutes at a retention time of ~4.5–6 min and the octanoylated $GSSFLC_{AcDan}$ product elutes at a retention time of ~12 min.
10. Integrate substrate and product peaks in the chromatogram using ChemStation software (Fig. 3b), and calculate % conversion using Eq. 1:

$$\%\text{conversion} = \frac{\text{Integrated fluorescence of octanoylated peptide}}{\text{Total integrated peptide fluorescence (octanoylated and non} - \text{octanoylated)}} \quad (1)$$

3.7 hGOAT Inhibition Assay

1. Perform hGOAT inhibition assays using the same reaction setup as for the titration assay in Subheading 3.5, with the desired concentration of inhibitor or vehicle control added during the preincubation step (Fig. 4a). Calculate % conversion at each inhibitor concentration as described in Subheading 3.5 (Fig. 4b, *see* **Note 15**).
2. Calculate % normalized activity at each inhibitor concentration using Eq. 2.

$$\%\text{normalized activity} = \frac{\%\text{conversion in presence of inhibitor}}{\%\text{conversion in presence of vehicle}} \quad (2)$$

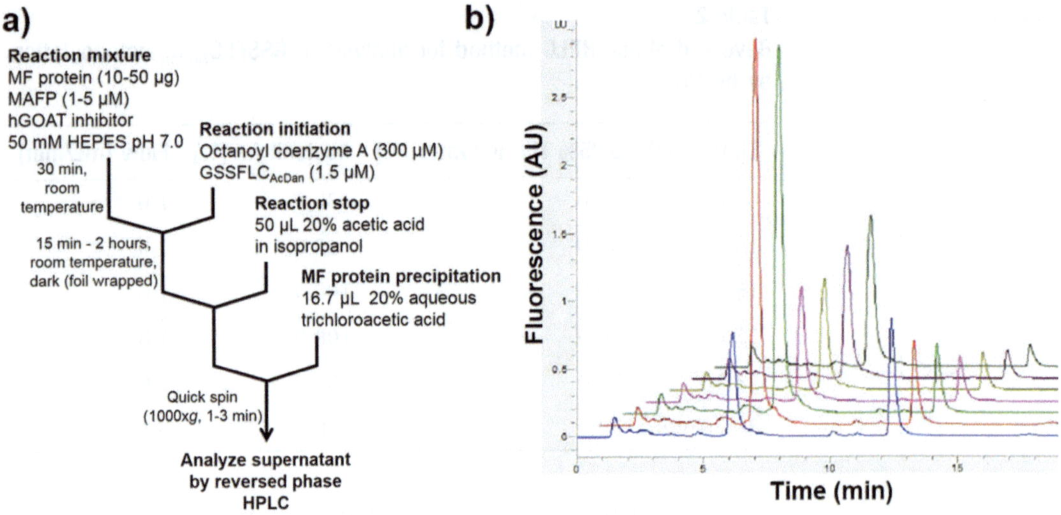

Fig. 4 Measuring inhibitor potency against hGOAT octanoylation activity. (**a**) Flow chart for hGOAT inhibition assays. (**b**) HPLC analysis of hGOAT inhibition reactions. The octanoylated $GSSFLC_{AcDan}$ peptide exhibits a retention time of 12.3 min, and reactions contained the following inhibitor concentrations: 0 μM (vehicle, blue trace), 0.5 μM (red trace), 1 μM (green trace), 1.5 μM (pink trace), 3 μM (yellow trace), 5 μM (purple trace), and 10 μM (black trace). Traces are horizontally offset by 0.3 min for visual clarity

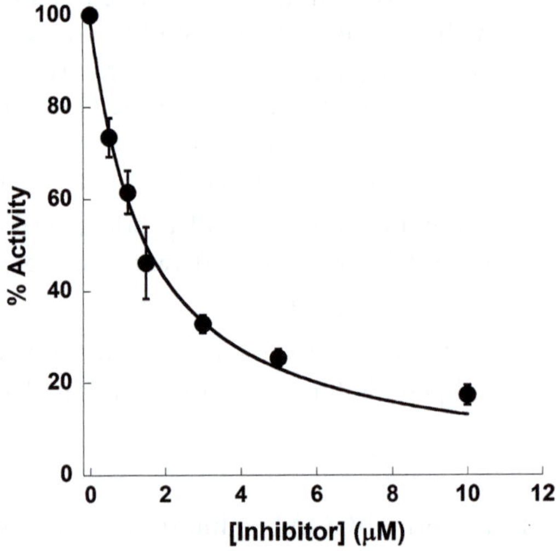

Fig. 5 Representative plot of inhibition of hGOAT-catalyzed $GSSFLC_{AcDan}$ octanoylation. Product fluorescence integrations from HPLC chromatograms were normalized to the vehicle (0 μM) and plotted against inhibitor concentrations, with error bars representing the standard deviation from a minimum of three replicates. Values for % activity and IC_{50} value were calculated as described in the text

3. To determine the IC_{50} value for an inhibitor, plot % normalized activity versus [inhibitor] and fit to Eq. 3, with % $activity_0$ denoting hGOAT activity in the presence of the vehicle alone (Fig. 5).

$$\%\text{normalized activity} = \%\text{activity}_0 \times \left(1 - \left(\frac{[\text{inhibitor}]}{[\text{inhibitor}] + \text{IC}_{50}}\right)\right) \quad (3)$$

4 Notes

1. The free acid has been increasingly hard to find from commercial sources, so we have tested both the Na^+ and K^+ salts of octanoyl coenzyme A and both exhibit similar reactivity to the free acid.
2. High-titer baculovirus stocks were prepared according to the manufacturer's protocol. We utilized the TagRFP red fluorescent protein in the second open reading frame of the pFastBac Dual plasmid to simplify determination of the optimal expression time, temperature and baculovirus multiplicity of infection (MOI) for hGOAT. We monitored the fluorescence of the supernatant and cell pellet ($\lambda_{ex} = 555$ nm, $\lambda_{em} = 585$ nm) at 6–12 h intervals to determine the length of infection that leads to cell lysis, releasing TagRFP into the supernatant. Incubation temperatures ranging from 26.5 °C to 28.5 °C were evaluated as well as MOIs ranging from 1 to 20.
3. Baculovirus titering was performed using a limiting dilution method in 96-well plates as described by Jordan and Crawford et al. [17].
4. All microsomal fraction manipulations should be performed at 4 °C or on ice.
5. We found that the concentration/density of the MF suspension affected the measured hGOAT activity. We tested multiple dilutions of the MF pellet into resuspension buffer based on pellet mass and determined activity level using our titration assay.
6. We tested the octanoylation activity of hGOAT after both passive freezing and flash freezing of the MF suspension stocks. There was no significant difference in hGOAT activity observed between the two methods. Enzyme activity has been tested up to 2 years after freezing with no significant decrease in activity.
7. Premade gels are available commercially or prepared in-house using the following: 10% running gel (25% (v/v) acrylamide 40% solution (19:1), 375 mM Tris–HCl pH 8.8, 0.1% (w/v) SDS,0.1% (w/v) ammonium persulfonate, 0.1% (v/v) TEMED in ultrapure water) and a 4% stacking gel (10% (v/v) acrylamide 40% solution (19:1), 125 mM Tris–HCl pH 6.8, 0.1% (w/v) SDS, 0.1% (w/v) ammonium persulfonate, 0.1% (v/v) TEMED in ultrapure water).

8. We found that hGOAT would aggregate when heated over 65 °C. We tested temperatures from 37–70 °C to determine the optimal temperature to heat the samples before loading them onto the gel.
9. We found that the primary Flag antibody can be reused up to 3 times after freezing and storing at −20 °C.
10. Although the calculated mass of hGOAT_3 × HA/FLAG/His_6 is 57.2 kDa, we and other groups have reported that it runs lower near 49 kDa [10].
11. DTNB, or Ellman's Reagent, undergoes disulfide exchange with the free sulfhydryl group of the cysteine of our peptide yielding 2-nitro-5-thiobenzoic acid (TNB), which can be quantitated colorimetrically at 412 nm using a molar extinction coefficient of 14,150 M^{-1} cm^{-1}.
12. Acrylodan concentration should be verified via absorbance at 393 nm (ε = 18,483 M^{-1} cm^{-1} per manufacturer specifications).
13. We aim for $GSSFLC_{AcDan}$ stock concentrations between 0.7 mM and1 mM.
14. Our lab has shown that the addition of alkyl fluorophosphonate to our MF reactions prevents esterase-catalyzed hydrolysis at the ghrelin octanoyl ester [15].
15. Each MF preparation is independently titrated for activity over time. We found that the optimal concentration of MF to use in inhibitor studies is that which gives us 50–60% peptide conversion over the chosen reaction time.

Acknowledgments

This work was supported by grants from the Foundation for Prader-Willi Research (FPWR) and the American Diabetes Association (1-16-JDF-042).

References

1. Müller TD, Nogueiras R, Andermann ML et al (2015) Ghrelin. Mol Metab 4:437–460
2. Kojima M, Hosoda H, Date Y et al (1999) Ghrelin is a growth-hormone-releasing acylated peptide from stomach. Nature 402:656–660
3. Hofmann K (2000) A superfamily of membrane-bound O-acyltransferases with implications for wnt signaling. Trends Biochem Sci 25:111–112
4. Gutierrez JA, Solenberg PJ, Perkins DR et al (2008) Ghrelin octanoylation mediated by an orphan lipid transferase. Proc Natl Acad Sci U S A 105:6320–6325
5. Yang J, Brown MS, Liang G et al (2008) Identification of the acyltransferase that octanoylates ghrelin, an appetite-stimulating peptide hormone. Cell 132:387–396
6. Darling JE, Zhao F, Loftus RJ et al (2015) Structure-activity analysis of human ghrelin O-acyltransferase reveals chemical

determinants of ghrelin selectivity and acyl group recognition. Biochemistry 54:1100–1110
7. Murtuza MI, Isokawa M (2018) Endogenous ghrelin-O-acyltransferase (GOAT) acylates local ghrelin in the hippocampus. J Neurochem 144:58–67
8. Hopkins AL, Nelson TAS, Guschina IA et al (2017) Unacylated ghrelin promotes adipogenesis in rodent bone marrow via ghrelin O-acyl transferase and GHS-R1a activity: evidence for target cell-induced acylation. Sci Rep. https://doi.org/10.1038/srep45541
9. Taylor MS, Ruch TR, Hsiao PY et al (2013) Architectural organization of the metabolic regulatory enzyme ghrelin O-acyltransferase. J Biol Chem 288:32211–32228
10. Taylor MS, Dempsey DR, Hwang Y et al (2015) Mechanistic analysis of ghrelin-O-acyltransferase using substrate analogs. Bioorg Chem 62:64–73
11. Yang J, Zhao TJ, Goldstein JL et al (2008) Inhibition of ghrelin O-acyltransferase (GOAT) by octanoylated pentapeptides. Proc Natl Acad Sci U S A 105:10750–10755
12. Ohgusu H, Shirouzu K, Nakamura Y et al (2009) Ghrelin O-acyltransferase (GOAT) has a preference for n-hexanoyl-CoA over n-octanoyl-CoA as an acyl donor. Biochem Biophys Res Commun 386:153–158
13. Darling JE, Prybolsky EP, Sieburg M et al (2013) A fluorescent peptide substrate facilitates investigation of ghrelin recognition and acylation by ghrelin O-acyltransferase. Anal Biochem 437:68–76
14. Barnett BP, Hwang Y, Taylor MS et al (2010) Glucose and weight control in mice with a designed ghrelin O-acyltransferase inhibitor. Science 330:1689–1692
15. McGovern-Gooch KR, Rodrigues T, Darling JE et al (2016) Ghrelin octanoylation is completely stabilized in biological samples by alkyl fluorophosphonates. Endocrinology 157:4330–4338
16. Riddles PW, Blakeley RL, Zerner B (1979) Ellman's reagent: 5,5′-dithiobis(2-nitrobenzoic acid)--a reexamination. Anal Biochem 94:75–81
17. Jordan KR, Crawford F, Kappler JW et al (2009) Vaccination of mice with baculovirus-infected insect cells expressing antigenic proteins. Curr Protoc Immunol 85:2.15.1–2.15.23

Chapter 19

In Vitro Analysis of Hedgehog Acyltransferase and Porcupine Fatty Acyltransferase Activities

James John Asciolla, Kalpana Rajanala, and Marilyn D. Resh

Abstract

Hedgehog and Wnt proteins are modified by covalent attachment of the fatty acids palmitate and palmitoleate, respectively. These lipid modifications are essential for Hedgehog and Wnt protein signaling activities and are catalyzed by related, but distinct fatty acyltransferases: Hedgehog acyltransferase (Hedgehog) and Porcupine (Wnt). In this chapter, we provide detailed methods to directly monitor Hedgehog and Wnt protein fatty acylation in vitro. Palmitoylation of Sonic hedgehog (Shh), a representative Hedgehog family member, is assayed using purified Hedgehog acyltransferase (Hhat) or Hhat-enriched membranes, a recombinant 19 kDa Shh protein or C-terminally biotinylated Shh 10-mer peptide, and ^{125}I-iodopalmitoyl CoA as the donor fatty acyl CoA substrate. The radiolabeled reaction products are quantified by SDS-PAGE and phosphorimaging or by γ-counting. To assay Wnt acylation, the reaction consists of a biotinylated, double disulfide-bonded Wnt peptide containing the sequence surrounding the Wnt3a acylation site, [^{125}I] iodo-cis-9-pentadecenoyl CoA, and Porcupine-enriched membranes. Radiolabeled, biotinylated Wnt3a peptide is captured on streptavidin coated beads and the reaction product is quantified by γ-counting.

Key words Hedgehog, Hedgehog acyltransferase, Porcupine, Wnt proteins, Fatty acylation

1 Introduction

Hedgehog and Wnt proteins are secreted signaling proteins that mediate growth, patterning, and differentiation during embryogenesis as well as tumorigenesis in adults [1, 2]. In order for these proteins to signal effectively, they must be modified by covalent attachment of a fatty acid. The 16-carbon saturated fatty acid palmitate is attached via amide bond to the N-terminal Cys of hedgehog proteins, a reaction catalyzed by the enzyme Hedgehog acyltransferase (Hhat) [3–5]. Wnt proteins contain the cis-Δ9-monounsaturated fatty acid, palmitoleate (C16:1Δ9), attached to a conserved, internal Ser; this reaction is catalyzed by Porcupine (Porcn) [6–8]. Both Hhat and Porcn are multipass membrane proteins that reside in the endoplasmic reticulum (ER), and are

Maurine E. Linder (ed.), *Protein Lipidation: Methods and Protocols*, Methods in Molecular Biology, vol. 2009,
https://doi.org/10.1007/978-1-4939-9532-5_19,

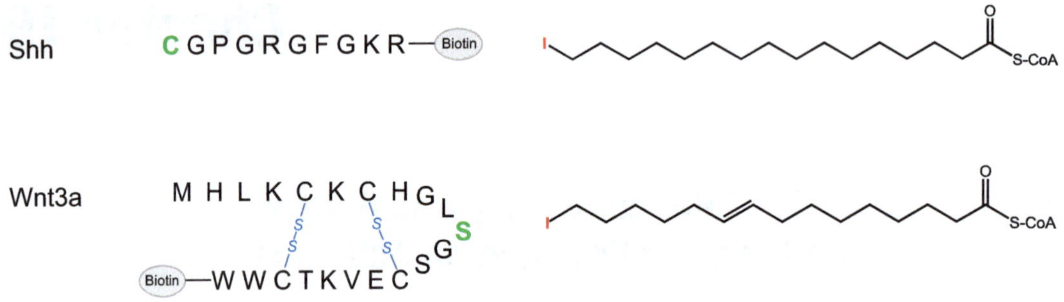

Fig. 1 Sequences of the Shh (top) and Wnt3a (bottom) biotinylated peptides

members of the MBOAT (membrane bound *O*-acyl transferase) family of acyltransferases [9, 10].

Cell-based assays have been developed to monitor fatty acylation of Hedgehog and Wnt proteins [3, 11, 12]. Cells are typically grown in medium supplemented with radiolabeled (^{3}H or ^{125}I) fatty acids or with fatty acid analogs amenable to click chemistry analysis. Fatty acids that enter the cell are converted to fatty acyl CoA, and then must enter the lumen of the ER, where Hhat and Porcn catalyze transfer of the fatty acid from the fatty acyl CoA to the protein substrate. Wnt protein acylation requires an additional step: Stearoyl CoA desaturase introduces a *cis*-double bond to convert palmitoyl CoA into palmitoleoyl CoA, the substrate for Porcn [13]. Given the essential role of fatty acylation in Hedgehog and Wnt signaling activities, it is important to understand the enzymatic mechanisms that underlie these reactions, and this requires the use of direct in vitro assays. Here we detail the reagents and methodology to monitor Hhat and Porcn fatty acylation activities in vitro. For Hhat, we outline several options for the source of enzyme (purified Hhat or Hhat in membranes from cells overexpressing Hhat) and protein substrate (recombinant protein or N-terminal peptide) [3, 14] (Fig.1). Sonic hedgehog (Shh), one of the 3 members of the mammalian family of hedgehog proteins, is used as it is the best studied of the family members. To date, there is only one option for monitoring Wnt acylation in vitro: membranes from cells overexpressing Porcn and a peptide modeled on the sequence surrounding the Wnt3a acylation site (Fig.1) [15].

2 Materials

2.1 Cell Growth and Transfection

1. 293FT cells.
2. Culture medium: DMEM, 10% fetal bovine serum, 50 units/mL penicillin, 50 μg/mL streptomycin, 500 μg/mL geneticin, 1 mM GlutaMAX, 1 mM sodium pyruvate, 0.1 mM nonessential amino acids.

3. Flag-tagged Hhat cDNA cloned into a mammalian expression vector (e.g., HA-Flag-his-Hhat [3]).
4. Lipofectamine 2000 (Invitrogen) transfection reagent (*see* **Note 1**).

2.2 P100 Membranes

1. NTE: 100 mM NaCl, 10 mM Tris–HCl pH 7.4, 1 mM EDTA. Store at 4 °C.
2. Hypotonic lysis buffer: 10 mM Tris–HCl (pH 7.4), 0.2 mM $MgCl_2$. Store at 4 °C.
3. 5× sucrose: 1.25 M. Add 21.39 g sucrose to 50 mL with dH_2O.
4. Sucrose–Tris–EDTA: 0.25 M sucrose, 10 mM Tris–HCl (pH 7.4), 1 mM EDTA.
5. Hypotonic Lysis Buffer/0.25 M sucrose/1 mM EDTA/ Thermo Scientific™ Halt™ Protease Inhibitor Single-Use Cocktail (AEBSF, aprotinin, bestatin, E64, leupeptin, pepstatin A).

2.3 Hhat Purification

1. NTE: 100 mM NaCl, 10 mM Tris–HCl pH 7.4, 1 mM EDTA. Store at 4 °C.
2. HEPES lysis buffer: 10 mM HEPES pH 7.3, 0.2 mM $MgCl_2$.
3. 5× sucrose: 1.25 M. Add 21.39 g sucrose to 50 mL with dH_2O.
4. Solubilization buffer: 350 mM NaCl, 20 mM HEPES 7.3, 1% octylglucoside, 1% glycerol.
5. FLAG-M2 beads (Sigma–Aldrich).
6. Elution buffer: Solubilization buffer supplemented with 300 ng/mL 3× Flag peptide.

2.4 Biotinylated Shh and Wnt Peptides

1. Shh peptides: A 10-mer peptide containing the first 10 amino acids of the mature human Shh sequence (CGPGRGFGKR) with a C-terminal PEG-biotin. A control peptide with the N-terminal cysteine replaced by alanine (AGPGRGFGKR) with a C-terminal PEG-biotin. Both peptides can be synthesized using standard methods by most in-house and commercial peptide synthesis facilities. Purity should be assessed by reversed phase HPLC and the sequence confirmed by LC-MS analysis.
2. Wnt3a peptides: C-terminal biotinylated (Biotin-PEG Nova-Tag™) containing 21 amino acids upstream and downstream of the serine acylation site of Wnt3a (MHL KC(S)KC(S-)HG LSG SC(S-)E VKT C(S-)WW). The peptides must contain two sets of disulfide-bonded cysteine residues as indicated in Fig.1. This mimics the two-dimensional conformation of Wnt proteins as revealed by the crystal structure [16]. Synthesis must be

performed by a lab with specific expertise in generating disulfide-bonded peptides. Purity should be assessed by reversed phase HPLC and the sequence confirmed by LC-MS analysis (*see* **Note 2**).

2.5 Iodo-Fatty Acyl CoA Synthesis ([^{125}I] IC16-CoA and [^{125}I] IC15:1-CoA)

1. Iodopalmitate (IC16) and iodo-cis-9-pentadecenoic acid (IC15:1): synthesized and radioiodinated with [^{125}I]NaI as described [3, 17] and in Subheading 3.4 (*see* **Note 3**).
2. 20 mM Lithium CoA, trilithium salt in dH_2O, pH to 5.0 with NaOH. Store in aliquots at −80 °C following reconstitution.
3. 50 mM ATP disodium salt hydrate in dH_2O, pH to 7.0 with NaOH. Store in aliquots at −80 °C following reconstitution.
4. 2 U/mL Acyl CoA Synthetase in 50 mM HEPES, pH 7.3. Store in aliquots at −80 °C following reconstitution.
5. 1 M HEPES pH 7.3.
6. 0.1% Triton X-100 (prepared from a 100% Triton X-100 stock).
7. 1 M DTT.
8. 1 M $MgCl_2$.
9. Grace™ Alltech™ Maxi-Clean™ Solid-Phase Extraction Cartridges, C18 (600 mg).
10. 50 mM ammonium acetate.
11. 25% acetonitrile in 50 mM ammonium acetate.
12. 45% acetonitrile in 50 mM ammonium acetate.

2.6 In Vitro Shh Peptide Fatty Acylation Assays

1. Reaction buffer: 167 mM MES, pH 6.5, 1.7 mM DTT, 0.083% Triton X-100.
2. RIPA buffer: 150 mM NaCl, 50 mM Tris–HCl pH 7.4, 1% Triton X-100, 0.5% sodium deoxycholate, 0.1% SDS, 1 mM EDTA.

2.7 Recombinant Shh Protein Expression

1. Shh cDNA encoding amino acids 24–197 in pET19b plasmid.
2. Competent BL21DE3pLysS E. coli cells.
3. SOC chloramphenicol–ampicillin agar plates: 2% tryptone, 0.5% yeast extract, 0.5% NaCl, 1.5% agar, 2.5 mM KCl, 10 mM $MgCl_2$, 20 mM glucose, 25 μg/mL chloramphenicol, 40 μg/mL ampicillin.
4. LB medium: 1% tryptone, 0.5% yeast extract, 0.5% NaCl.
5. LB chloramphenicol–ampicillin medium: LB, 25 μg/mL chloramphenicol, 40 μg/mL ampicillin.
6. 1 M IPTG.
7. Resuspension Buffer: 50 mM Tris–HCl pH 8, 20% sucrose.

8. 5 M NaCl.
9. 1.43 M β-mercaptoethanol (1:10 dilution of the liquid stock supplied by the manufacturer).
10. 2 M imidazole .
11. 10% IGEPAL [0.1%].
12. Wash Buffer: 350 mM NaCl, 20 mM Tris–HCl pH 8.0, 10–20 mM Imidazole, 1 mM βME.
13. Elution Buffer: 350 mM NaCl, 250 mM Imidazole, 20 mM Tris–HCl pH 8.0, 1 mM βME.
14. Dialysis Buffer: 350 mM NaCl, 20 mM Tris–HCl pH 8.0, 1 mM βME.
15. Storage buffer: 20 mM Tris–HCl pH 8.0, 1 mM βME.
16. Enterokinase.

3 Methods

3.1 Cell Transfection

1. Grow cells in a 100 mm tissue culture dish until they are 70–80% confluent.
2. Add 6 μg of either Hhat or Porcupine cDNA plasmid to 15 μL Lipofectamine 2000 and incubate for 20 min at room temperature. Add DNA-lipid complex dropwise to cells and incubate overnight at 37 °C.
3. The following day, split the cells 1:2 and grow for 24–48 h.

3.2 P100 Membrane Preparation (Volumes per 1 × 100 mm Tissue Culture Plate)

1. Aspirate media from tissue culture plate.
2. Rinse plate with 5–10 mL cold NTE.
3. Scrape or pipette cells off the dish with 5–10 mL NTE. Centrifuge cells at 1000 × *g* for 5 min at 4 °C.
4. Add 0.8 mL Hypotonic lysis buffer to the cell pellet. Gently vortex and incubate on ice for 15 min.
5. Break open cells by 30 up and down strokes in a Dounce homogenizer, being careful not to cause foaming (*see* **Note 4**).
6. Add 200 μL 5× sucrose and 2 μL 0.5 M EDTA.
7. Centrifuge cell suspension at 1000 × *g* for 10 min at 4 °C. Use a glass Corex tube so you can see the pellet clearly.
8. Carefully remove the supernatant (S1) and transfer to an ultracentrifuge tube. Avoid pulling up any of the nuclear pellet.
9. Resuspend the pellet in 1 mL Sucrose/Tris/EDTA and Dounce homogenize 10 strokes.
10. Centrifuge at 1000 × *g* for 10 min at 4 °C. Remove supernatant and combine with the S1 supernatant from **step 8** above.

11. Centrifuge the combined S1 fractions at 100,000 × *g* for 45 min at 4 °C.
12. Remove the supernatant. Resuspend the pellet (P100) in Hypotonic Lysis Buffer/0.25 M sucrose/1 mM EDTA/protease inhibitor, (Thermo Scientific™ Halt™ Protease Inhibitor Single-Use Cocktail [AEBSF, aprotinin, bestatin, E64, leupeptin, pepstatin A]) and Dounce homogenize.
13. Aliquot, quick freeze on dry ice, and store at −80 °C. Membrane aliquots should be thawed only once.

3.3 Hhat Purification

1. Start with 20 × 100 mm plates of 293FT cells transfected with Hhat-HA-Flag-His in pcDNA3.1. Remove plates from incubator and place on ice. All manipulations from this point on are performed on ice or in the cold room and all solutions and buffers are at 4 °C.
2. Aspirate media and wash each plate twice with 5 mL of NTE. 293FT are weakly adherent so wash as gently as possible.
3. Add 5 mL of NTE to each plate and scrape off cells with a rubber policeman. Collect and combine the suspensions into two 50 mL tubes.
4. Centrifuge at 1000 × *g* for 10 min at 4 °C.
5. Aspirate supernatant and resuspend each cell pellet in 4 mL of HEPES Lysis buffer. Keep tubes on ice for 15–30 min.
6. Combine both suspensions and lyse in a Dounce homogenizer, using 30–40 up and down strokes. Add 2 mL of 5× Sucrose and homogenize 10–15 more strokes. Transfer homogenate to Beckman Ti 70.1 ultracentrifuge tubes. Spin in ultracentrifuge for 45 min at 45,000 rpm (186,000 × *g*), 4 °C.
7. Aspirate supernatant. Resuspend pellet in 8 mL Hypotonic Lysis Buffer +2 mL 5× Sucrose and collect membranes by ultracentrifugation as above.
8. Aspirate supernatant. Resuspend pellet in 10 mL of solubilization buffer. If desired, the lysate can be stored at this point by freezing in liquid N_2 and storing overnight at −80 °C. The following day, thaw on ice and proceed to **step 9**.
9. Rock in cold room for 1 h to solubilize Hhat. Spin as in **step 6** to pellet insoluble material.
10. While the tube is spinning, prepare FLAG M2 Resin. Remove 1 mL from the 50% slurry and spin down at 850 × *g* in a 15-mL conical tube. Aspirate the supernatant (packing buffer). Resuspend the beads in 500 μL Solubilization buffer, mix, then collect resin by centrifugation. Repeat this wash **step 2** more times.

11. Transfer the supernatant containing solubilized Hhat from **step 9** to a 15 mL conical tube containing 1 mL of the prewashed resin (packed beads) in solubilization buffer.
12. Rock in cold room (4 °C) for 1 h.
13. Spin down the beads at 850 × *g*. Remove and save the supernatant as a precaution.
14. Wash beads 4 times with 5 mL of Solubilization buffer. Save washes as above.
15. After the last wash, remove as much wash buffer as possible without disturbing the beads. Then add 500 μL of Elution Buffer and mix and spin as in **step 13**.
16. Remove and save supernatant. Repeat **step 15** two additional times for a total elution volume of 1.5 mL.
17. Remove a 50 μL aliquot for Western blotting analysis.
18. Aliquot the remaining eluant, freeze in liquid nitrogen and store at −80 °C (*see* **Note 5**).

3.4 Iodo-Fatty Acid Synthesis

1. Prepare a reaction vial by melting the tip of a Pasteur pipette. Set the temperature of a heating block to 55 °C and place in a fume hood used for iodination. Place a crystallization dish containing mineral oil onto the heating block and insert a thermometer to monitor the temperature.
2. Add 200 μL of IC16 or IC15:1 (10 μM) resuspended in acetone into the reaction vial, dry under nitrogen and add 10 μL of glacial acetic acid.
3. From this point on, work behind a lead-embedded plexiglass shield. Resuspend the NaI^{125} in 200 μL of acetone and add to reaction vial. Rinse the container with another 200 μL acetone and add to reaction vial.
4. Seal the pipette reaction vial with an 8 mm rubber septum. Insert a charcoal trap (sandwich activated charcoal between glass wool inside a 5 cc syringe with a 23 gauge needle) and incubate in the 55 °C oil bath overnight or for 18 h.
5. The next day, remove the septum and trap and dispose. Add 0.5 mL water, and extract the aqueous layer twice with 1.2 mL chloroform. Transfer the combined chloroform (lower) layers to a glass 1-dram vial and dry under air or nitrogen. Dissolve the residue in 500 μL ethanol.

3.5 Iodo-Fatty Acyl CoA Synthesis

1. Set up a reaction with the following components: 100 μL 1 M HEPES pH 7.3, 1 μL Triton X-100, 250 μL 20 mM Lithium CoA, 200 μL 50 mM ATP, 1 μL 1 M DTT, 100 μL [^{125}I]IC16 or [^{125}I]IC15:1, 125 μL 2 units/mL Acyl CoA synthetase, 10 μL 1 M $MgCl_2$, 213 μL dH_2O. Incubate at room temperature for 3 h with stirring.

2. Preactivate a C18 reversed-phase column with 10 mL methanol, followed by a 5 mL H_2O wash and a 5 mL 50 mM ammonium acetate wash (*see* **Note 6**).
3. Add 5 mL Ammonium acetate to the column, and then add the reaction mix.
4. Wash the column with 15 mL 50 mM ammonium acetate.
5. Wash the column with 25 mL 25% acetonitrile in 50 mM ammonium acetate.
6. Elute in 5–8 mL 45% acetonitrile in 50 mM ammonium acetate.
7. Aliquot the eluate and dry down by rotary evaporating for 2.5–3 h.
8. Resuspend in 400–500 μL dH_2O.
9. Prepare a 1:1000 dilution in water and measure the activity of the preparation by determining counts per minute using a gamma counter.
10. Prepare a 1:200 dilution and measure the concentration of the preparation, using the extinction coefficient of palmitoyl CoA (*see* **Note 7**).

3.6 Recombinant Shh Protein Expression

1. Prepare SOC agar plates that contain both chloramphenicol [25 μg/mL] and ampicillin [100 μg/mL] (*see* **Note 8**).
2. Add 1 μL of plasmid DNA encoding Shh 24–197 in pET19b to 20 μL of competent BL21DE3pLysS *E. coli* cells.
3. Incubate on ice for 5 min.
4. Heat-shock for 30 s in a water bath set to 42 °C.
5. Place cells on ice for 2 min.
6. Add 80 μL of SOC or LB media to each tube of cells.
7. Incubate cells for 1 h in a shaker set to 250 rpm and 37 °C.
8. Plate the entire 80 μL of transformed cells and incubate plates for >16 h at 37 °C (*see* **Note 9**).
9. Pick one colony and grow overnight in 4 mL of LB containing 25 μg/mL chloramphenicol and 100 μg/mL ampicillin.
10. The next day, add 2.5 mL of the overnight culture to 100 mL of LB containing 25 μg/mL chloramphenicol and 100 μg/mL ampicillin.
11. Grow the culture in a shaker at 37 °C, rpm ~250. Monitor the OD at 600 nm periodically.
12. When the culture reaches an OD of 0.8, induce protein expression by adding 100 μL of 1 M IPTG (Final concentration = 10 mM IPTG). Incubate for ~3 h in shaker.

13. Pour culture in a 50 mL falcon tube and spin down @ 1000 × *g* for 10 min. Discard the supernatant and spin the rest of the culture.
14. Remove the supernatant and flash-freeze the pellet in liquid nitrogen. Store at −80 °C.
15. Put the pellet on ice and allow it to thaw.
16. Resuspend the pellet in ~ 1 mL of Resuspension Buffer, and add: 150 μL of 5 M NaCl (final concentration, 0.5 M), 1 μL of ten-fold dilution βME (1 mM), 15 μL of 100 mM PMSF (final conc., 1 mM), 7.5 μL of 2 M imidazole (final concentration, 10 mM), 15 μL of 10% IGEPAL (final concentration, 0.1%). Mix by pipetting up and down. AVOID BUBBLES.
17. Keeping the suspension on ice at all times, dissociate with six, 30 s pulses at 20% using a microtip on a sonicator. Allow the mixture to cool (for ~30 s) between pulses. Keep the sample on ice and make sure it does not get too hot (*see* **Note 10**).
18. Transfer the sonicated suspension to a clean tube.
19. Save a 10 μL sample (label it "prespin"). Place it on ice.
20. Centrifuge the remainder of the sample for 30 min, 19,000 × *g*, at 4 °C.
21. Transfer the supernatant to a clean tube. Take a 10 μL sample for the gel and label it "sup."
22. Prepare the nickel resin by taking 100 μL of Ni slurry using a wide-bore pipette tip, place it in a new microfuge tube, and spin in a microfuge (6500 × *g*, 5 min, 4 °C). Remove the ethanol from the Ni beads, and add 1 mL of Wash Buffer. Vortex and spin in a microfuge (6500 × *g*, 5 min, 4 °C). Repeat two more times.
23. Add 100 μL of Ni slurry to the supernatant from **step 21**.
24. Place the sample on a nutator in the cold room and rock for 10 min (*see* **Note 11**).
25. Remove the sample from the nutator and spin down in a microfuge (6500 × *g*, 5 min, 4 °C).
26. Remove supernatant. Save the supernatant and keep it on ice. Take a sample for the gel and label it "unbound."
27. Wash the pellet with 1 mL Wash Buffer and centrifuge (8000 rpm, 5 min, 4 °C). Repeat two more times. Take gel samples from each wash, label each "Wash 1, 2, or 3." Keep each wash and put them on ice.
28. Elute the protein from Ni beads: Add 100 μL of Elution Buffer to the beads (1,1 ratio of elution buffer to Ni beads). Place sample on the nutator for 10 min. Spin sample down 6500 × *g*, 5 min, 4 °C. Remove supernatant from beads. Repeat the

Elution step one more time and pool both Elution Fractions and label the pooled sample "Eluate." Take a 10 μL sample from the "Eluate" for the gel.

29. Flash-freeze the sample and store at −80 °C. All other samples that were taken from this prep can also be flash-frozen and stored at −80 °C.
30. Prepare Dialysis Buffer. Place buffer in the cold room in a beaker. Soak the dialysis cassette in dialysis buffer.
31. Defrost the eluate sample. Add ~8–10 units of enterokinase directly to the Eluate sample and mix by pipetting up and down (*see* **Note 12**).
32. Add the eluate plus enterokinase sample into the dialysis cassette using a syringe. Dialyze O/N in the cold room (4 °C).
33. The next day, remove the fraction from the dialysis cassette (take a sample for the gel). Filter the fraction through a 2 μm filter (take a sample for the gel) (*see* **Note 13**).
34. Place the filtered sample over the Ni beads again (*see* **steps 24–29**).
35. Take the supernatant (which contains your cleaved purified protein) keep on ice (take sample for gel).
36. Elute the remaining protein off the beads using elution buffer (This elution contains uncleaved protein) (*see* **step 28**) (take sample for gel). Analyze all fractions by SDS-PAGE with Coomassie blue staining. The total yield of Shh protein from a 100 mL culture is 40–50 μg. The culture can be scaled up accordingly for larger protein yield.

    Concentrating Protein into Storage Buffer (Only do this for Large Scale Preps) (*see* **Note 14**).
37. Soak the membrane containing portion of a Micon Centriprep YM-10 10,000 MWCO concentrator in storage buffer.
38. Place the protein solution and the storage buffer in the designated compartments and spin (*see* **Note 15**).
39. Concentrate the protein down to ~3 mg/mL. Determine the concentration with the Bradford reagent or other protein determination kit.
40. Aliquot the protein into eppendorf tubes (~100 μL/tube). Analyze by either Coomassie staining of an SDS-PAGE gel, or by Western blotting with anti-Shh antibody.

3.7 In Vitro Shh Peptide Fatty Acylation Assay

1. In a 100 μL Eppendorf tube, add 10 μg of P100 membranes prepared from 293FT cells expressing Hhat, 100 μM Shh peptide, 167 μM [^{125}I]IC16 CoA, and 30 μL reaction buffer.
2. Incubate for 1 h at room temperature.

3. Add 400 μL RIPA buffer and 50 mL streptavidin-agarose beads. Incubate for 1 h at 4 °C with continuous mixing.
4. Centrifuge at 1000 × *g* for 5 min at 4 °C.
5. Wash the beads (pellet) three times with 500 μL RIPA buffer.
6. Count ^{125}I cpm in a γ counter.

3.8 In Vitro Shh Protein Fatty Acylation Assay with Purified Hhat

1. Combine the following in a final volume of 50 μL: 10 μL Hhat-HA-Flag-His in 20 mM HEPES, pH 7.3, 350 mM NaCl, 1% octylglucoside and 1% glycerol, 10 μL recombinant Shh (dilute purified prep to 0.2–0.4 mg/mL in 20 mM MES, pH 6.5, 1 mM EDTA, 1 mM DTT) and 30 μL of reaction buffer containing 167 μM [^{125}I]IC16 CoA (Subheading 2.6, **item 1**).
2. Incubate for 30 min at room temperature.
3. Stop with 50 μL 2× SDS-PAGE sample buffer.
4. Run samples on 12.5% SDS-PAGE; dry gel; expose to phosphorimager. Quantify ^{125}I cpm incorporation into the 19 kDa Shh protein band.

3.9 In Vitro Wnt Peptide Fatty Acylation Assay [15]

1. In a 100 μL Eppendorf tube, add 10 μg of P100 membranes prepared from 293FT cells expressing Porcupine, 100 μM Wnt peptide, 167 μM [^{125}I]IC15:1 CoA, and 30 μL reaction buffer.
2. Incubate for 1 h at 37 °C.
3. Add 400 μL RIPA buffer and 50 mL streptavidin-agarose beads and incubate for 1 h at 4 °C with continuous mixing.
4. Follow **steps 4–6** as described for Subheading 3.7 above.

4 Notes

1. Any transfection reagent suitable for protein expression in mammalian cells can be used.
2. Synthesis and purification of a double disulfide bonded peptide is technically challenging, time-consuming, and expensive. We used AnaSpec (Fremont, CA) to synthesize C-terminal biotinylated (Biotin–PEG NovaTag™), disulfide-bonded Wnt3a WT (MHL KC(S-)KC(S-)HG LSG SC(S-)E VKT C(S-)WW) and C-terminal biotinylated (Biotin–PEG NovaTag™) disulfide-bonded Wnt3a S209A (MHL KC(S-)K C(S-)HG LAG SC(S-)E VKT C(S-)WW) and then analyzed by HPLC to assess purity to be greater than 95%. The handling of the peptides is a critical aspect of the procedure as the peptides must be used immediately after reconstitution for optimal results. Additionally, because of the limited quantity that can be synthesized at one time, great care must be taken to use the

peptides as efficiently as possible as synthesis can take up to 2 months.

3. The iodination procedure should be conducted in a fume hood designated for iodinations, and all manipulations should be behind lead-embedded shielding. Monitor the surrounding areas at all times using a Geiger counter. The temperature of the oil should be checked at least one additional time prior to overnight incubation. Personnel should have thyroid scans prior to and after the iodination reaction.
4. When lysing cells by Dounce homogenization, check for cell lysis under the microscope.
5. To check the purity of purified Hhat by silver-stained gel, the final Flag eluate fraction should be concentrated to a final volume of ~100–200 μL.
6. We use Grace™ Alltech™ Maxi-Clean™ Solid-Phase Extraction Cartridges, C18 (5122340). Push solutions through the column slowly and remove column when removing the plunger to add solutions as to avoid damaging the column.
7. The concentration of Palmitoyl-CoA can be determined using an extinction coefficient of 15.4×10^3 mM^{-1} cm^{-1} at 260 nm.
8. Prepare plates at least 1 day before you plan to transform.
9. To avoid over growth, split the 80 μL and plate 40 μL on two plates.
10. The suspension should become less viscous after sonication. If there is no change in viscosity, use a stronger or longer pulse. Be careful not to oversonicate the suspension, as this may lead to denaturation of the protein.
11. The Ni beads should remain suspended at all times during the incubation.
12. One unit is defined as the amount of enterokinase required to cleave ~25 μg of the fusion protein at 25 °C for 16 h.
13. Affinity based removal of the residual enterokinase from the cleavage reaction mixture can be done at this stage. We use the enterokinase removal kit from Sigma and follow their protocol.
14. The purpose of this step is to bring the protein to a concentration suitable for the in vitro reaction, and to lower the salt concentration to ~80 mM–125 mM.
15. This step depends on the type of concentrator that you will use. See the directions for your concentrator of choice.

References

1. Briscoe J, Therond PP (2013) The mechanisms of hedgehog signalling and its roles in development and disease. Nat Rev Mol Cell Biol 14 (7):418–431. https://doi.org/10.1038/nrm3598
2. Nusse R, Clevers H (2017) Wnt/beta-catenin signaling, disease, and emerging therapeutic modalities. Cell 169(6):985–999
3. Buglino JA, Resh MD (2008) Hhat is a palmitoyl acyltransferase with specificity for N-palmitoylation of sonic hedgehog. J Biol Chem 283:22076–22088
4. Pepinsky RB, Zeng C, Wen D, Rayhorn P, Baker DP, Williams KP, Bixler SA, Ambrose CM, Garber EA, Miatkowski K, Taylor FR, Wang EA, Galdes A (1998) Identification of a palmitic acid-modified form of human sonic hedgehog. J Biol Chem 273 (22):14037–14045
5. Chen MH, Li YJ, Kawakami T, Xu SM, Chuang PT (2004) Palmitoylation is required for the production of a soluble multimeric hedgehog protein complex and long-range signaling in vertebrates. Genes Dev 18 (6):641–659
6. Takada R, Satomi Y, Kurata T, Ueno N, Norioka S, Kondoh H, Takao T, Takada S (2006) Monounsaturated fatty acid modification of Wnt protein: its role in Wnt secretion. Dev Cell 11(6):791–801
7. Galli LM, Barnes TL, Secrest SS, Kadowaki T, Burrus LW (2007) Porcupine-mediated lipid-modification regulates the activity and distribution of Wnt proteins in the chick neural tube. Development 134(18):3339–3348
8. Biechele S, Cox BJ, Rossant J (2011) Porcupine homolog is required for canonical Wnt signaling and gastrulation in mouse embryos. Dev Biol 355(2):275–285
9. Shindou H, Eto M, Morimoto R, Shimizu T (2009) Identification of membrane O-acyltransferase family motifs. Biochem Biophys Res Commun 383(3):320–325
10. Resh MD (2016) Fatty acylation of proteins: the long and the short of it. Prog Lipid Res 63:120–131
11. Miranda M, Galli LM, Enriquez M, Szabo LA, Gao X, Hannoush RN, Burrus LW (2014) Identification of the WNT1 residues required for palmitoylation by porcupine. FEBS Lett 588(24):4815–4824
12. Gao X, Hannoush RN (2014) Single-cell imaging of Wnt palmitoylation by the acyltransferase porcupine. Nat Chem Biol 10(1):61–68
13. Rios-Esteves J, Resh MD (2013) Stearoyl CoA desaturase is required to produce active, lipid-modified Wnt proteins. Cell Rep 4 (6):1072–1081
14. Petrova E, Rios-Esteves J, Ouerfelli O, Glickman JF, Resh MD (2013) Inhibitors of hedgehog acyltransferase block sonic hedgehog signaling. Nat Chem Biol 9:247–249
15. Asciolla JJ, Miele MM, Hendrickson RC, Resh MD (2017) An in vitro fatty acylation assay reveals a mechanism for Wnt recognition by the acyltransferase porcupine. J Biol Chem 292(33):13507–13513
16. Janda CY, Waghray D, Levin AM, Thomas C, Garcia KC (2012) Structural basis of Wnt recognition by frizzled. Science 337(6090):59–64
17. Berthiaume L, Peseckis SM, Resh MD (1995) Synthesis and use of iodo-fatty acid analogs. Methods Enzymol 250:454–466

Part VI

Prenylation and Post-Prenylation Processing

Chapter 20

Production of Farnesylated and Methylated Proteins in an Engineered Insect Cell System

William Gillette, Peter Frank, Shelley Perkins, Matthew Drew, Carissa Grose, and Dominic Esposito

Abstract

Protein prenylation is a common posttranslational modification that enhances the ability of proteins to interact with membrane components within the cell. In many cases, these prenylated proteins are involved in important human diseases, including aging-related disorders and cancer. To effectively study these proteins or develop therapeutics, large quantities of properly modified proteins are required. Historically, production of fully modified farnesylated and methylated proteins at high yield has been challenging. Recently, an engineered insect cell system which is capable of producing authentically modified KRAS protein was used to generate material for structural studies and assay development. Here we describe protocols for extending this work to other farnesylated and methylated substrates.

Key words Prenylation, Baculovirus, Insect cells, Protein expression, Farnesylation, Carboxylmethylation, Ras proteins, Farnesyl transferase, Protein engineering

1 Introduction

More than 100 human proteins have been shown to be farnesylated during posttranslational modification, including a number of proteins relevant to human disease [1]. The processing required for proper modification of farnesylated proteins include three main steps: first, the nascent polypeptide is recognized by farnesyltransferase (FNT) through a C-terminal CaaX motif (a = any aliphatic amino acid, X = usually M, S, A, or E) [2]. The cysteine residue of the CaaX motif is then farnesylated using farnesyl pyrophosphate as a donor. The farnesylated protein is then localized to the outer face of the endoplasmic reticulum where it is acted upon by a metalloprotease which removes the final three amino acids after the farnesylcysteine. There are multiple proteases in this family of proteins which seem to have different specificities for target proteins. In the case of the RAS superfamily, the protein most commonly involved is RCE1 [3]. The modified protein is then a substrate for the final

Maurine E. Linder (ed.), *Protein Lipidation: Methods and Protocols*, Methods in Molecular Biology, vol. 2009,
https://doi.org/10.1007/978-1-4939-9532-5_20,

protein in the processing pathway, isoprenylcysteine methyltransferase (ICMT) that converts the C-terminal hydroxyl of the farnesylcysteine residue to a methyl ester to create the final farnesylated and methylated protein [4].

The proper processing of farnesylated proteins is often essential for ensuring their correct subcellular localization and function, in many cases through protein–protein interactions at the relevant membrane location. Among these proteins are a number of small GTPases of the RAS superfamily, which are responsible for a wide variety of human cancers [5]. Many early studies on RAS proteins were carried out using soluble mutant proteins which did not properly localize to the plasma membrane, thus providing an incomplete understanding of the authentic activity of the protein in its proper cellular compartment. While many efforts were made to produce properly farnesylated and methylated RAS proteins, low yields and incomplete processing (either mixed with unprocessed protein or with species that failed to carboxymethylated) inhibited production of proteins in relevant amounts for drug screening and structural biology [6]. Attempts were subsequently made to generate analogs of farnesyl groups by protein ligation or chemical synthesis, but in all cases, the proteins produced failed to accurately recapitulate the authentic in vivo protein [7, 8].

In general, the major deficiency in insect cell systems used to initially generate small amounts of processed RAS proteins may have been the inability of insect farnesyltransferases (FNT) to properly process the amount of protein produced in a baculovirus-driven overexpression experiment. Alternatively, insect FNTs may have a lower affinity or activity toward human protein substrates, or an altered specificity of the CAAX prenylation signal. To overcome both of these limitations, we developed a system in which human farnesyltransferase was overexpressed in insect cells to permit high-level processing of farnesylated proteins [9]. Using a recombineering approach, we generated a bacmid DNA called DE35 (Fig. 1) which coexpresses human FNTA using the baculovirus p10 promoter, and FNTB using the baculovirus polyhedrin promoter. These genes were placed into the baculovirus genome at the site of a deletion of the viral chitinase and cathepsin proteins, removal of which leads to the production of higher quality proteins [10]. This DE35 bacmid strain can be used for standard baculovirus production using the Bac-to-Bac system [11] or newer versions of a similar Tn7-based transposition system [12]. Using this system, we have been able to generate more than 5 mg of purified, farnesylated and methylated native KRAS4b protein per liter of insect cell culture, and have utilized this protein to generate a high-resolution crystal structure of this protein in its fully processed form for the first time [13].

Here we describe the details of the process for production of human KRAS4b-FMe (farnesylated and methylated), but the same

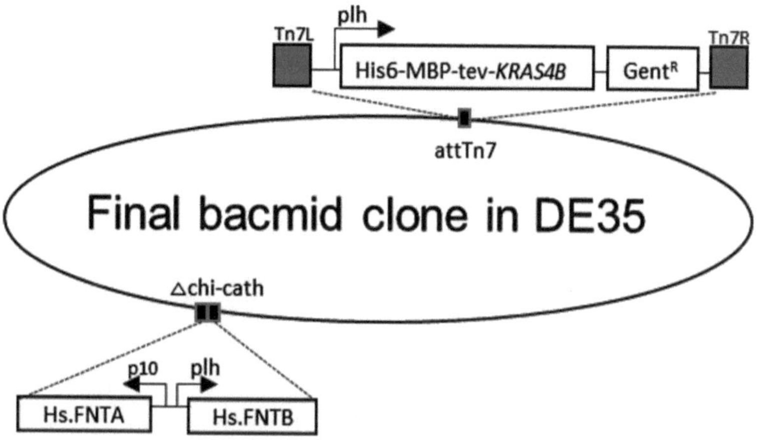

Fig. 1 Map of DE35 bacmid with a KRAS4b insert. Shown is a schematic map of a DE35 bacmid, identifying the insertion of the farnesyltransferase cassette and KRAS4b coding sequence. The dual cassette with baculovirus p10 promoter driving human FNTA and polyhedrin promoter driving human FNTB is inserted in place of the viral chitinase and cathepsin genes. The His6-MBP fusion to KRAS4b is driven from a polyhedrin promoter inserted into the attTn7 locus in the DE35 bacmid using standard Bac-to-Bac style technology. Details on the construction of DE35 can be found in [9]

process can be used for production of any farnesylated protein with the assumption that slight additional modifications to the purification may be necessary for different proteins. The system utilizes (1) a donor plasmid encoding a fusion protein consisting of an N-terminal hexahistidine sequence, enabling purification by immobilized metal affinity chromatography (IMAC), followed by maltose-binding protein (MBP) to increase expression and protein solubility [14], a Tobacco Etch Virus (TEV) protease site to permit cleavage of the solubility tag during the purification scheme, and the protein of interest; (2) the modified DE35 bacmid strain for FNT coexpression (Fig. 1), and (3) the Tni-FNL cell line, a modified *Trichoplusia ni* cell line which can produce higher levels of processed protein than other cell lines tested.

The purification scheme (Fig. 2) employs sequential chromatography over IMAC and cation exchange (CEX) resins, followed by cleavage of the tev sequence with His6-tagged TEV protease. A final IMAC chromatography step resolves purified KRAS4b-FMe from the hexahistidine-tagged protease and hexahistidine-tagged MBP.

2 Materials

2.1 Generation of Bacmids

1. Donor plasmid (e.g., pFastBac) encoding the His6-MBP-tev-protein of interest (*see* **Note 1**).
2. *E. coli* DE35 competent cells (Kerafast). Store at −80 °C; do not reuse open vials.

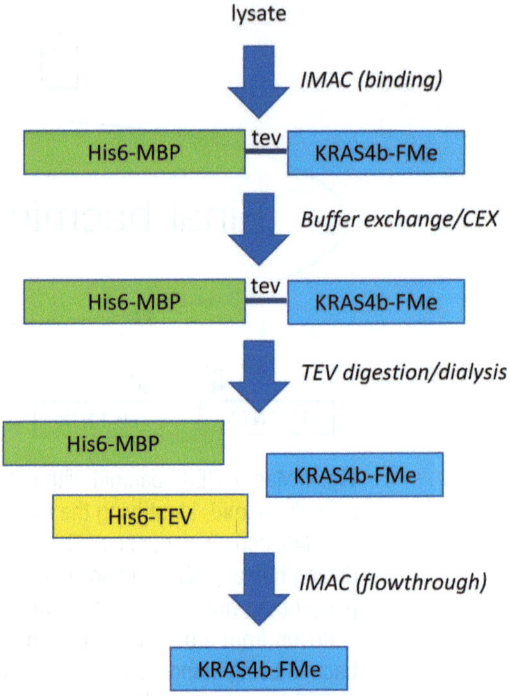

Fig. 2 Chromatographic scheme used to purify KRAS4b. The target protein is captured from the clarified lysate utilizing the N-terminal hexahistidine (His6) tag on the maltose binding protein (MBP)-KRAS4b fusion protein via Immobilized Metal Affinity Chromatography (IMAC). This step is followed by dialysis to remove imidazole, followed by cation exchange (CEX) chromatography. Pooled fractions are digested at the tev (tobacco etch virus) site using His6-TEV protease to release untagged KRAS4b-FMe and buffer exchanged by dialysis. A second IMAC step serves to separate His6-MBP and His6-TEV from KRAS4b-FMe

3. Lysogeny broth (LB): 10 g/L tryptone, 5 g/L yeast extract, 5 g/L sodium chloride.
4. LB-agar petri plates: LB, 1.5% agar.
5. LB-KGTXI agar plates: LB-agar petri plates with 7 μg/mL gentamycin, 50 μg/mL kanamycin, 10 μg/mL tetracycline, 40 μg/mL IPTG, and 100 μg/mL Bluo-gal (Teknova).
6. LB-KG medium: LB, 50 μg/mL kanamycin, 7 μg/mL gentamycin; autoclave for 20 min, then add antibiotics.
7. Plasmid Prep Buffer A: 25 mM Tris–HCl, pH 8.0, 10 mM EDTA, 0.9% (w/v) D-glucose. Store at room temperature.
8. Plasmid Prep Buffer B: 0.20 M NaOH, 1.33% (w/v) sodium dodecyl sulfate (SDS). Store at room temperature. SDS may precipitate at low temperatures—redissolve before use by heating at 37 °C. Make fresh after 6 months and be sure to cap immediately after use.

9. Plasmid Prep Buffer C: 7.5 M ammonium acetate. Store at 4 °C.
10. Bacmid Resuspension Buffer: 10 mM Tris–HCl, pH 8.0, 0.1 mM EDTA, 0.1 μg/mL ribonuclease A (Sigma). Store at room temperature for up to 6 months.

2.2 Baculovirus Production

1. Sf-9 insect cell line (ThermoFisher Scientific).
2. Sf-9 culture medium: Sf-900 III Serum-Free Media (ThermoFisher Scientific).
3. Fetal bovine serum (FBS).
4. Cellfectin II (ThermoFisher Scientific).
5. 250 mL Optimum Growth Flasks (Thomson Instrument Company).
6. Innova-44 incubating, refrigerated shaker with 2″ orbit (Eppendorf).
7. CedEx HiRes (Roche Diagnostics)—automated image-based cell analyzer.
8. ViroCyt 3100 Virus Counter (Sartorius AG).
9. ViroCyt Combo Dye Kit (Sartorius AG).
10. 2 mL sterile tubes.

2.3 Insect Cell Expression

1. Tni.FNL cells (Kerafast or licensed through the NIH Office of Technology Transfer) (*see* **Note 2**).
2. Sf-900 III Serum-Free Media (ThermoFisher Scientific).

2.4 Protein Purification

1. Buffer A: 20 mM HEPES, pH 7.3, 300 mM NaCl, 5 mM $MgCl_2$, 1 mM TCEP.
2. Buffer B: 20 mM HEPES, pH 7.3, 300 mM NaCl, 35 mM imidazole, 5 mM $MgCl_2$, 1 mM TCEP.
3. Buffer C: 20 mM HEPES, pH 7.3, 300 mM NaCl, 500 mM imidazole, 5 mM $MgCl_2$, 1 mM TCEP.
4. Buffer D: 20 mM MES, pH 6.0, 200 mM NaCl, 5 mM $MgCl_2$, 1 mM TCEP.
5. Buffer E: 20 mM MES, pH 6.0, 5 mM $MgCl_2$, 1 mM TCEP.
6. Buffer F: 20 mM MES, pH 6.0, 100 mM NaCl, 5 mM $MgCl_2$, 1 mM TCEP.
7. Buffer G: 20 mM MES, pH 6.0, 1 M NaCl, 5 mM $MgCl_2$, 1 mM TCEP.
8. Protease Inhibitor Cocktail without EDTA or other chelators (e.g., Sigma P8849).
9. IMAC column (GE HisPrep FF 16/10).

10. CEX column (20 mL HiPrep SP Sepharose High Performance).
11. Whatman GD/XP PES 0.45 μm syringe filter.
12. 2 M imidazole, pH 7.5.
13. Ultracentrifuge capable of 100,000 × *g*.
14. High-speed/benchtop centrifuge up to 4000 × *g*.
15. Spectrophotometer capable of reading at 280 nm.
16. Tobacco Etch Virus (TEV) protease (purchased or purified in the lab using protocols in [15]).

3 Methods

3.1 Generation of Bacmids

1. Transform the donor plasmid encoding the protein of interest into *E. coli* DE35 cells using the following protocol: mix 5 ng of donor plasmid DNA and 50 μL of competent DE35 cells (*see* **Note 3**) in a Falcon 2059 tube, and incubate on ice for 5 min. Heat-shock samples at 42 °C for 45 s and add 450 μL LB to the tube. Incubate tubes at 37 °C for 4 h in a shaking incubator at 200 rpm.
2. Plate 5 μL of the culture on LB-KGTXI plates with appropriate antibiotics (*see* **Note 4**) and incubate the plates overnight at 37 °C. Restreaking of positive clones may be required on the following day (*see* **Note 5**).
3. Pick a white colony and inoculate 3 mL LB-KG medium; grow overnight at 37 °C in a shaking incubator at 200 rpm. These cultures can be used to prepare bacmid DNA using a standard alkaline lysis plasmid preparation procedure as follows (*see* **Note 6**).
4. Centrifuge 2 mL of culture in a microcentrifuge tube at maximum speed for 1 min, and aspirate the supernatant.
5. Add 250 μL Plasmid Prep buffer A and mix by pipetting.
6. Add 250 μL Plasmid Prep buffer B and mix by gentle inversion.
7. Add 250 μL Plasmid Prep buffer C and mix by gentle inversion. A white flocculent precipitate will form.
8. Centrifuge for 10 min at maximum speed and transfer 700 μL sample to a new microcentrifuge tube.
9. Centrifuge for an additional 10 min at maximum speed and transfer 600 μL to a new 2- mL microcentrifuge tube.
10. Add 1.2 mL of 100% ethanol and mix by inversion.
11. Centrifuge for 20 min at maximum speed, decant supernatant, centrifuge briefly, and pipet out the remaining liquid.

12. Dry the tubes containing the pellets at 37 °C for 5 min with the caps open to evaporate any remaining ethanol (*see* **Note 7**).
13. Resuspend in 100 μL Bacmid Resuspension buffer and incubate for 1 h at 37 °C in benchtop heat block with shaking at 1100 × *g* (*see* **Note 8**).
14. Centrifuge at maximum speed for 2 min and transfer the supernatant (~95 μL) to a new tube, avoiding transfer of any solid pellet or particulates.
15. If desired, PCR analysis can be performed to verify the bacmid DNA integrity (*see* **Note 9**).

3.2 Baculovirus Generation and Amplification

The preferred insect host for the creation of high-titer baculovirus stock is the Sf-9 cell line, which can be grown in either adherent or suspension culture in serum-free media. Below we outline (1) a method for transfecting Sf-9 cells in suspension culture that will produce ~100 mL of high-titer baculovirus stock from an initial transfection [16] and (2) a large-scale viral amplification (*see* **Note 10**).

1. Culture Sf-9 cells at 1.5×10^6 viable cells/mL in 100 mL of Sf-900 III + 1% (v/v) FBS Final in a 250-mL Optimum Growth Flask. Set culture to shake at 27 °C, 125 rpm.
2. Add 0.5 mL of Sf-900 III to 2 × 2 mL sterile tubes.
3. Label one sterile tube "B" for bacmid and one "C" for Cellfectin II (*see* **Note 11**).
4. Add 0.05 mL of bacmid to "B" tube and 0.250 mL Cellfectin II to tube labeled "C." Incubate for 5 min at room temperature.
5. Add contents of "B" tube dropwise to "C" tube and invert 5 times to mix. Let stand for 20 min at room temperature.
6. Add bacmid/Cellfectin complex dropwise to 100 mL Sf-9 culture.
7. Set culture to shake for 5 days at 27 °C, 125 rpm.
8. Remove a 0.3 mL aliquot and measure cell count, viability, and cell size.
9. Collect the Sf-9 cells by centrifugation at 1100 × *g* for 15 min at 4 °C.
10. Decant supernatant into a sterile 125 mL bottle and store at 4 °C.
11. To titer the virus stock using ViroCyt, dilute the virus 1:500 in ViroCyt Sample Dilution Buffer into a total volume of 0.25 mL.
12. Aliquot 0.125 mL of diluted virus into two ViroCyt sample vials.

13. Add 0.06 mL of ViroCyt Combo Dye (prepared according to ViroCyt Instructions) to each sample vial.
14. Cap vials and vortex to mix. Incubate at room temperature in the dark for 30 min.
15. Determine total viral titer according to ViroCyt 3100 protocol. Each vial will be enough sample for four readings.
16. Average all readings per sample and divide value by 15 to determine infectious titer (*see* **Note 12**).
17. Add FBS, final concentration of 3% (v/v), to the harvested supernatant to preserve an aliquot of the virus for long-term storage at 4 °C.
18. The remaining virus stock can be used for expression (Subheading 3.3) and/or amplification as described next in **steps 19–28** below (*see* **Note 13**).
19. For large-scale virus amplification, set up 900 mL of Sf-9 cells at 1.5×10^6 viable cells per mL in a 1.6 L Optimum Growth Flask.
20. Set flask to shake at 125 rpm at 27 °C for 24 h.
21. After 24 h, remove a 0.3 mL sample to determine cell count, cell viability and cell size.
22. Use the calculation below to determine the amount of titered baculovirus stock (from Subheading 3.2, **step 16**) to add to infect the culture at a multiple of infection (MOI) of 0.1.

    ((viable cells per mL) × (Volume) × 0.1)/baculovirus titer
23. After adding the required amount of baculovirus stock, set the culture to shake at 27 °C at 125 rpm for 72 h.
24. After 72 h, remove 0.3 mL sample to determine cell count, cell viability, and cell size (*see* **Note 14**).
25. Divide the culture evenly between two 500-mL centrifuge bottles and spin at 1100 × *g* for 10 min at 4 °C.
26. Collect supernatant in a sterile 1 L container and carry out titer determination.
27. Once baculovirus stock titer has been determined, add FBS to a final concentration of 3% (v/v) as a preservative.
28. Store virus in the dark at 4 °C for longer shelf life (*see* **Note 15**).

3.3 Large Scale Culture for Protein Purification

1. Set 900 mL of Tni-FNL cells at 8×10^5 viable cells/mL in a 1.6-L Optimum Growth Flask (for larger expressions set multiple flasks at 900 mL each).
2. Incubate flasks at 27 °C, shaking at 125 rpm for 24 h.
3. After 24 h remove a 0.3 mL aliquot from each flask and determine cell count, cell viability and cell size.

4. Determine amount of baculovirus stock to add using the following calculation: (viable cells per mL × volume × 3)/baculovirus titer.
5. Add calculated amount of virus to each flask and shake at 21 °C at 125 rpm for 72 h (*see* **Note 16**).
6. After 72 h, remove 0.3 mL aliquot from each flask to determine cell count, cell viability and cell size (*see* **Note 14**).
7. Harvest cells by centrifugation at 1100 × *g* for 15 min at 4 °C in 500-mL conical centrifuge bottles.
8. If the sample is not to be processed immediately, freeze the pellet on dry ice and store at −80 °C.

3.4 Protein Purification

1. For purification from 1.8 L of insect expression culture, resuspend cells in 200 mL Buffer A with added 1:200 v/v protease inhibitor per liter of insect cell culture in a beaker with a stir bar at room temperature (*see* **Note 17**). Stir sample until it is fully in suspension and homogeneous.
2. Lyse cell pellet in a M110-EH microfluidizer (Microfluidics) at 10,000 psi for 2 passes. Flush the system with 20–30 mL of buffer A to get all material out of the lines (*see* **Note 18**).
3. Clarify lysed cells by centrifugation at 100,000 × *g* in an ultracentrifuge for 30 min. Decant supernatant and filter through a 0.45 μM PES filter (*see* **Note 19**).
4. Determine volume of the clarified lysate and adjust the sample to 35 mM imidazole by the addition of 2 M imidazole stock.
5. Load the sample at 3 mL/min onto a 20 mL IMAC (*see* **Note 20**) column (10 mL of resin per liter of culture) which has been preequilibrated in Buffer B for 3 column volumes (CVs). Collect column flow-through in batch. This and all subsequent chromatography steps should be performed at room temperature.
6. Wash the column with Buffer B at 3 mL/min for 20 mL (1 CV) then increase the flow rate to 4 mL/min for the remainder of the wash step until the Abs_{280} reaches a baseline (preferably lower than 100 mAU). This usually takes 3–4 CVs of Buffer B.
7. Elute the protein with a 400 mL (20 CV) gradient from Buffer B to Buffer C while collecting 8 mL fractions (*see* **Note 21**).
8. Analyze protein fractions using SDS-PAGE. Peak fractions should then be pooled and dialyzed overnight against 2 L of Buffer D at 4 °C (*see* **Note 22**).
9. Remove the sample from dialysis and centrifuge at 4000 × *g* for 10 min to remove any precipitate (*see* **Note 23**).
10. Load the clarified supernatant onto a 20 mL HiPrep SP Sepharose High Performance column (GE Healthcare, *see*

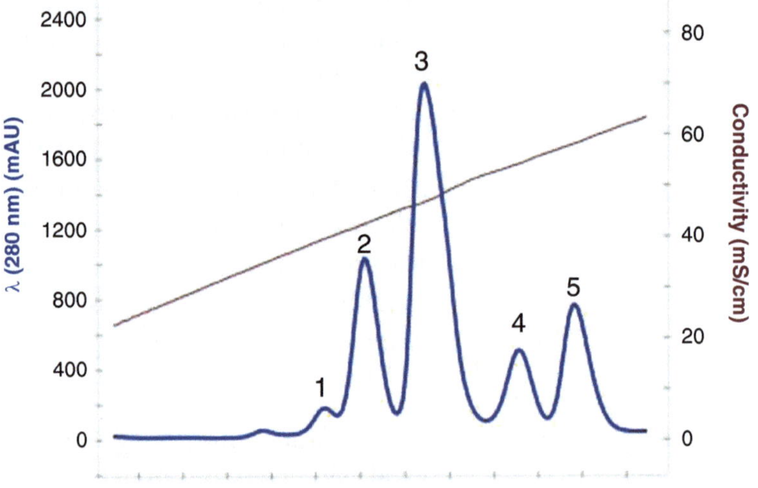

Fig. 3 CEX chromatography elution profile. The typical CEX elution profile has four to five major peaks. Aside from peak 1 which typically is a proteolyzed form of unprocessed KRAS lacking the four C-terminal amino acids, the four peaks numbered 2 through 5 in the panel result from the variability in the processing of the N and C-termini of the protein, with progressively more positively charged molecules eluting later in the gradient (*see* **Notes 27–29**)

Note 24). Immediately prior to loading, dilute the sample to 100 mM NaCl with Buffer E (*see* **Note 25**).

11. Wash the column with Buffer F for ~3 CV until a baseline absorbance is reached.
12. Elute the protein with a 400 mL (20 CV) gradient from Buffer F to 65% Buffer G. Collect 6 mL fractions throughout the elution gradient (*see* **Note 26**). When the gradient is finished, wash with 100% Buffer G for 1.5 CV.
13. Analyze the fractions by SDS-PAGE. Pool fractions based on the gel analysis and the cation exchange chromatogram (*see* **Notes 27–29** and Fig. 3).
14. Add ~200 μg of His6-tagged TEV protease (*see* **Note 30**) per mL of pool.
15. Dialyze pooled sample against >40 mL Buffer A per mL eluent pool (*see* **Note 31**).
16. After digestion and dialysis, load the protein directly to a 20 mL IMAC (*see* **Note 22**) column. The loading and first CV of wash should be done at 2 mL/min.
17. Following the column load, wash the column at 4 mL/min with Buffer A for a total of 3 CVs or until a baseline absorbance is reached. Collect 7 mL fractions during this process.
18. Elute the target protein with a gradient from 0% to 10% Buffer C for 5 CVs, continuing to collect 7 mL fractions throughout.

19. Elute remaining proteins with 100% Buffer C, again collecting 7 mL fractions, and analyze all fractions by SDS-PAGE (*see* **Note 32**).
20. Pool all fractions containing target protein in the FT or column wash.
21. Protein concentration should be determined by spectrophotometry at 280 nm (*see* **Note 33**).
22. Validate the proper posttranslational processing of the purified proteins using electrospray ionization mass spectrometry (ESI-MS). This method allows for accurate determination of the mass of the protein, which should be modified by the addition of 208 Da for the farnesyl group, and subtraction of the molecular weight of the three residues removed from the C-terminus during processing. If the protein is carboxymethylated, an addition of another 14 kDa for the methyl group will also be observed. ESI-MS is the best way to accurately quantitate the percentage of properly processed protein. In the case of KRAS, we routinely see >90% FMe protein in these final fractions. If further verification is required, MS/MS analysis can be used to identify the actual modified peptide as was done for KRAS [9].
23. Analyze samples from each of the purification steps by SDS-PAGE and visualize using Coomassie Blue staining (Fig. 4).

4 Notes

1. To effectively utilize the Bac-to-Bac-based DE35 strain, the expression vector must contain the left and right Tn7 transposon binding sequences as found in the pFastBac series of vectors from ThermoFisher, or analogous vectors. More recent modifications of this system such as the Bac-2-the-Future system [12] can also be used.
2. The Tni-FNL cell line is a derivative of *Trichoplusia ni* BTI-Tn-5B1-4 and is the preferred cell line for high-level protein expression. It is available for nonprofit organizations from Kerafast (www.kerafast.com) or can be licensed through the NIH Office of Technology Transfer. This cell line comes suspension- and serum-free growth adapted and is ready to use. It can consistently produce 5–10 times the amount of protein produced in Sf-9 cells and reproducibly better than the commercially available High Five™ cell line from ThermoFisher.
3. To generate competent DE35 cells, follow the procedures in [17]. This process regularly generates *E. coli* K-12 strains (such as DE35) at a transformation efficiency of at least 1×10^8 cfu/

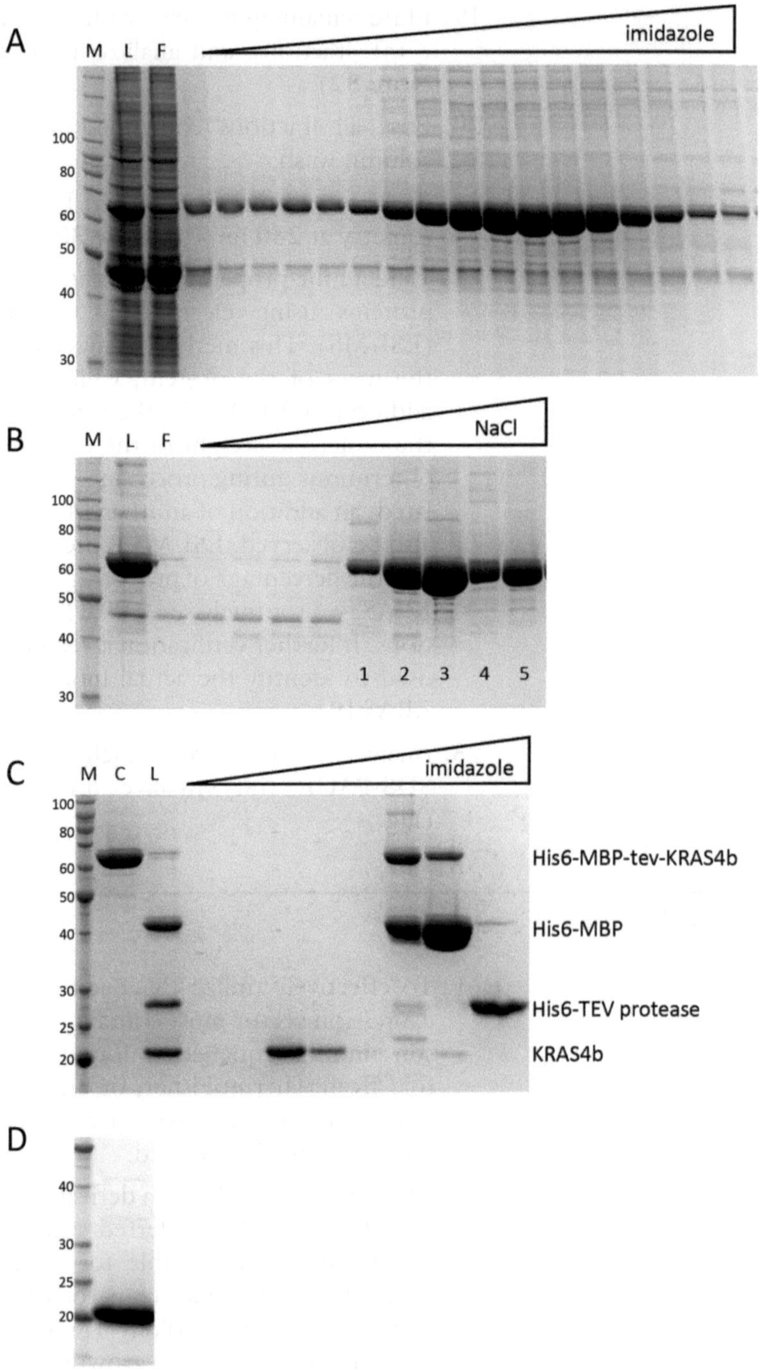

Fig. 4 SDS-PAGE gels of the purification process. (**a**) IMAC capture from lysate. FNTA and FNTB are the strong bands migrating at ~48 kDa, and the His6-MBP-tev-KRAS4b is the strong band migrating at ~67 kDa. M, protein molecular weight ladder; L, clarified lysate/column load; F, column flow-through. Fractions are eluted with an increasing imidazole concentration gradient. (**b**) Cation exchange (CEX). Fractions labeled 1–5 correspond to the peak fractions from

μg. This level of competence is more than sufficient for bacmid production under normal circumstances. If higher levels are required, for instance for expression from libraries of constructs, we suggest using electroporation instead of chemical transformation.

4. Appropriate antibiotics for plating the bacmid clones will depend on the system used. For Bac-to-Bac expression clones, plates should contain 7 μg/mL gentamycin, 50 μg/mL kanamycin, 10 μg/mL tetracycline, 40 μg/mL IPTG, and 100 μg/mL Bluo-gal.
5. In a standard Bac-to-Bac transformation, two types of colonies will be observed—blue colonies are nonrecombinant, while white colonies contain the proper recombinant baculovirus. Some white colonies can contain small amounts of nonrecombinant bacmid DNA that can affect downstream success. To ensure minimal background, white colonies should be restreaked onto fresh plates and grown for an additional day at 37 °C. Colonies which produce streaks with any blue color should be discarded. Blue colonies are smaller in size—for this reason, the largest white colonies should be picked where possible. To avoid the issue of contaminating nonrecombinant DNA, the Bac-2-the-Future system can be used, which eliminates the need for additional screening of positive colonies.
6. DE35 cells harbor a 130-kB single copy plasmid, known as a bacmid, which contains the entire baculovirus genome. This bacmid DNA cannot be readily purified by many commercially available kits due to its size and low copy number. However, a classical alkaline lysis preparation will produce enough material for subsequent insect cell transfections and is both simple and inexpensive.
7. Do not over dry the DNA at this step—excess drying will make resuspension of the bacmid DNA very challenging and reduce yields.
8. We find this additional incubation critical to getting high yields of bacmid DNA. Shortening the time of the resuspension will significantly affect yield as some bacmid DNA will not be recovered from the tube walls.
9. Analysis of bacmid DNA on agarose gels is difficult due to the size of the bacmid, interference from residual plasmid DNA, and low DNA concentration. We prefer PCR verification of

Fig. 4 (continued) the elution peaks shown in the chromatogram of Fig. 3. (**c**) TEV digestion and second IMAC. C—pool from CEX step, pre-TEV cleavage; L—load sample post-TEV cleavage. Various species of interest are labeled. (**d**) Final KRAS4b-FMe protein

bacmids using the suggestions in the manufacturer's Bac-to-Bac kit or as found in [12].

10. There are numerous methods for determining the titer of a baculovirus stock. We recommend using the ViroCyt System for its ease of use and shortened assay time but other more traditional cell-based systems such as the Sf-9 Easy Titer cells [18] provide the same results. Sf-9 Easy Titer cells are available from Kerafast (www.kerafast.com).
11. Other transfection reagents such as linear polyethylenimine (PEI) (Polysciences, Inc) or Insect GeneJuice (EMD Millipore) can be used for the initial transfection, but the ratio of transfection reagent to bacmid DNA and optimal harvest time will need to be optimized. For PEI we have found the optimal ratio is 0.720 mL PEI to 0.070 mL bacmid and 7 days of growth to give optimal virus titers. For Insect GeneJuice we have found that the same ratio and harvest time as Cellfectin II works well.
12. The correction factor (divide ViroCyt titer by 15) was determined empirically. The ViroCyt measures total viral particles whereas the Sf-9 Easy Titer (end-point dilution assay) is selective for infectious particles only. We compared multiple baculovirus stocks tittered in parallel with both methods and arrived at a standard adjustment factor of 15.
13. Amplification of baculovirus may be required if the desired scale of the expression is large. Given an average baculovirus titer of 1×10^8 pfu/mL, one can generally express more than 4 L of culture from the initial baculovirus stock, so frequently amplification is not necessary.
14. Cell counts should be between 3×10^6 and 8×10^6 viable cells per mL with cell viability between 75% and 85% at harvest. If parameters are outside of this range, it could mean that the quality of virus produced is not optimal, possibly leading to reduced protein expression.
15. It is best to use virus stocks that are less than 6–9 months old. After this time, virus stability can be reduced and protein expression levels dramatically lower. If more than 6–9 months have passed, produce a fresh virus stock starting from bacmid DNA.
16. Expressing the protein for 72 h at 21 °C will help to minimize any protein insolubility, misfolding, aggregation, or proteolysis. Using shorter expression times can potentially lead to a larger amount of farnesylated protein lacking the carboxyterminal methylation, as opposed to the desired fully processed FME protein.

17. Lysis volume can be scaled for any culture volume. Generally, cultures are resuspended in 10% of the total volume of the culture. Note also that the protease inhibitor must be EDTA-free to avoid chelating metal from the IMAC column and inhibiting protein binding.
18. We have had the best success using microfluidic lysis; however, many other lysis methods should be suitable for insect cells, including Emulsiflex-style pressure lysis or sonication. The manufacturer's standard conditions should be used for these alternative methods, but some optimization may be required.
19. After clarification, the sample often has a white lipid residue on the surface of the supernatant; therefore, multiple filters are often required as the lipids tend to block the filters quickly.
20. Specific choice of IMAC resins may vary—we have had good success with GE HisPrep FF 16/10 columns, but it is likely other IMAC resins would work as well. Flow rates and other process variables may need to be altered depending on the resin used.
21. For small GTPases, the target proteins usually elute from the IMAC column early in the gradient and it is often not necessary to complete the entire gradient. However, there is no guarantee that this early elution behavior will be similar for other farnesylated proteins. Accordingly, initial chromatography experiments should run the complete gradient to enable better understanding of protein-specific phenomena.
22. The pH change during this dialysis occurs very slowly, and it has been found that a shorter dialysis time is usually not sufficient to fully exchange this buffer. A small amount of precipitation is normal during this step; the lower the NaCl, the more precipitation there is. This is why Buffer D used for dialysis contains 200 mM NaCl, which is twice the concentration used for the subsequent CEX chromatography. Accordingly, the NaCl concentration must be reduced prior to loading the CEX column. It is unclear if this precipitation phenomenon is specific to small GTPases or may occur with other farnesylated proteins, but the majority of the precipitate that we observe is entirely processed protein. For this reason, we suggest using the higher NaCl concentration in the dialysis buffer for any protein of interest.
23. The final dialyzed sample at this step is still often hazy even after centrifugation, but can be applied as is onto the subsequent CEX column.
24. We have used SP Sepharose High Performance resin (GE Healthcare) for the cation exchange step and shown that it can resolve the differently charged species that arise from incomplete N-terminal and C-terminal processing (*see* **Note**

27 and Fig. 3). The choice of resin for the CEX may affect whether these species are resolved. In our initial optimization experiments, Source 15S (GE Healthcare) resin failed to produce the distinct multiple peaks seen in Fig. 3. Also, colleagues have reported that SP Sepharose Fast Flow (GE Healthcare) in their hands failed to produce the pattern seen in Fig. 3.

25. Due to the potential precipitation of the protein at lower NaCl concentration, the sample is diluted a small amount at a time. Generally, we mix 10 mL of sample with 10 mL of buffer E and immediately load that onto the column at 3 mL/min. When that portion is nearly finished loading, another 20 mL is prepared and added to the column inlet. This pattern is continued until the entire sample has been loaded.

26. The smaller the fraction size at this point, the easier it is to refine your peak separation when pooling fractions. This may permit finer resolution of fully processed and partially processed proteins.

27. In the case of small GTPases with the standard MBP vectors, there are usually five main peaks observed (Fig. 3). The first peak is generally unprocessed and/or proteolyzed protein; the second is a farnesylated intermediate (FARN); the third is fully processed FMe protein (*see* **Note 28**). These are followed by two more peaks that represent proteins that are not N-acetylated at the amino terminus (*see* **Note 29**): a second FARN peak (4) and an FMe peak (5). As peaks 2/4 and 3/5 will produce identical proteins after tag removal, these can be pooled at this stage if desired.

28. N-terminal processing is affected by the presence of a glycine residue immediately after the starting methionine in most commercially available MBP fusion vectors. In cells, this combination results in the removal of the initiator methionine nearly 100% of the time, and acetylation of the new N-terminal glycine residue at some percentage. The acetylation is often variable and leads to a combination of species which are either acetylated (no charge) or nonacetylated (positively charged), resulting in alterations in mobility in ion exchange chromatography. The presence of multiple peaks driven by the incomplete acetylation of the MBP N-terminus can also be addressed by modifications of the MBP protein sequence. As the N-terminus of this protein is not essential for folding or solubility enhancement, we have recently constructed vectors which replace the Met-Gly residues at the N-terminus with a Met-Pro. Proline at the +2 position abrogates removal of the initiator methionine and prohibits acetylation. In this case, we see collapse of peaks 2 and 4 into a single peak, and collapse of peaks 3 and 5 into a single peak, allowing for a more simplified

purification process. We have not yet clearly demonstrated whether these vectors produce the same yield of protein or how they work for other proteins, but this has the potential to be a useful mutation to aid purification of prenylated proteins. We have seen similar chromatographic results for other small GTPases, and predict that a similar pattern will likely arise for other farnesylated proteins. However, a detailed mass spectrometric analysis of individual peak fractions should be performed for any new purification process to identify which modifications are present on a particular protein of interest. It may be necessary to slightly modify the ion exchange process for a given protein to maximize separation of the desired species.

29. The final carboxymethylation step of processing seems to be incomplete in insect cells, leading to the presence of a small but variable amount of protein which has been farnesylated and proteolytically processed, but not methylated. This FARN-only species has lower hydrophobicity due to the lack of the methyl group and also carries an additional negative charge from the hydroxyl at the C-terminus, leading to potential mobility changes during ion exchange. In the case of small GTPases, there typically is an accumulation of some percentage of the FARN peak. Whether this pool accumulates and can be resolved by CEX for other farnesylated proteins is unknown. However, purification of the FARN protein can be useful in teasing apart the contributions of the farnesyl and methyl groups, as was done from KRAS4b [13].
30. In our hands, TEV protease is the best option for tag removal. TEV protease is available commercially, but can also easily be made in the laboratory using plasmids obtainable from Addgene (https://www.addgene.org/8827).
31. TEV protease digestion occurs during dialysis and requires at least 2 h at room temperature. For best results, the digestion/dialysis step is usually done overnight at 4 °C.
32. The elution pattern of the target protein in this second IMAC step can be variable and depends on the purification conditions and the protein of interest. Most of the time, the target protein released from His6-MBP binds to the column with low affinity and elutes in the 6% Buffer C wash at 30 mM imidazole. However, the target protein is occasionally present in the flow-through. For this reason, it is vital to collect fractions throughout the process. The high imidazole elution gradient should contain only cleaved His6-MBP, His6-tagged TEV protease, and any uncleaved fusion proteins.
33. For some proteins, Bradford or other protein assay quantitation will suffice for measuring protein concentration. For small

GTPases, the potential presence of bound nucleotide may hinder accurate quantitation with these methods, so we suggest calculating protein extinction coefficients (being sure to add the contribution of nucleotide) and use 280 nm absorbance to quantitate protein.

Acknowledgments

The authors thank Jennifer Mehalko, Vanessa Wall, Zhaojing Meng, and Oleg Chertov for assistance in the development of this technology. This work has been funded in whole or in part with federal funds from the National Cancer Institute, National Institutes of Health, under contract HHSN261200800001E. The content of this publication does not necessarily reflect the views or policies of the Department of Health and Human Services, nor does the mention of trade names, commercial products, or organizations imply an endorsement by the US government.

References

1. Novelli G, D'Apice MR (2012) Protein farnesylation and disease. J Inherit Metab Dis 35 (5):917–926. https://doi.org/10.1007/s10545-011-9445-y
2. Krzysiak AJ, Rawat DS, Scott SA, Pais JE, Handley M, Harrison ML, Fierke CA, Gibbs RA (2007) Combinatorial modulation of protein prenylation. ACS Chem Biol 2 (6):385–389. https://doi.org/10.1021/cb700062b
3. Manolaridis I, Kulkarni K, Dodd RB, Ogasawara S, Zhang Z, Bineva G, O'Reilly N, Hanrahan SJ, Thompson AJ, Cronin N, Iwata S, Barford D (2013) Mechanism of farnesylated CAAX protein processing by the intramembrane protease Rce1. Nature 504 (7479):301–305. https://doi.org/10.1038/nature12754
4. Yang J, Kulkarni K, Manolaridis I, Zhang Z, Dodd RB, Mas-Droux C, Barford D (2011) Mechanism of isoprenylcysteine carboxyl methylation from the crystal structure of the integral membrane methyltransferase ICMT. Mol Cell 44(6):997–1004. https://doi.org/10.1016/j.molcel.2011.10.020
5. Stephen AG, Esposito D, Bagni RK, McCormick F (2014) Dragging ras back in the ring. Cancer Cell 25(3):272–281. https://doi.org/10.1016/j.ccr.2014.02.017
6. Page MJ, Hall A, Rhodes S, Skinner RH, Murphy V, Sydenham M, Lowe PN (1989) Expression and characterization of the ha-ras p21 protein produced at high levels in the insect/baculovirus system. J Biol Chem 264 (32):19147–19154
7. Chavan TS, Meyer JO, Chisholm L, Dobosz-Bartoszek M, Gaponenko V (2014) A novel method for the production of fully modified K-Ras 4B. Methods Mol Biol 1120:19–32. https://doi.org/10.1007/978-1-62703-791-4_2
8. Lowe PN, Page MJ, Bradley S, Rhodes S, Sydenham M, Paterson H, Skinner RH (1991) Characterization of recombinant human Kirsten-ras (4B) p21 produced at high levels in Escherichia coli and insect baculovirus expression systems. J Biol Chem 266 (3):1672–1678
9. Gillette WK, Esposito D, Abreu Blanco M, Alexander P, Bindu L, Bittner C, Chertov O, Frank PH, Grose C, Jones JE, Meng Z, Perkins S, Van Q, Ghirlando R, Fivash M, Nissley DV, McCormick F, Holderfield M, Stephen AG (2015) Farnesylated and methylated KRAS4b: high yield production of protein suitable for biophysical studies of prenylated protein-lipid interactions. Sci Rep 5:15916. https://doi.org/10.1038/srep15916
10. Kaba SA, Salcedo AM, Wafula PO, Vlak JM, van Oers MM (2004) Development of a chitinase and v-cathepsin negative bacmid for improved integrity of secreted recombinant proteins. J Virol Methods 122(1):113–118. https://doi.org/10.1016/j.jviromet.2004.07.006

11. Luckow VA, Lee SC, Barry GF, Olins PO (1993) Efficient generation of infectious recombinant baculoviruses by site-specific transposon-mediated insertion of foreign genes into a baculovirus genome propagated in *Escherichia coli*. J Virol 67(8):4566–4579
12. Mehalko JL, Esposito D (2016) Engineering the transposition-based baculovirus expression vector system for higher efficiency protein production from insect cells. J Biotechnol 238:1–8. https://doi.org/10.1016/j.jbiotec.2016.09.002
13. Dharmaiah S, Bindu L, Tran TH, Gillette WK, Frank PH, Ghirlando R, Nissley DV, Esposito D, McCormick F, Stephen AG, Simanshu DK (2016) Structural basis of recognition of farnesylated and methylated KRAS4b by PDEdelta. Proc Natl Acad Sci U S A 113 (44):E6766–E6775. https://doi.org/10.1073/pnas.1615316113
14. Kapust RB, Waugh DS (1999) *Escherichia coli* maltose-binding protein is uncommonly effective at promoting the solubility of polypeptides to which it is fused. Protein Sci 8 (8):1668–1674. https://doi.org/10.1110/ps.8.8.1668
15. Tropea JE, Cherry S, Waugh DS (2009) Expression and purification of soluble his(6)-tagged TEV protease. Methods Mol Biol 498:297–307. https://doi.org/10.1007/978-1-59745-196-3_19
16. Gillette WK, Esposito D, Taylor TE, Hopkins RF, Bagni RK, Hartley JL (2011) Purify first: rapid expression and purification of proteins from XMRV. Protein Expr Purif 76 (2):238–247. https://doi.org/10.1016/j.pep.2010.12.003
17. Hanahan D, Jessee J, Bloom FR (1991) Plasmid transformation of Escherichia coli and other bacteria. Methods Enzymol 204:63–113
18. Hopkins R, Esposito D (2009) A rapid method for titrating baculovirus stocks using the Sf-9 easy Titer cell line. BioTechniques 47 (3):785–788. https://doi.org/10.2144/000113238

Chapter 21

A Quantitative FRET Assay for the Upstream Cleavage Activity of the Integral Membrane Proteases Human ZMPSTE24 and Yeast Ste24

Erh-Ting Hsu, Jeffrey S. Vervacke, Mark D. Distefano, and Christine A. Hrycyna

Abstract

The integral membrane protease ZMPSTE24 plays an important role in the lamin A maturation pathway. ZMPSTE24 is the only known enzyme to cleave the last 15 residues from the C-terminus of prelamin A, including a farnesylated and carboxyl methylated cysteine. Mutations in ZMPSTE24 lead to progeroid diseases with abnormal prelamin A accumulation in the nucleus. Ste24 is the yeast functional homolog of ZMPSTE24 and similarly cleaves the **a**-factor pheromone precursor during its posttranslational maturation. To complement established qualitative techniques used to detect the upstream enzymatic cleavage by ZMPSTE24 and Ste24, including gel-shift assays and mass spectrometry analyses, we developed an enzymatic in vitro FRET-based assay to quantitatively measure the upstream cleavage activities of these two enzymes. This assay uses either purified enzyme or enzyme in crude membrane preparations and a 33-amino acid **a**-factor analog peptide that is a substrate for both Ste24 and ZMPSTE24. This peptide contains a fluorophore (2-aminobenzoic acid—Abz) at its N-terminus and a quencher moiety (dinitrophenol—DNP) positioned four residues downstream from the cleavage site. Upon cleavage, a fluorescent signal is generated in real time at 420 nm that is proportional to cleavage of the peptide and these kinetic data are used to quantify activity. This assay should provide a useful tool for kinetic analysis and for studying the catalytic mechanism of both ZMPSTE24 and Ste24.

Key words ZMPSTE24/Ste24, Metalloproteases, Membrane proteins, Fluorescence quenching, Fluorescence resonance energy transfer (FRET), Assay

1 Introduction

Human ZMPSTE24 is a unique intramembrane zinc metalloprotease that is localized to both the endoplasmic reticulum and inner nuclear membranes. This enzyme plays important dual roles in the maturation of the nuclear scaffold protein lamin A. Like all CaaX proteins, the precursor of lamin A, prelamin A, undergoes a series of posttranslational modifications including farnesylation by farnesyltransferase, endoproteolysis of the AAX residues by ZMPSTE24

Maurine E. Linder (ed.), *Protein Lipidation: Methods and Protocols*, Methods in Molecular Biology, vol. 2009, https://doi.org/10.1007/978-1-4939-9532-5_21, © Springer Science+Business Media, LLC, part of Springer Nature 2019

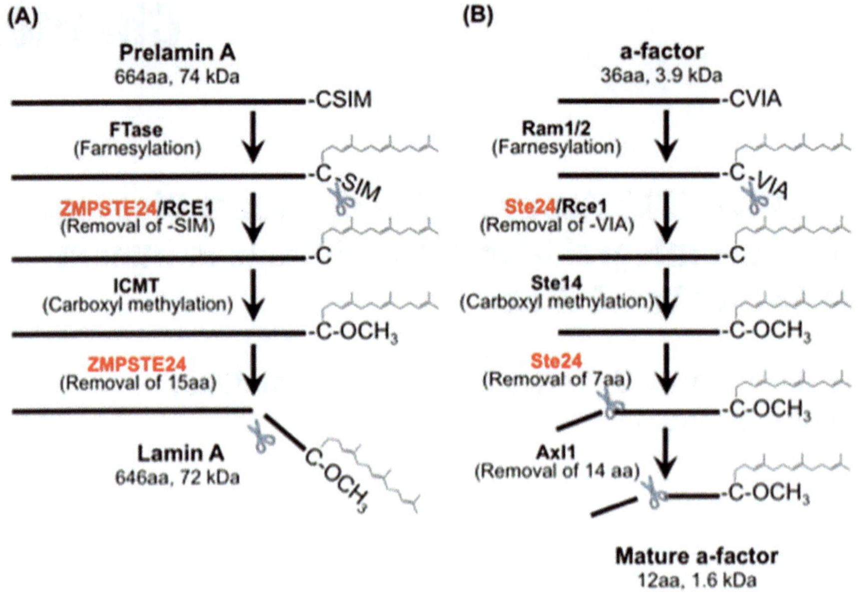

Fig. 1 Comparison of the maturation pathways for lamin A and **a**-factor. (**a**) Prelamin A processing pathway to form mature lamin A. Prelamin A is a 74 kDa protein with a CSIM motif at the C-terminus. The pathway includes a series of posttranslational modifications, called CAAX processing, which includes farnesylation, endoproteolysis, and carboxyl methylation. Furthermore, an additional cleavage event is required to remove the last 15 residues, which is performed by ZMPSTE24 only. (**b**) Yeast **a**-factor biogenesis pathway. The **a**-factor precursor is a 36-mer peptide. Like prelamin A and other CAAX proteins, the CVIA motif at the C-terminus triggers the CAAX processing, followed by two additional cleavages by Ste24 and Axl1 to form mature **a**-factor (Adapted from [1])

or RCE1, and carboxyl methylation by isoprenylcysteine carboxyl methyltransferase (ICMT). Prelamin A then undergoes a discrete site-specific upstream cleavage mediated only by ZMPSTE24. This step removes 15 residues from the C-terminus, including the farnesylated and carboxyl methylated cysteine, releasing mature lamin A into the nucleoplasm (Fig. 1a) [1, 2]. Ste24, which is the functional homolog of ZMPSTE24 from *S. cerevisiae*, cleaves the mating pheromone **a**-factor during its posttranslational maturation process. Like ZMPSTE24, Ste24 catalyzes two distinct cleavages of the **a**-factor peptide precursor. The first cleavage is the endoproteolysis of the AAX tail. The second cleavage is to remove the first 7 residues from the N-terminus (Fig. 1b) [1, 3].

Improper prelamin A processing caused by mutations in the gene encoding ZMPSTE24 results in progeroid diseases [4, 5]. Since ZMPSTE24 and another novel CAAX protease, RCE1, could both efficiently cleave the CAAX motif from prelamin A, the ability of ZMPSTE24 to perform the upstream cleavage determines the level of accumulated uncleaved prelamin A in progeroid diseases. Therefore, it is critical to develop effective tools to detect the upstream cleavage of ZMPSTE24. Currently, immunoblot analysis is commonly used to detect the upstream cleavage by ZMPSTE24 in mammalian cells, since the mature prelamin A is 2 kDa smaller than its precursor [6, 7]. A mass spectrometry binding assay has been reported to monitor the proteolysis by ZMPSTE24 [8]. However, both methods are laborious, technically demanding and difficult for quantification. Similar assays have been utilized for **a**-factor cleavage in yeast [9–11].

Herein, we introduce a FRET-based assay to quantify the upstream cleavage activity of ZMPSTE24 and Ste24. The method is adapted from a previously reported FRET assay for CAAX proteases that measures the fluorescence from a dequenched peptide [12, 13]. We utilize an **a**-factor sequence-based peptide as the probe because **a**-factor can be processed by both ZMPSTE24 and Ste24 in yeast [4, 10, 11].

The 33-mer **a**-factor analog (Peptide 1) has a 2-aminobenzoic (Abz) fluorophore at the N-terminus and a dinitrophenol (DNP) quencher located on the opposite side of the cleavage site. This peptide was synthesized using methods previously described by Distefano and coworkers [14, 15]. The peptide does not contain the AAX tail, but instead contains a C-terminal *S*-farnesyl-cysteine methyl ester, which mimics the processed substrate for the upstream cleavage. Because of internal quenching from the DNP group, the fluorescence of the Abz group in the intact peptide is significantly diminished. Following site-specific proteolysis by ZMPSTE24 or Ste24, increased fluorescence is observed over time due to the separation of the Abz/DNP fluorophore–quencher pair and the signal can be detected continuously in real time using a fluorimeter. These data are used to quantitatively determine the enzymatic activities of each enzyme.

This chapter first describes the purification process of the integral membrane proteases ZMPSTE24 and Ste24 expressed in the yeast *S. cerevisiae*. Each enzyme is tagged at the N-terminus with ten histidine residues and three tandem repeats of the hemagglutinin epitope (HA). Crude membranes derived from these cells containing over-expressed enzymes are prepared and subsequently solubilized in buffer containing detergent. TALON® cobalt resin is used for enzyme purification, exploiting the presence of the His-tags [16]. These enzymes can also be expressed in insect cells and purified by other chromatographic methods [17–19].

The second part of the chapter discusses the preparation of the FRET peptide, how to calibrate the assay and how to use the assay to determine enzyme activity and kinetic parameters. For peptide preparation, lyophilized peptides might contain from 10% to as much as 70% water and salts by weight. Therefore, it is difficult to determine the actual peptide concentration based on its mass. Using the Beer–Lambert law and the known extinction coefficient for DNP-lysine, the actual peptide concentration can be calculated [12]. Next, before performing the assay for further studies, a standard curve (Peptide 2) for conversion of RFU into specific activity is constructed. It is also important to generate another calibration curve (Peptide 2 + 3) for fluorescence correction since this Abz/DNP pair exhibits a strong inner filter effect, which leads to decreases in fluorescence when using a high concentration of fluorogenic substrate [12, 20].

The reaction setup in this FRET assay enables real-time monitoring of the peptide cleavage and direct determination of kinetic parameters. Together, this assay provides a convenient way for studying the upstream cleavage of both ZMPSTE24 and Ste24, which will aid in the elucidation of the precise catalytic mechanism of these novel enzymes.

2 Materials

2.1 Purification of ZMPSTE24 and Ste24

1. *Saccharomyces cerevisiae* strain (*Δste24Δrce1*) overexpressing His-tagged ZMPSTE24 or Ste24 (*2 μ URA3 P_{PGK}-His_{10}-HA_3-ZMPSTE24* or *STE24*) [4, 16].
2. Synthetic complete medium without uracil (SC-URA): For 1 L SC-URA, mix 20 g of glucose, 5 g of ammonium sulfate, 1.7 g of yeast nitrogen base, and 0.7 g of SC-URA powder in a 2-L flask. Add deionized water to 1 L then autoclave the medium.
3. Lysis buffer: 300 mM sorbitol, 100 mM NaCl, 6 mM $MgCl_2$, 10 mM Tris–HCl pH 7.5, 10 μg/mL aprotinin, 2 mM AEBSF, and 1 mM DTT. Add protease inhibitors and DTT freshly. Keep on ice.
4. Liquid nitrogen.
5. French press (SLM Aminco).
6. Disposable 1-mL syringe and 18, 20, 22, and 25 gauge needles.
7. Buffer S: 300 mM sorbitol, 100 mM NaCl, 6 mM $MgCl_2$, 10 mM Tris–HCl, pH 7.5, 10% glycerol, 10 μg/mL aprotinin, and 2 mM AEBSF. Add protease inhibitors and glycerol fresh. Keep on ice.
8. 10% dodecyl maltoside (DDM): Dissolve 250 μg DDM in 2.5 mL of autoclaved water (*see* **Note 1**). Always prepare fresh just before use and keep on ice.
9. 5 M Imidazole: Dissolve 17.2 g of imidazole in about 25 mL of autoclaved water. Adjust pH to 7.5 by adding HCl. Add autoclaved water to bring the final volume to 50 mL. Store at 4 °C and avoid exposure to light.
10. TALON® cobalt resin (Clontech). Store at 4 °C.
11. Buffer A: 40 mM imidazole and 1% (w/v) DDM in buffer S. Always prepare fresh buffer and keep on ice.

12. Buffer B: 500 mM KCl, 40 mM imidazole, and 1% (w/v) DDM in buffer S. Always prepare fresh buffer and keep on ice.
13. Buffer C: 500 mM KCl, 40 mM imidazole, and 0.1% (w/v) DDM in buffer S. Always prepare fresh buffer and keep on ice.
14. Buffer E: 250 mM imidazole and 0.1% (w/v) DDM in buffer S. Always prepare fresh buffer and keep on ice.
15. Disposable 10-mL chromatography column.
16. Amicon Ultra-15 centrifugal filter (30-kDa): Purchased from EMD Millipore.
17. Lab armor beads: Store at −80 °C.
18. 5× SDS-PAGE buffer: 0.5 M Tris–HCl, pH 6.8, 25% β-mercaptoethanol, 0.1% bromophenol blue, 30% sucrose, and 10% SDS. Store at −20 °C.

2.2 Preparation of FRET Peptides Stock Solution

1. FRET peptides: Abz-MQPSTATAAPK(DNP)EKTSSEKKD-NYIIKGVFWDPAC(Fr)-OMe (Peptide 1), Abz-MQPSTAT (Peptide 2) and AAPK(DNP)EKTSSEKKDN-YIIKGVFWD-PAC(Fr)-OMe (Peptide 3) (Fig. 2). Synthesized in the Distefano lab (unpublished data) using previously described methods [14, 15].
2. Dimethylformamide (DMF): Store at room temperature in supplied container.

2.3 FRET Assay

1. FRET peptides: Stock solution of Peptides 1, 2 and 3 prepared as in Subheading 3.2. Dilute to 1 mM in 100% DMF. Store at −80 °C. Incubate at room temperature for 10 min before usage and avoid exposure to light.

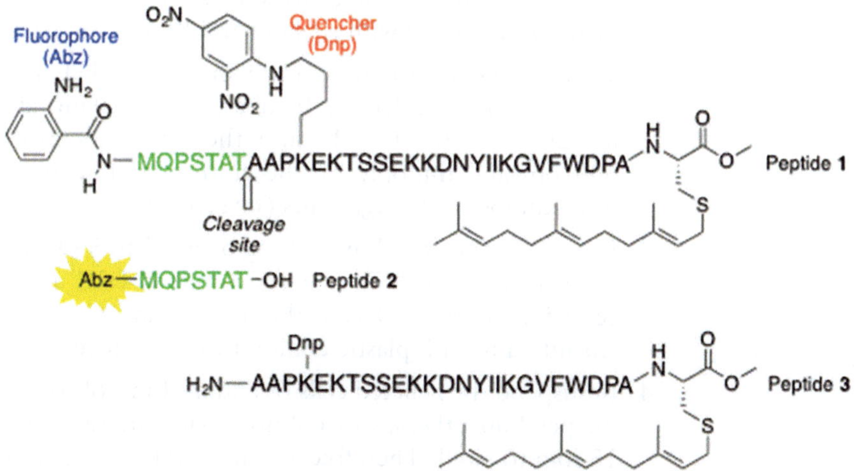

Fig. 2 Schematic representation of the FRET assay for ZMPSTE24 and Ste24 cleavage. The internally quenched FRET substrate, Peptide 1, contains an Abz fluorophore and a DNP quencher positioned at the N-terminus, and 4 residues downstream from the cleavage site, respectively. Cleavage by ZMPSTE24 or Ste24 between the T7 and A8 residues results in two cleavage products, Peptides 2 and 3. Fluorescence released from the dequenched Abz fluorophore can be measured at 420 nm

2. Assay buffer: 150 mM Tris–HCl, pH 7.5. Dissolve 18.171 g of Tris base in about 800 mL of deionized water. Adjust the pH to 7.5 with HCl. Add deionized water to 1 L. Store at 4 °C.
3. 625 μg/μL *E. coli* polar lipid extract (Avanti Polar Lipids). For 50 mg/mL lipid stock, add 2 mL of autoclaved water and 4 μL of 1 M β-mercaptoethanol (final: 2 μM) to 100 mg of lipid extract (one ampule). Gently resuspend by repeated pipetting on ice. Aliquot 500 μL to new eppendorf tubes. To prepare 625 μg/μL lipid stock, mix 50 μL of 50 mg/mL stock with 3950 μL of autoclaved water. Aliquot 500 μL to new eppendorf tubes. Purge the tubes with nitrogen and store at −80 °C. Thaw and keep on ice before use (*see* **Note 2**).
4. 0.15 μg/μL ZMPSTE24 and Ste24 protein: Prepared as in Subheading 3.1. Freshly dilute concentrated proteins with 10 mM Tris–HCl, pH 7.5 to 0.15 μg/μL stock. Keep on ice.
5. Black polystyrene 96-well half-area plate with flat bottom: purchased from Corning.
6. Synergy™ H4 hybrid microplate reader (BioTek). All experiments described in Subheadings 3.3 to 3.5 are designed using this microplate reader. Other multiwell fluorimeters with an excitation wavelength around 320 nm and capable of measuring emission around 420 nm may also be suitable.

3 Methods

3.1 Purification of Ste24 and ZMPSTE24

1. Grow *Δste24Δrce1* strain of *Saccharomyces cerevisiae* overexpressing His-tagged ZMPSTE24 or Ste24 from fresh (less than 2-months old) working stock plate in 15 mL of SC-URA medium. Incubate with shaking at 220 rpm for 20 h at 30 °C.
2. Transfer the overnight culture into fresh 1 L of SC-URA medium for inoculation. Incubate with shaking at 220 rpm for 20 h at 30 °C. Measure the optical density (OD) at 600 nm and multiply by the culture volume to obtain the total number of OD_{600} units (*see* **Note 3**).
3. Harvest the yeast cells in late-log phase (3 to 5 OD_{600}/mL) by centrifugation at 3750 × *g* for 5 min at 4 °C using 500-mL centrifuge bottles. Decant the supernatant. Combine the pellets into a 50-mL plastic conical tube (*see* **Note 4**).
4. Resuspend the pelleted cells by adding 1 mL of cold lysis buffer for per 800 OD_{600}'s of culture. Place the sample on ice for 15 min to swell. Then freeze and thaw the sample twice using liquid nitrogen.
5. Lyse the cells using a French Press. Prechill the French press cell on ice. Apply the sample to the French Press cell and bring

the cell to 12,000 psi. While maintaining the pressure, adjust the outlet flow rate to about 1 drop/s. Collect cell lysate in a 50-mL plastic conical tube that is kept on ice (*see* **Note 5**). Pass the lysate through the chilled French Press as described one additional time.

6. Centrifuge the lysate at 500 × *g* for 10 min at 4 °C. Transfer the supernatants to new conical tubes and centrifuge again to completely remove cell debris and unbroken cells. Transfer the supernatants to ultracentrifugation tubes. Ultracentrifuge at 100,000 × *g* for 1 h at 4 °C to obtain a pellet that includes the crude membranes containing the overexpressed ZMPSTE24 or Ste24. Remove the supernatant by decanting.
7. Add 300 μL of prechilled 10 mM Tris–HCl, pH 7.5 to the pellet. Resuspend the pellet with 1-mL syringe by passing through bent needles (18-, 20-, 22-, and 25-gauge) on ice. Then transfer the resuspended crude membranes to a new eppendorf tube using the 25-gauge needle (*see* **Note 6**). A 1-L yeast culture could yield about 10 mg crude membranes.
8. Solubilize 10 mg of crude membrane in 200 μL of 10% DDM (final: 1%), 8 μL of 5 M imidazole (final: 20 mM) and add buffer S to final volume of 2 mL. Incubate the sample on a rotary mixer for 1 h at 4 °C. Remove the insoluble fraction by ultracentrifugation at 100,000 × *g* for 45 min at 4 °C.
9. Prepare the cobalt resin by adding 2 mL of a 50% suspension of the resin into a 15-mL plastic conical tube. Spin down at 350 × *g* for 2 min at 4 °C then remove the supernatant. Wash the resin twice by adding 2 mL of buffer S to the resin. Gently mix, then spin down at 350 × *g* for 2 min at 4 °C. Decant the supernatant (*see* **Note** 7).
10. Transfer the supernatant fraction containing the solubilized ZMPSTE24 or Ste24 from **step 8** to the preequilibrated cobalt resin. Incubate the sample on a rotary mixer for 1 h at 4 °C. His-tagged ZMPSTE24 or Ste24 should bind and be retained on the resin.
11. Centrifuge at 350 × *g* for 2 min at 4 °C then remove the unbound fraction. Then wash the resin by adding 5 mL of buffer A. Incubate the sample on a rotary mixer for 10 min at 4 °C. Spin down at 350 × *g* for 2 min at 4 °C then remove the supernatant. Repeat washing with another 5 mL of buffer A, B, and C.
12. Transfer the resin to a 10-mL chromatography column. Elute the protein with 5 mL of buffer E and collect the elution at 4 °C.
13. Transfer the elution sample to a prerinsed 30-kDa concentrator and perform centrifugation at 5000 × *g* for 20 min at 4 °C.

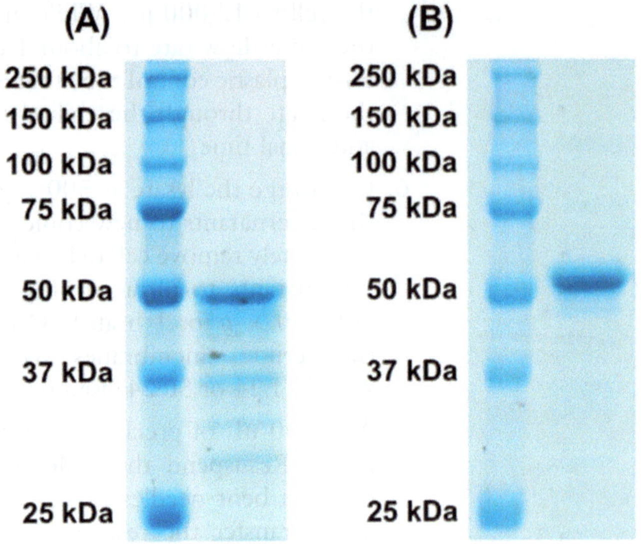

Fig. 3 Purification of ZMPSTE24 and Ste24 proteins. Coomassie blue stained SDS-PAGE shows all proteins were >90% pure. (**a**) Purified His-HA-ZMPSTE24 runs at approximate 50 kDa. (**b**) Purified His-HA-Ste24 runs at the expected size of 58 kDa

Aliquot 10 μL of concentrated protein to new eppendorf tubes, then flash-freeze on −80 °C lab armor beads (*see* **Note 8**). 10 mg of crude membrane should yield about 50 μg pure ZMPSTE24 or Ste24 protein.

14. Determine the protein concentration using Amido Black protein assay (*see* **Note 9**) [21]. To prepare SDS-PAGE samples, dilute purified protein to 0.5 μg/μL. Take 10 μL of diluted protein and mix with 2.5 μL of 5× SDS-PAGE buffer. Incubate at room temperature for 30 min and vortex every 5 min. Evaluate the purity of purified ZMPSTE24 or Ste24 by SDS-PAGE, followed by Coomassie blue staining (Fig. 3).

3.2 Preparation of FRET Peptide Stock Solutions

1. Dissolve the Peptides (1, 2 and 3) in 100% DMF to prepare about 20 μg/μL stock solutions.
2. Measure the absorbance of Peptides 1 and 3 at 360 nm to determine the molar concentration using the molar extinction coefficient of $\varepsilon = 17{,}530\ M^{-1}\ cm^{-1}$ for the DNP group. To determine the molar concentration of Peptide 2, measure the absorbance at 320 nm and use the molar extinction coefficient of $\varepsilon = 21{,}870\ M^{-1}\ cm^{-1}$ for the Abz group for conversion (*see* **Note 10**) [22].
3. Avoid exposure to light. Store the stock solutions at −80 °C for up to 1 year.

3.3 Conversion and Calibration of Relative Fluorescence Units (RFUs) into Specific Activity

1. For the standard curve, prepare a serial dilution of Peptide 2. Dilute Peptide 2 in assay buffer to 250, 150, 75, 37.5, 18.75, and 9.375 μM with a final volume of 30 μL. Use 30 μL of assay buffer for the 0 μM stock. Keep at room temperature and avoid exposure to light.
2. For the calibration curve, prepare a serial dilution of Peptides 2 and 3 mixture. Dilute both peptides in assay buffer to 500, 300, 150, 75, 37.5, and 18.75 μM with a final volume of 15 μL. Then combine the solutions with the same concentrations in new eppendorf tubes. Those tubes should contain 30 μL of 250 to 9.375 μM of both Peptides 2 and 3. Use 30 μL of assay buffer for the 0 μM stock. Keep at room temperature and avoid exposure to light.
3. Transfer the serial dilutions of Peptide 2 (from **step 1**) and the serial dilutions of the mixed Peptides 2 and 3 (**step 2**) to the 96-well plate. Cover with lid to prevent light exposure. Keep at room temperature.
4. Prepare master mix. Rapidly reconstitute 80 μL of 0.15 μg/μL purified ZMPSTE24 or Ste24 protein into 160 μL of 625 μg/μ L *E. coli* polar lipid extract in a 15-mL plastic conical tube. Add 1040 μL of assay buffer into the tube. Gently vortex and incubate on ice for 10 min.
5. Warm up the master mix in a 30 °C water bath for 5 min. Transfer 80 μL of master mix to 12 wells in the 96-well plate.
6. Place the 96-well plate into the microplate reader and wait about 3 min for the temperature to reach 30 °C.
7. Use an 8-channel pipette to transfer 20 μL of FRET peptide prepared from **step 3** to each well containing 80 μL of master mix (*see* **Note 11**). Thus, the total volume of each sample is 100 μL; final peptide concentrations range from 50 to 0 μM.
8. Mix by shaking at medium speed for 30 s. Measure the fluorescence readings at excitation and emission wavelengths of 320 ± 9 nm and 420 ± 9 nm at 30 °C (*see* **Note 12**).
9. Plot the fluorescence reading against final peptide concentration (Fig. 4). The extinction coefficient (ε) obtained from the standard curve (Peptide 2) can be used for converting from RFU to concentration of dequenched peptide product.
10. A correction factor (C) for the inner filter effect is calculated at each concentration of peptide used in the assay, as given by the following relationship:

$$C = \frac{F(\text{Peptide 2})}{F(\text{Peptide 2} + 3)}$$

where F(Peptide 2) is the fluorescence reading of Peptide 2 and F(Peptide 2 + 3) is the fluorescence reading of Peptide 2 in the

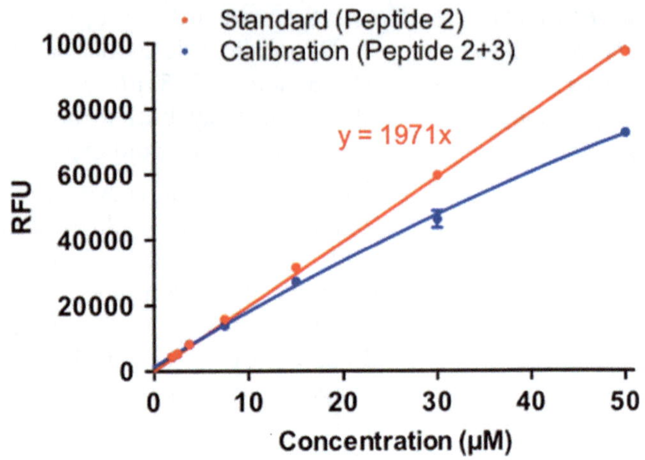

Fig. 4 Assay calibration. The standard curve (*red*) correlates the relative fluorescence unit (RFU) to a range of concentrations of the dequenched product, Peptide 2. The extinction coefficient (ε) was determined as 1971 RFU/μM from the slope of the standard curve. The calibration curve (*blue*) was measured using an equimolar mixture of Peptides 2 and 3 over the same range of concentrations. All fluorescence readings were taken under standard assay conditions with purified Ste24

presence of equimolar amount of Peptide 3. C will be applied to the same concentrations of FRET peptides to yield the corrected fluorescence value.

3.4 Determination of Upstream Cleavage Activity

1. Freshly dilute Peptide 1 in assay buffer to 150 μM with a final volume of 30 μL. Keep at room temperature and avoid exposure to light.
2. Rapid reconstitute 7.5 μL of 0.15 μg/μL purified ZMPSTE24 or Ste24 protein into 15 μL of 625 μg/μL *E. coli* polar lipid extract in an eppendorf tube. Add 97.5 μL of assay buffer into the tube. Gently vortex and incubate on ice for 10 min.
3. Warm up the protein/liposome sample in a 30 °C water bath for 5 min. Transfer 80 μL to a well in the 96-well plate.
4. Place the 96-well plate into the microplate reader and wait about 3 min for the temperature to reach 30 °C.
5. Initiate the assay by adding 20 μL of 150 μM Peptide 1 prepared from **step 1**. Final volume in each well is 100 μL. Each reaction contains 0.75 μg of purified enzyme reconstituted in 6.25 mg of *E. coli* lipid and 30 μM substrate in 100 mM Tris–HCl, pH 7.5 (*see* **Note 13**). Mix by shaking at medium for 30 s. Measure the fluorescence readings at excitation and emission wavelengths of 320 ± 9 nm and 420 ± 9 nm at 30 °C every 30 s (*see* **Note 14**). Constantly record the increase in fluorescence with time (Fig. 5).

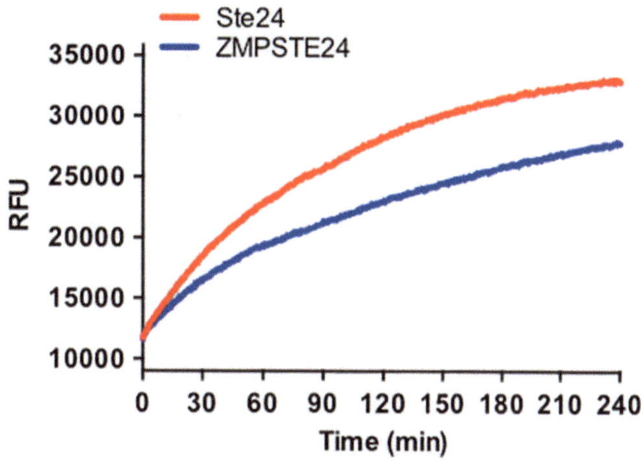

Fig. 5 Real-time monitoring of the upstream cleavage of the FRET substrate by ZMPSTE24 and Ste24. Fluorescence readings over a time course of 4 h were detected using an excitation wavelength of 320 nm and an emission wavelength of 420 nm. The fluorescence increased remarkably during the cleavage by purified ZMPSTE24 (*blue*) and Ste24 (*red*) proteins. Initial rates were found to be within the first 10 min of reaction

6. Determine the cleavage rate of ZMPSTE24/Ste24 using the slope from the first linear region (usually the first 10 min) (*see* **Note 15**). Convert the cleavage rate to the specific activity using the following equation:

$$\text{Specific activity (pmol/mg/min)} = \text{Rate (RFU/min)} \times C \times \frac{1}{\varepsilon(\text{RFU}/\mu\text{M})} \times \frac{10^6(\text{pmol}/\mu\text{mol})}{10^4\,(\text{L})} \times \frac{1}{0.0075\,(\text{mg})}$$

where the correction factor (C) and the extinction coefficient (ε) are defined in Subheading 3.3. The specific activity is expressed as pmol of substrate cleaved per min by per mg of enzyme.

3.5 Kinetic Analysis

1. Freshly dilute Peptide 1 in assay buffer to 150, 75, 37.5, 18.75, and 9.375 μM with a final volume of 30 μL. Use 30 μL of assay buffer for the 0 μM stock. Keep at room temperature and avoid exposure to light.
2. Transfer solutions made in **step 1** to the 96-well plate. Cover with lid to prevent light exposure. Keep at room temperature.
3. Prepare master mix. Rapid reconstitute 40 μL of 0.15 μg/μL purified ZMPSTE24 or Ste24 protein into 80 μL of 625 μg/μL *E. coli* polar lipid extract in a 15-mL plastic conical tube. Add 520 μL of assay buffer into the tube. Gently vortex and incubate on ice for 10 min.

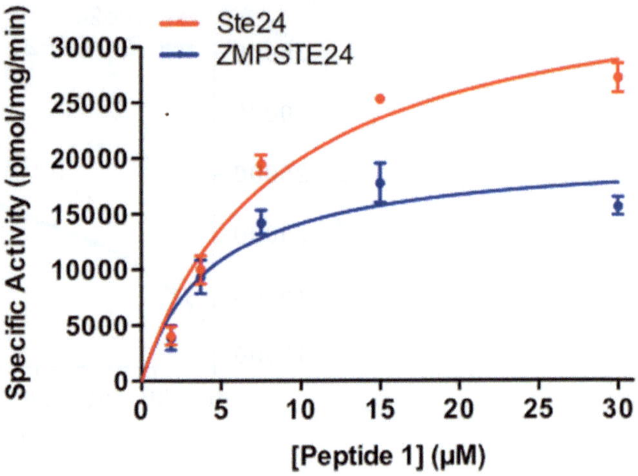

Fig. 6 Kinetic analysis of the upstream cleavage of the FRET substrate by ZMPSTE24 and Ste24. Specific activities and substrate concentrations were plotted and fitted to the Michaelis-Menten equation using the Prism 6 software. From these data, the values of kinetic parameters were determined. For ZMPSTE24 (*blue*), V_{max} is 27,000 pmol/mg/min and K_m is 6.7 μM. For Ste24 (*red*), V_{max} is 38,000 pmol/mg/min and K_m is 9.1 μM

4. Warm up the master mix in a 30 °C water bath for 5 min. Transfer 80 μL of master mix to 6 wells in the 96-well plate.
5. Place the 96-well plate into the microplate reader and wait about 3 min for the temperature to reach 30 °C.
6. Use 8-channel pipette to transfer 20 μL of FRET peptide solution (Peptide 1) prepared from **step 2** to each well containing 80 μL of master mix. Thus, total volume of each sample is 100 μL. Final peptide concentrations are from 30 to 0 μM.
7. Mix by shaking at medium speed for 30 s. Constantly measure the fluorescence readings at excitation and emission wavelengths of 320 ± 9 nm and 420 ± 9 nm at 30 °C for at least 10 min until seeing the first linear region.
8. Calculate specific activity at each peptide concentration as Subheading 3.4 description. Plot the specific activities against the peptide concentration. Fit the curve to the Michaelis–Menten equation using GraphPrism software to determine kinetic parameters (Fig. 6).

4 Notes

1. Always prepare DDM solution freshly to avoid hydrolysis and oxidation. Gently invert the solution or use rotator until DDM goes into solution. This will prevent bubble formation.

2. After each use, repurge the tube with nitrogen and store at −80 °C immediately to protect lipid from oxidation.
3. Generally speaking, the density of yeast overexpressing wild-type Ste24 or ZMPSTE24 will reach about 4 OD_{600} after 20 h incubation. Some mutant strains grow slower and might to longer incubation time.
4. The pellet can be stored at −80 °C for long-term storage.
5. If the volume of cell lysate is more than 40 mL, separate the sample into multiple tubes. Confirm the suggested minimum and maximum volume of the French Press cell before use.
6. The crude membranes can be stored at −80 °C for up to 2 years. Protein concentration of the crude membrane can be determined by the Bradford assay and should be between 25 and 35 mg/mL.
7. Do not allow the resin to dry after adding the lysate. It is also critical to avoid using strong reducing agents, such as DTT, or chelators, such as EDTA, which will disrupt the function of the cobalt resin.
8. Proteins may also be flash-frozen in liquid nitrogen with additional glycerol to a final concentration of 10–20%.
9. The Amido Black protein assay can be used to accurately determine low amounts of protein in the presence of high levels of both ionic and nonionic detergents, including DDM used for purification. Measuring the absorbance at 280 nm by NanoDrop spectrophotometer can also be used to determine the protein concentration. However, it tends to have a higher error at very low protein concentrations.
10. Depending on the spectrophotometer used, peptides might need to be diluted or prepared in larger volumes to construct a standard curve to determine the concentrations. The final concentration of FRET peptides should be around 5 mM for Peptides 1 and 3 and 20 mM for Peptide 2.
11. Since the samples contain detergent, it is easy to add bubbles to the wells. To avoid air bubbles, dispense the liquid while touching the bottom corner of the well and only pressing the pipette to its first stop. If there are still bubbles, use a pipette top to remove them, or perform a quick spin using a suitable centrifuge.
12. Changing assay conditions might result in emission intensity and spectral shifts. Recording an emission spectrum to confirm the maximum peak will help to select the best wavelength for the assay.
13. We do not suggest using more than 50 μM substrate in each reaction. This leads to a strong decrease in the fluorescence

intensity due to the increase in the inner filter effect. Moreover, the FRET peptide might not be able to dissolve in the assay buffer because of its limited solubility.

14. The time interval should be selected according to the reaction rate, the total sample number and the microplate reader model. We generally use 10 s to 1 min for measurement.

15. It is important to initially run the reactions to completion to determine the linear region. Since the 96-well plates are uncovered during analysis, evaporation can cause sample concentrations to increase, which affects data collection. Use plate sealing film if needed. If seeing an "S" shape curve, incubate the sample at 30 °C or shake in the microplate reader for a longer time.

References

1. Barrowman J, Michaelis S (2009) ZMPSTE24, an integral membrane zinc metalloprotease with a connection to progeroid disorders. Biol Chem 390:761–773
2. Michaelis S, Hrycyna CA (2013) A protease for the ages. Science 339:1529–1530
3. Michaelis S, Barrowman J (2012) Biogenesis of the *Saccharomyces cerevisiae* pheromone **a**-factor, from yeast mating to human disease. Microbiol and Mol Biol Rev 76:626–651
4. Barrowman J, Wiley PA, Hudon-Miller SE, Hrycyna CA, Michaelis S (2012) Human ZMPSTE24 disease mutations: residual proteolytic activity correlates with disease severity. Human Mol Genet 21:4084–4093
5. Davies BS, Fong LG, Yang SH, Coffinier C, Young SG (2009) The posttranslational processing of prelamin a and disease. Annu Rev Genomics Hum Genet 10:153–174
6. Barrowman J, Hamblet C, Kane MS, Michaelis S (2012) Requirements for efficient proteolytic cleavage of prelamin a by ZMPSTE24. PLoS One 7:e32120
7. Bergo MO, Gavino B, Ross J, Schmidt WK, Hong C, Kendall LV, Mohr A, Meta M, Genant H, Jiang Y, Wisner ER, Van Bruggen N, Carano RA, Michaelis S, Griffey SM, Young SG (2002) Zmpste24 deficiency in mice causes spontaneous bone fractures, muscle weakness, and a prelamin a processing defect. Proc Natl Acad Sci U S A 99:13049–13054
8. Mehmood S, Marcoux J, Gault J, Quigley A, Michaelis S, Young SG, Carpenter EP, Robinson CV (2016) Mass spectrometry captures off-target drug binding and provides mechanistic insights into the human metalloprotease ZMPSTE24. Nat Chem 8:1152–1158
9. Chen P, Sapperstein SK, Choi JD, Michaelis S (1997) Biogenesis of the *Saccharomyces cerevisiae* mating pheromone **a**-factor. J Cell Biol 136:251–269
10. Schmidt WK, Tam A, Michaelis S (2000) Reconstitution of the Ste24p-dependent N-terminal proteolytic step in yeast **a**-factor biogenesis. J Biol Chem 275:6227–6233
11. Tam A, Schmidt WK, Michaelis S (2001) The multispanning membrane protein Ste24p catalyzes CAAX proteolysis and NH2-terminal processing of the yeast **a**-factor precursor. J Biol Chem 276:46798–46806
12. Arachea BT, Wiener MC (2017) Acquisition of accurate data from intramolecular quenched fluorescence protease assays. Anal Biochem 522:30–36
13. Hollander I, Frommer E, Mallon R (2000) Human ras-converting enzyme (hRCE1) endoproteolytic activity on K-ras-derived peptides. Anal Biochem 286:129–137
14. Diaz-Rodriguez V, Ganusova E, Rappe TM, Becker JM, Distefano MD (2015) Synthesis of peptides containing C-terminal esters using trityl side-chain anchoring: applications to the synthesis of C-terminal ester analogs of the *Saccharomyces cerevisiae* mating pheromone **a**-factor. J Org Chem 80:11266–11274
15. Diaz-Rodriguez V, Mullen DG, Ganusova E, Becker JM, Distefano MD (2012) Synthesis of peptides containing C-terminal methyl esters using trityl side-chain anchoring: application to the synthesis of **a**-factor and **a**-factor analogs. Org Lett 14:5648–5651

16. Hudon SE, Coffinier C, Michaelis S, Fong LG, Young SG, Hrycyna CA (2008) HIV-protease inhibitors block the enzymatic activity of purified Ste24p. Biochem Biophys Res Commun 374:365–368

17. Clark KM, Jenkins JL, Fedoriw N, Dumont ME (2017) Human CaaX protease ZMPSTE24 expressed in yeast: structure and inhibition by HIV protease inhibitors. Protein Sci 26:242–257

18. Pryor EE Jr, Horanyi PS, Clark KM, Fedoriw N, Connelly SM, Koszelak-Rosenblum M, Zhu G, Malkowski MG, Wiener MC, Dumont ME (2013) Structure of the integral membrane protein CAAX protease Ste24p. Science 339:1600–1604

19. Quigley A, Dong YY, Pike AC, Dong L, Shrestha L, Berridge G, Stansfeld PJ, Sansom MS, Edwards AM, Bountra C, von Delft F, Bullock AN, Burgess-Brown NA, Carpenter EP (2013) The structural basis of ZMPSTE24-dependent laminopathies. Science 339:1604–1607

20. Hildebrandt ER, Arachea BT, Wiener MC, Schmidt WK (2016) Ste24p mediates proteolysis of both isoprenylated and non-prenylated oligopeptides. J Biol Chem 291:14185–14198

21. Schaffner W, Weissmann C (1973) A rapid, sensitive, and specific method for the determination of protein in dilute solution. Anal Biochem 56:502–514

22. Arellano M, Coll PM, Yang W, Duran A, Tamanoi F, Perez P (1998) Characterization of the geranylgeranyl transferase type I from *Schizosaccharomyces pombe*. Mol Microbiol 29:1357–1367

Part VII

Biochemistry of Protein Lipidation

Chapter 22

Monitoring RhoGDI Extraction of Lipid-Modified Rho GTPases from Membranes Using Click Chemistry

Akiyuki Nishimura and Maurine E. Linder

Abstract

The posttranslational lipid modification of Rho GTPases is important for their proper subcellular localization and signal transduction. Rho GTPases terminate in a CaaX motif, in which the cysteine residue is modified with either a farnesyl or geranylgeranyl isoprenoid. RhoGDI renders Rho GTPases soluble by masking their lipid moieties. We recently identified that the brain-specific splice variant of Cdc42 (bCdc42) containing a noncanonical CCaX motif harbors a dual prenyl–palmitoyl modification that prevents its binding to RhoGDI. This chapter describes a method to analyze RhoGDI extraction of Rho GTPases containing different lipid modifications from membranes using a liposome reconstitution assay and click chemistry.

Key words Rho GTPase, Rho guanine nucleotide dissociation inhibitor, Prenylation, Palmitoylation, Click chemistry

1 Introduction

Rho-family small GTPases play key roles in many physiological processes by regulating cytoskeletal organization. The subcellular localization of Rho GTPases is important for their interaction with regulators and effectors and proper signal transduction [1, 2]. Posttranslational lipid modifications of the C-terminal CaaX motif control the subcellular localization of Rho GTPases [3, 4]. In canonical CaaX processing, the cysteine residue of the CaaX motif is modified by a farnesyl or geranylgeranyl isoprenoid, followed by proteolytic cleavage of the -aaX tripeptide and carboxyl methylation of the prenylated cysteine [5]. Prenylation with either a farnesyl or geranylgeranyl group provides Rho GTPases with a membrane anchor.

Rho guanine nucleotide dissociation inhibitors (RhoGDIs) play a regulatory role in the activity and subcellular localization of Rho GTPases [6]. These regulatory proteins were initially identified as cytosolic proteins that bind to the GDP-bound form of Rho GTPases and inhibit their GDP dissociation [7]. However,

Maurine E. Linder (ed.), *Protein Lipidation: Methods and Protocols*, Methods in Molecular Biology, vol. 2009,
https://doi.org/10.1007/978-1-4939-9532-5_22,

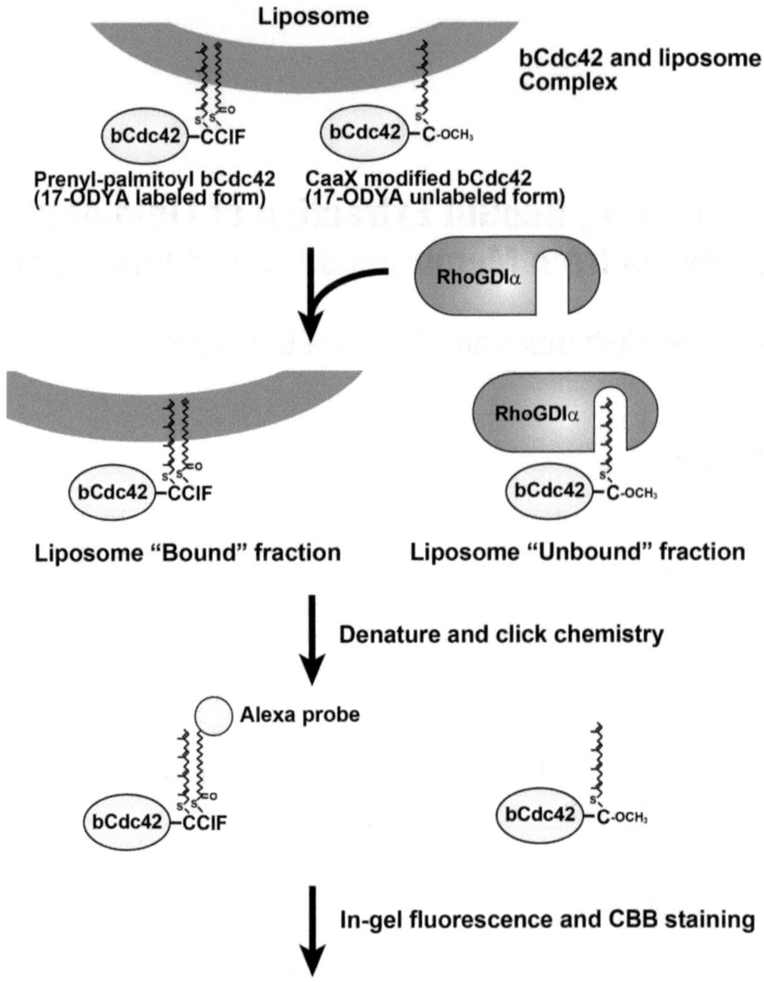

Fig. 1 Schematic outline of RhoGDIα extraction of proteoliposomes

RhoGDIs also dissociate Rho GTPases from membranes by binding and sequestering the prenyl group within a hydrophobic pocket, forming a stable cytosolic complex [8]. This ability of RhoGDI to extract Rho GTPases from membranes can be demonstrated using a reconstituted system with purified components in which proteoliposomes containing lipid-modified GTPases are incubated with RhoGDI, then fractionated by centrifugation [9, 10]. Rho proteins that are associated with RhoGDI are detected in the supernatant fraction, whereas those that remain associated with liposomes are detected in the pellet (Fig. 1).

Metabolic labeling with radioactive lipid is a classical assay to detect protein lipidation. Because of potential environmental and health hazards in the handling and disposal of radioisotopes, non-isotopic assays to detect protein lipidation have been developed. Click chemistry involves the copper-catalyzed azide and alkyne cycloaddition (CuAAC) reaction forming a covalent bond between azide and alkyne groups [11, 12]. We previously analyzed the lipid modification of the brain-specific splice variant of Cdc42 (bCdc42) terminating in the CCaX motif using click chemistry-based lipid probes and identified that the CCaX motif of bCdc42 undergoes either canonical CaaX processing or a tandem prenyl–palmitoyl modification [13] (Fig. 1). We found that whereas canonically processed bCdc42 can be extracted from membranes with RhoGDI, the tandem prenyl–palmitoyl form of bCdc42 does not associate with RhoGDI and remains in the membrane. This chapter describes the analysis of RhoGDI extraction of Rho GTPases from membranes using an adaptation of the liposome-based reconstitution assay [9]. The palmitoylated form of the RhoGTPase is labeled with a palmitate analog, 17-octadecynoic acid (17-ODYA) and detected by click chemistry using a fluorescent azide reporter.

2 Materials

2.1 Purification of bCdc42 Labeled with a Clickable Fatty Acid Analogue

1. TriEX Sf9 insect cells (Novagen).
2. TriEX serum-free medium (Novagen).
3. Dialyzed fetal bovine serum (FBS).
4. Baculovirus expressing His-bCdc42 or protein of interest (*see* **Note 1**).
5. 25 mM 17-octadecynoic acid (17-ODYA) in DMSO.
6. Lysis Buffer: 40 mM Hepes–NaOH, pH 7.4, 100 mM NaCl, 5 mM $MgCl_2$, 10 μM GDP.
7. Protease inhibitors: 100 mM PMSF in ethanol (100×), 10 mg/mL Leupeptin in DMSO (1000×).
8. 20% sodium cholate.
9. High Salt Buffer A: 20 mM Hepes–NaOH, pH 7.4, 500 mM NaCl, 5 mM $MgCl_2$, 10 μM GDP, 20 mM imidazole, 1% sodium cholate.
10. Elution Buffer A: 20 mM Hepes–NaOH, pH 7.4, 100 mM NaCl, 5 mM $MgCl_2$, 10 μM GDP, 200 mM imidazole, 11 mM CHAPS.
11. Storage Buffer: 20 mM Hepes–NaOH, pH 7.4, 100 mM NaCl, 5 mM $MgCl_2$, 10 μM GDP, 11 mM CHAPS.
12. Cell disruption vessel for nitrogen cavitation (Parr Instrument Co.).
13. Ni-NTA agarose (Qiagen).
14. Amicon Ultra-15, molecular weight cutoff - 10,000.

2.2 Purification of GST-Tagged RhoGDIα from E. coli

1. Bacterial strain *E. coli* BL21 (DE3) (Novagen).
2. Bacterial expression plasmid for GST-RhoGDIα.
3. 1 M isopropyl 1-thio-β-D-galactopyranoside (IPTG).
4. Extraction Buffer: 40 mM Tris–HCl, pH 7.4, 150 mM NaCl, 1 mM EDTA, 1 mM DTT.
5. High Salt Buffer B: 20 mM Tris–HCl, pH 7.4, 500 mM NaCl, 1 mM EDTA, 1 mM DTT.
6. Elution Buffer B: 100 mM Tris–HCl, pH 8.0, 1 mM EDTA, 1 mM DTT, 20 mM glutathione.
7. Dialysis Buffer: 20 mM Hepes–NaOH, pH 7.4, 1 mM EDTA.
8. 10% NP-40,
9. 1 M $MgCl_2$
10. 10 mg/mL lysozyme from chicken egg white in distilled water.
11. 1 mg/mL DNase I in distilled water.
12. Glutathione Sepharose 4B.
13. Dialysis tubing, molecular weight cutoff—10,000.

2.3 Liposome Reconstitution of bCdc42

1. 10 mM 1,2-dioleoyl-*sn*-glycero-3-phosphoethanolamine (DOPE) in chloroform.
2. 10 mM 1,2-dioleoyl-*sn*-glycero-3-phospho-L-serine (DOPS) in chloroform.
3. 10 mM L-α-phosphatidylinositol (PI) (bovine liver) in chloroform.
4. 25 mM cholesterol (ovine) in chloroform.
5. Reconstitution Buffer: 50 mM Hepes–NaOH, pH 7.4, 150 mM NaCl, 5 mM $MgCl_2$.
6. Mini-Extruder equipped with 0.2 μm pore size filter.
7. Vacuum freeze dryer.

2.4 RhoGDIα Extraction and Click Chemistry

1. 4 mM Alexa Fluor 488 Azide (Invitrogen) in DMSO.
2. 10 mM Tris[(1-benzyl-1*H*-1,2,3-triazol-4-yl)methyl]amine (TBTA) in DMSO.
3. 50 mM copper(II) sulfate in water, prepared just before use.
4. 50 mM TCEP HCl in water, prepared just before use.

3 Methods

3.1 Purification of bCdc42 Labeled with Clickable Fatty Acid

1. Culture TriEX Sf9 cells at 27 °C at a density of 0.8×10^6 cells/mL in 100 mL TriEX medium using a 250 mL sterile Erlenmeyer flask (*see* **Note 2**).

2. The next day, infect cells with 2.5 mL baculovirus stock encoding His-bCdc42 and culture for an additional 48 h (*see* **Note 3**).
3. Before cells are harvested, label cells with 17-ODYA. Mix 400 μL of 25 mM 17-ODYA with 5 mL of dialyzed FBS. Add this mixture to the cell culture (final 100 μM 17-ODYA), and culture cells for an additional 6 h (*see* **Note 4**).
4. Harvest cells by centrifugation (5000 × *g* for 10 min); store cell pellets at −80 °C if purification will be performed later.
5. Resuspend the cell pellet in 40 mL Lysis Buffer with protease inhibitors (*see* **Note 5**).
6. Disrupt cells by nitrogen cavitation at 500 psi for 30 min (*see* **Note 6**).
7. Centrifuge the homogenate at 500 × *g* for 5 min to pellet nuclei, unbroken cells, and large debris. Transfer the supernatant into ultracentrifugation tubes.
8. Resuspend the pellet in 40 mL of Lysis Buffer with protease inhibitors and repeat **steps 6** and 7.
9. Collect membranes by ultracentrifugation at 100,000 × *g* for 45 min (*see* **Note** 7). Recover the pellets and resuspend in 19 mL of Lysis Buffer using a Dounce homogenizer.
10. Add 1 mL of 20% sodium cholate to the resuspended membranes and rotate for 1 h to extract lipidated bCdc42.
11. Centrifuge at 100,000 × *g* for 45 min. The supernatant contains the detergent-extracted bCdc42. Check the protein concentration of the supernatant by Bradford assay and adjust to 1 mg/mL.
12. Save an aliquot of the supernatant as Input for later analysis.
13. Mix the supernatant with 250 μL of Ni-NTA agarose (preequilibrated in Lysis Buffer) and 300 μL of 1 M imidazole (final 15 mM), and incubate for 2 h at 4 °C with gentle rotation.
14. Transfer the sample into a column. Collect and save the flow-through for later analysis.
15. Wash the column with 10 mL High-Salt Buffer A.
16. Add 3 mL Elution Buffer A and collect the eluate in 0.5 mL fractions.
17. Analyze the input, flow-through, washes, and elutions by SDS-PAGE. Detect by Coomassie Brilliant Blue (CBB) staining or western blot. Concentrate the peak elution fractions containing His-bCdc42 and buffer exchange with Storage Buffer using an Amicon Ultra-15. Divide the final pool into 20 μL aliquots, snap-freeze in liquid nitrogen, and store at −80 °C.

18. Serially dilute samples and BSA standards and analyze by SDS-PAGE. Stain with CBB and determine sample concentration by densitometric analysis (*see* **Note 8**).

3.2 Purification of GST-Tagged RhoGDIα from E. coli

1. Culture *E. coli* BL21 (DE3) transformed with a GST-RhoGDIα expression plasmid in 300 mL LB medium at 30 °C. When the culture reaches an OD_{600} of 0.5, induce protein expression by adding IPTG to a final concentration of 300 μM. Culture the bacteria for an additional 4 h at 30 °C.
2. Harvest the bacteria by centrifugation (8000 × *g* for 10 min).
3. Store the pellets at −80 °C until purification.
4. Resuspend the cell pellet in 40 mL of Extraction Buffer with protease inhibitors (*see* **Note 9**).
5. Mix the homogenate with 2 mL of 10 mg/mL lysozyme and incubate for 10 min with gentle rotation (*see* **Note 10**).
6. Add 4.2 mL of 10% NP-40 (final 1%), 140 μL of 1 M $MgCl_2$ (final 3 mM), and 400 μL of 1 mg/mL DNase I, and incubate for 10 min with gentle rotation.
7. Disrupt the bacteria by passing the mixture through a 23G needle 15 times.
8. Centrifuge at 100,000 × *g* for 45 min and save the supernatant.
9. Save an aliquot of the supernatant as Input for later analysis.
10. Mix the supernatant with 1 mL of Glutathione Sepharose 4B (preequilibrated in Extraction Buffer) and incubate for 1 h with gentle rotation.
11. Load the sample into the column; collect and save the flow-through for later analysis.
12. Wash the column resin with 10 mL High Salt Buffer B.
13. Add 6 mL Elution Buffer B and collect the eluate in 1 mL fractions.
14. Analyze the input, flow-through, and fractions by SDS-PAGE by CBB staining.
15. Pool peak elution fractions containing GST-RhoGDIα; dialyze three times for a minimum of 4 h against 500 mL of Dialysis Buffer.
16. Divide the sample into 150 μL aliquots, snap-freeze in liquid nitrogen, and store at −80 °C (*see* **Note 11**).

3.3 Liposome Reconstitution of bCdc42

Liposomes are prepared with a composition of 35% PE, 25% PS, 5% PI, and 35% cholesterol to mimic the plasma membrane [9], then reconstituted with purified bCdc42. The proteoliposomes are recovered by centrifugation.

1. Add 100 μL of chloroform to a clean glass tube.

2. To make 700 μL of 1 mM liposomes, add 24.5 μL of 10 mM DOPE, 17.5 μL of 10 mM DOPS, 3.5 μL of 10 mM PI, and 9.8 μL of 25 mM cholesterol using a glass syringe.
3. Dry the lipid mixture with a gentle stream of argon gas for 10 min. While under the gas stream, rotate the tube slowly to form a thin lipid film on the wall.
4. Freeze-dry the lipid mixture using a vacuum freeze dryer for 1 h.
5. Resuspend the dry lipid film in 700 μL of Reconstitution Buffer by pipetting up and down.
6. Freeze the hydrated liposome solution in liquid nitrogen and thaw it in a 37 °C water bath. Repeat this freeze–thaw cycle five times to disrupt large multilamellar vesicles.
7. Make unilamellar vesicles using a mini-extruder equipped with a 0.2 μm pore size filter. Pass the liposome solution through the filter 21 times, and collect extruded liposomes in the syringe opposite of the sample loading syringe.
8. Mix 1 μM 17ODYA-labeled bCdc42 with 700 μL of liposome solution, and incubate for 30 min at room temperature with gentle agitation.
9. Centrifuge the samples at 16,000 × *g* for 20 min at room temperature, and discard the supernatant to remove liposome-unbound bCdc42.
10. Resuspend the liposome pellet in 700 μL of Reconstitution buffer.
11. Divide the liposome solution into 6 × 100 μL aliquots.

3.4 RhoGDIα Extraction and Click Chemistry

1. Add 20 μL of increasing amounts of GST-RhoGDIα (0, 30, 60, 90, 120, and 150 pmol) to each tube containing 100 μL of proteoliposomes. Incubate the samples for 30 min at room temperature with gentle agitation.
2. Centrifuge the samples at 16,000 × *g* for 20 min at room temperature. Transfer 84 μL of supernatants into new tubes, add 10 μL of 10% SDS, and save as "Unbound" samples.
3. Resuspend the pellet in 120 μL of Reconstitution Buffer. Transfer 84 μL into new tubes, add 10 μL of 10% SDS, and save as "Bound" samples (*see* **Note 12**).
4. Prepare click chemistry premix: for a 6 μL premix/sample, mix 1 μL of 4 mM Alexa 488 azide, 1 μL of 10 mM TBTA, 2 μL of freshly prepared 50 mM TCEP, and 2 μL of freshly prepared 50 mM $CuSO_4$.
5. To perform click chemistry, mix 94 μL of "Unbound" or "Bound" samples with 6 μL of click chemistry premix.

6. Incubate the samples for 1 h at room temperature in the dark with gentle agitation.
7. Perform a methanol–chloroform precipitation to remove the free Alexa probe (*see* **Note 13**). Add 400 μL of methanol and vortex. Then add 100 μL of chloroform and vortex.
8. Add 300 μL of water and vortex. Centrifuge at 16,000 × *g* for 5 min at room temperature.
9. Carefully discard the upper aqueous phase (*see* **Note 14**).
10. Add 300 μL of methanol and vortex. Centrifuge again.
11. Discard supernatant as completely as possible and air-dry the pellet for 10 min.
12. Resuspend the pellet in 50 μL Binding Buffer.
13. Add 12.5 μL of 5× sample buffer, and boil the samples.
14. Apply the samples to duplicate SDS polyacrylamide gels; one for in-gel fluorescence imaging (Fig. 2a) and the second for CBB staining (Fig. 2b).

The results are shown in Fig. 2. Total bCdc42 that is detected by CBB staining is moved from the "Bound" fraction to "Unbound" fraction in a RhoGDIα concentration-dependent manner. On the other hand, palmitoylated (17-ODYA labeled) bCdc42 is only detected in "Bound" fraction, consistent with palmitoylation inhibiting bCdc42 binding to RhoGDIα.

4 Notes

1. In our study of bCdc42 [13], the mouse cDNA (NM_001243769) gene is inserted into the baculovirus expression vector pFastBac HT B vector (Invitrogen) to generate N-terminal His-tagged bCdc42. Recombinant baculovirus is generated using the Bac-to-Bac baculovirus expression system (Invitrogen) according to the manufacturer's instructions.
2. Culture medium volume should be less than 40% of flask size to give sufficient aeration.
3. The volume of virus to add is dependent on the titer of the virus stock. A small-scale expression test should be performed in advance to determine the volume of virus stock to be added.
4. The concentration and incubation time for ODYA labeling is dependent on the protein of interest. A small-scale expression test should be performed in advance.
5. All purification steps are performed at 4 °C after the cell harvest.

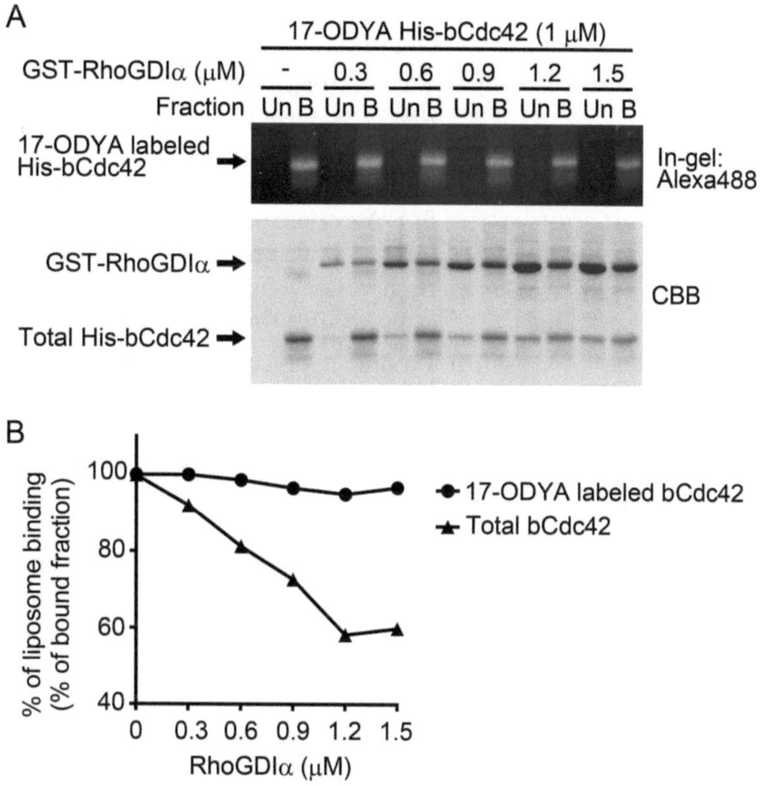

Fig. 2 (**a**) Total bCdc42 detected by CBB staining is extracted from the liposome-bound fraction to the unbound (Un) fraction in a RhoGDIα concentration-dependent manner. On the other hand, palmitoylated (17-ODYA-labeled) bCdc42 is detected by in-gel fluorescence only in the liposome bound fraction, suggesting that palmitoylation inhibits bCdc42 binding to RhoGDIα. (**b**) Quantification of liposome-bound bCdc42. (Figure adapted from Ref. [13])

6. Nitrogen cavitation is a gentle and efficient method to disrupt mammalian cells. Mechanical homogenization methods using Dounce or Potter-Elvehjem homogenizers can also be used.
7. The supernatant contains nonlipidated bCdc42, which can be purified by Ni-NTA affinity chromatography if desired.
8. We purified more than 300 μg His-bCdc42 from a 100-mL culture.
9. All purification steps are performed at 4 °C after cell harvest.
10. In **steps 4–6**, bacterial cell walls and DNA are digested using lysozyme and DNase I. Sonication can also be used for bacterial lysis.
11. We purified more than 2.5 mg GST-RhoGDIα from a 300-mL culture.
12. Buffer composition may affect the efficiency of click chemistry. Accordingly, we added SDS to a final concentration of 1% to

the "Unbound" samples to make them equivalent to the buffer composition of the "Bound" samples.

13. There is a large fluorescence signal from free Alexa 488 azide that is detected at around 15 kDa on the gel. This signal interferes with detection of the 25 kDa GTPase labeled with ODYA. The methanol–chloroform precipitation removes the free Alexa probe to reduce the background. If the protein of interest is well resolved from the free probe, this step may not be necessary and 5× sample buffer can be added directly to samples after the click chemistry reaction.
14. Protein is precipitated at the interface of upper and bottom layer. Be careful not to disturb the interface.

Acknowledgements

This work was supported by the National Institutes of Health (R01 GM121540).

References

1. Jaffe AB, Hall A (2005) RHO GTPASES: biochemistry and biology. Annu Rev Cell Dev Biol 21:247–269
2. Heasman SJ, Ridley AJ (2008) Mammalian Rho GTPases: new insights into their functions from in vivo studies. Nat Rev Mol Cell Biol 9:690–701
3. Winter-Vann AM, Casey PJ (2005) Post-prenylation-processing enzymes as new targets in oncogenesis. Nat Rev Cancer 5:405–412
4. Roberts PJ, Mitin N, Keller PJ et al (2008) Rho Family GTPase modification and dependence on CAAX motif-signaled posttranslational modification. J Biol Chem 283:25150–25163
5. Adamson P, Marshall CJ, Hall A, Tilbrook PA (1992) Post-translational modifications of p21rho proteins. J Biol Chem 267:20033–20038
6. Garcia-Mata R, Boulter E, Burridge K (2011) The 'invisible hand': regulation of RHO GTPases by RHOGDIs. Nat Rev Mol Cell Biol 12(8):493–504
7. Ueda T, Kikuchi A, Ohga N, Yamamoto J, Takai Y (1990) Purification and characterization from bovine brain cytosol of a novel regulatory protein inhibiting the dissociation of GDP from and the subsequent binding of GTP to rhoB p20, a ras p21-like GTP-binding protein. J Biol Chem 265:9373–9380
8. Hoffman GR, Nassar N, Cerione RA (2000) Structure of the Rho family GTP-binding protein Cdc42 in complex with the multifunctional regulator RhoGDI. Cell 100:345–356
9. Johnson JL, Erickson JW, Cerione RA (2009) New insights into how the Rho guanine nucleotide dissociation inhibitor regulates the interaction of Cdc42 with membranes. J Biol Chem 284(35):23860–23871
10. Johnson J, Cerione RA, Erickson JW (2012) A quantitative fluorometric approach for measuring the interaction of RhoGDI with membranes and Rho GTPases. Methods Mol Biol 827:107–119
11. Charron G, Zhang MM, Young JS et al (2009) Robust fluorescent detection of protein fatty-acylation with chemical reporters. J Am Chem Soc 131:4967–4975
12. Martin BR, Cravatt BF (2009) Large-scale profiling of protein palmitoylation in mammalian cells. Nat Methods 6:135–138
13. Nishimura A, Linder ME (2013) Identification of a novel prenyl and palmitoyl modification at the CaaX motif of Cdc42 that regulates RhoGDI binding. Mol Cell Biol 33:1417–1429

Chapter 23

Purification of the Rhodopsin–Transducin Complex for Structural Studies

Yang Gao, Jon W. Erickson, Richard A. Cerione, and Sekar Ramachandran

Abstract

G protein-coupled receptors (GPCRs) comprise the largest family of transmembrane receptors and are targets for over 30% of all drugs on the market. Structural information of GPCRs and more importantly that of the complex between GPCRs and their signaling partner heterotrimeric G proteins is of great importance. Here we present a method for the large-scale purification of the rhodopsin–transducin complex, the GPCR–G protein signaling complex in visual phototransduction, directly from their native retinal membrane using native proteins purified from bovine retinae. Formation of the complex on native membrane is orchestrated in part by the proper engagement of lipid-modified rhodopsin and transducin (i.e., palmitoylation of the rhodopsin C-terminus, myristoylation and farnesylation of the α_T and γ_1, respectively). The resulting complex is of high purity and stability and has proved suitable for further biophysical and structural studies. The methods described here should be applicable to other recombinantly expressed receptors from insect cells or mamalian cells by forming stable, functional complexes directly on purified cell membranes.

Key words G protein-coupled receptor (GPCR), G protein, Rhodopsin, Transducin, GPCR–G protein complex

1 Introduction

G protein-coupled receptors (GPCRs) represent the largest family of transmembrane proteins in metazoan biology with about 800 members [1] and, in human medicine, are targets for over 30% of all drugs on the market [2]. They share a signature structural motif of seven transmembrane helixes and transmit signals from a vast array of extracellular stimuli, including hormones, neurotransmitters, odorants, and photons, across the plasma membrane [3] and thus modulate a wide range of cellular responses. This process is achieved mostly through the activation of their canonical signaling partners, heterotrimeric G proteins, which are composed of three subunits, the nucleotide-binding α subunit and two constitutively associated subunits β and γ. And they can be

Maurine E. Linder (ed.), *Protein Lipidation: Methods and Protocols*, Methods in Molecular Biology, vol. 2009,
https://doi.org/10.1007/978-1-4939-9532-5_23,

classified into four different types based on the amino acid sequence similarity of the α subunits: G_S, G_i, G_q, and G_{12} [4].

In contrast to the striking diversity of the GPCR family, in humans there are only 16 genes encoding 21 different α subunits [5] which implies that each G protein isoform must be able to interact with a large number of different GPCRs. Furthermore, there is good evidence demonstrating that some GPCRs, for example the β_2 adrenergic receptor [6], can couple to more than one G protein. Therefore, proper regulation regarding the specificity between GPCRs and G proteins is vital for signal transduction inside cells. However, due to the intrinsic flexibility and instability of GPCRs and the GPCR–G protein complexes, it has been challenging to extract and purify component proteins and elucidate the structural mechanism underlying this specificity.

In recent years, with the development of a wide range of techniques, including novel protein engineering [7], in meso crystallization [8], microfocus beamlines [9] at synchrotron facilities, and cryo-electron microscopy [10], the number of resolved GPCR structures have been growing almost exponentially, among which are three high-resolution structures of GPCR–G protein complexes: the β_2 adrenergic receptor–G_S protein complex (solved by X-ray crystallography) [11], the calcitonin receptor–G_S protein complex [12], and the glucagon-like peptide 1 receptor–G_S protein complex [13] (the last two solved with cryo-electron microscopy). As all these complex structures are of the same G protein G_S, structural information of complexes formed with different classes of G proteins will be crucial for understanding the specificity between GPCRs and G proteins. To this end, the visual phototransduction cascade is an ideal system, as its G protein transducin belongs to the G_i family.

The visual phototransduction system is a prototypical GPCR signaling system. Rhodopsin, the GPCR, has a covalently bound inverse agonist 11-cis retinal which photoisomerizes into all-trans retinal and becomes an agonist for rhodopsin upon light-activation. Light-activated rhodopsin then binds and catalyzes the exchange of GDP for GTP in the heterotrimeric G protein transducin (G_T, subunits designated as α_T, β_1, and γ_1). The GTP-bound α_T can then activate cGMP phosphodiesterase (PDE) which hydrolyzes cGMP to GMP, eventually leading to visual neuron signals [14]. This system offers certain advantages for obtaining structural insights into GPCR-signaling, as each of the principal components can be purified from native tissue in large quantities. As a result, rhodopsin represents the first and only GPCR to date for which X-ray crystal structures [15–19] have been solved in the native form.

Here we present a method for extracting and purifying the complex between light-activated rhodopsin and nucleotide-free transducin directly from the native retinal membranes [20]

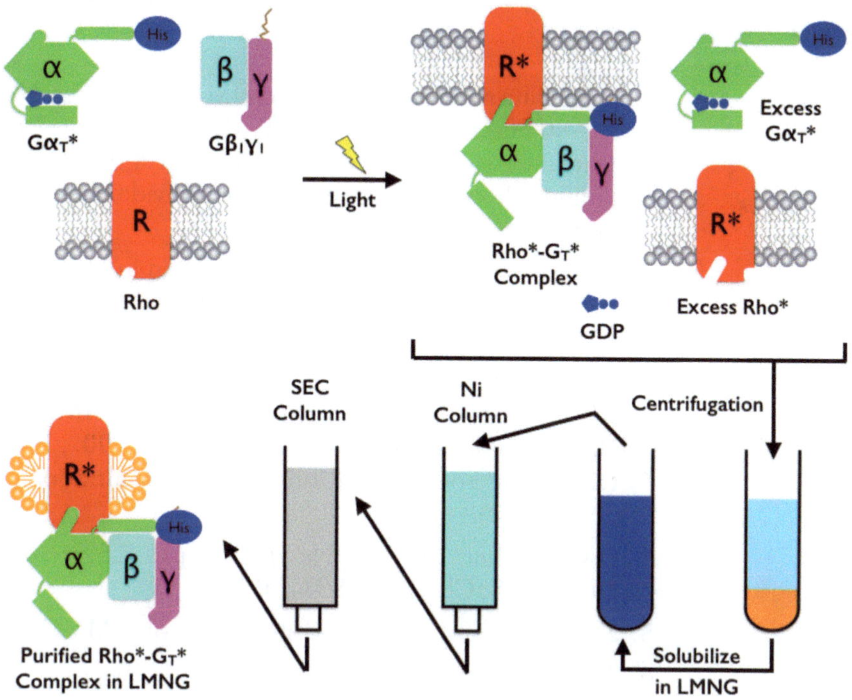

Fig. 1 Rhodopsin–transducin complex purification scheme

(as shown in Fig. 1) using protein components purified from bovine retinae and an α_T/α_{i1} chimera (α_T*), which can be expressed in *E. coli* and can undergo rhodopsin catalyzed nucleotide exchange, resulting in an activation of PDE that rivals native α_T [21]. Forming the complex on the membrane allows for easy separation of the proteins by centrifugation from free GDP (which is destabilizing for the complex) dissociated from G_T upon activation. Moreover, as both rhodopsin and G_T are lipid-modified (rhodopsin is palmitoylated at its C-terminus [22] and γ_1 is farnesylated at the C-terminus [23]), formation of the complex on membranes prior to detergent-extraction enables proper engagement using these lipid modifications. The resulting complex can then be solubilized with a mild detergent (lauryl maltose neopentyl glycol (LMNG)), affinity purified by utilizing a His_6-tag on α_T* and further purified with size-exclusion chromatography (SEC). The purified complex has a 1:1 ratio of rhodopsin to transducin, which can be verified with UV-Vis spectroscopy, as the ligand all-trans retinal has a distinct absorbance at 380 nm. The complex is very stable and can be stored at 4 °C in the dark for over a week without undergoing significant dissociation. The approach outlined here is applicable to any recombinantly expressed receptors from insect cells or mammalian cells by forming complexes directly on purified cell membranes.

2 Materials

2.1 Formation and Extraction of the Rhodopsin–Transducin Complex

1. Dark-state rhodopsin, stored in HMN buffer with 10% glycerol at −80 °C, aliquoted into 500 μL aliquots in 1.7 mL microcentrifuge tubes at a concentration of 140 μM. Rhodopsin aliquots should be wrapped in foil individually to avoid exposure to light (*see* **Note 1**).
2. Purified transducin $\beta_1\gamma_1$ subunits, stored in HMN buffer with 10% glycerol at −80 °C, aliquoted into 500 μL aliquots in 1.7 mL microcentrifuge tubes at a concentration of 20 μM (*see* **Note 1**).
3. His_6-tagged α_T/α_{i1} chimera (α_T*) stored in HMN buffer with 10% glycerol at −80 °C aliquoted into 500 μL aliquots in 1.7 mL microcentrifuge tubes at a concentration of 22 μM (*see* **Note 2**).
4. HMN buffer: 20 mM Na-HEPES pH 7.5, 2 mM $MgCl_2$, 100 mM NaCl, 100 μM TCEP.
5. 2% (w/v) lauryl maltose neopentyl glycol (LMNG) (Anatrace).
6. Aluminum foil.
7. A tabletop microcentrifuge and an end-to-end rocker kept in a cold room at 4 °C.
8. 1.7 mL microcentrifuge tubes and 15 mL conical tubes.
9. A desk lamp with standard UV filter.
10. Floor lamps covered with 3 M 616 lithographer's tape.

2.2 Purification and Characterization of the Rhodopsin–Transducin Complex

1. HisTrap HP column (1 mL) (GE Life Sciences).
2. An FPLC equipped with a UV absorbance monitor set at 280 nm and a fraction collector and kept in a cold box at 4 °C. A desk lamp covered with 3 M 616 lithographer's tape.
3. HisTrap buffer A: 20 mM HEPES pH 7.5, 2 mM $MgCl_2$, 100 mM NaCl, 100 μM TCEP, 0.02% LMNG.
4. HisTrap buffer B: HisTrap buffer A, 500 mM imidazole pH 7.5.
5. Superdex 200 10/300 GL column (GE Life Sciences).
6. SEC buffer: 20 mM HEPES pH 7.5, 2 mM $MgCl_2$, 100 mM NaCl, 100 μM TCEP, 0.003% LMNG.
7. Amicon Ultra-0.5 mL centrifugal filters with 100 kD molecular weight cutoff (EMD Millipore).
8. UV-Vis spectrophotometer.

3 Methods

3.1 Formation and Extraction of the Rhodopsin–Transducin Complex

1. Thaw one 500 μL aliquot each of rhodopsin, α_T* and $\beta_1\gamma_1$ on ice and mix them in a 1.7 mL microcentrifuge tube, resulting in 7:1.1:1 molar ratio of rhodopsin–α_T*–$\beta_1\gamma_1$. Incubate mixture on ice in the dark for 5 min (*see* **Notes 3** and **4**).
2. In a cold room, lay the tube containing the mixture on an end-to-end rocker and place a desktop lamp about 10 cm above the tube. Turn on both the rocker and the lamp and expose the mixture to light for about 20 min (*see* **Note 5**).
3. *All of the following steps should be carried out in the dark with red lights.*
4. Centrifuge the tube at 16,000 × *g* in a tabletop microcentrifuge for 30 min in the cold room.
5. Carefully aspirate and discard the supernatant. The rhodopsin–transducin complex and excess rhodopsin will be in the membrane pellet.
6. Resuspend the pellet with 2 mL cold HMN buffer with 1% LMNG and transfer the suspension into a 15 mL conical tube. Wrap the tube in foil and gently rock it in the cold room for 30 min (*see* **Note 6**).
7. Add 8 mL cold HMN buffer to the tube, wrap the tube in foil again, and gently rock the tube in the cold room for an additional 60 min.

3.2 Purification of the Rhodopsin–Transducin Complex

1. Equilibrate a 1 mL HisTrap HP column with 10 mL HisTrap buffer A at 1 mL/min on an FPLC equipped with a UV absorbance monitor set at 280 nm (*see* **Note 7**).
2. Add 0.2 mL HisTrap buffer B to the solubilized complex and load the mixture onto the column at 1 mL/min (*see* **Note 8**).
3. Wash the column with 20 mL 4% HisTrap buffer B at 1 mL/min (*see* **Note 9**).
4. Elute with a 10 mL 4–40% B gradient and an additional 10 mL 40% B step gradient at 1 mL/min. Collect 1 mL fractions (*see* **Note 10**).
5. Concentrate the fractions containing the complex with two 0.5 mL 100 kD molecular weight cutoff Amicon concentrators by centrifugation at 14,000 × *g* for 3 min in a tabletop microcentrifuge kept in a dark cold room with red lights. Dilute the concentrated complex back to 0.5 mL three times with HisTrap buffer A in the concentrators to remove imidazole, centrifuging for 3 min each time. The final volume of the complex should be about 200 μL. At this stage the complex

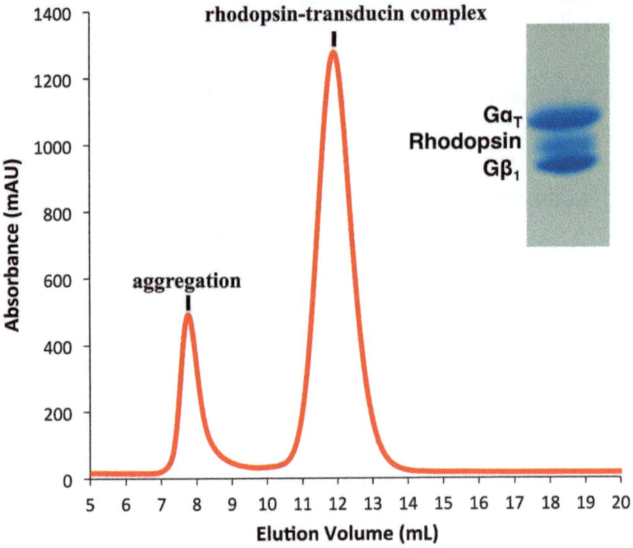

Fig. 2 SEC profile of rhodopsin–transducin complex purification and SDS-PAGE gel of concentrated peak fractions

can be either stored wrapped in foil on ice overnight or used directly for the next step (*see* **Note 11**).

6. Equilibrate a Superdex 200 10/300 GL column with 60 mL SEC buffer at 0.4 mL/min.
7. Inject the complex from **step 5** onto the column at a flow rate of 0.4 mL/min with SEC buffer. Collect 0.4 mL fractions. A typical SEC profile is shown in Fig. 2.
8. Pool fractions containing the complex from **step 7** and concentrate with a 0.5 mL 100 kD molecular weight cutoff Amicon concentrator to about 100 μL. This final purified complex should be at a concentration of ~10 mg/mL (*see* **Note 12**).

3.3 Characterization of the Purified Complex with UV-Vis Spectroscopy and Analytical SEC

1. Measure the concentration of the complex with a UV-Vis spectrophotometer at 280 nm ($A_{280\ nm}$) using the extinction coefficient $\varepsilon_{280\ nm} = 155{,}070\ M^{-1}\ cm^{-1}$, and molecular weight $M_w = 126{,}040$ Da (*see* **Note 13**).
2. Assess the purity of the complex with UV-Vis spectroscopy by measuring the absorbance at 280 nm ($A_{280\ nm}$, the absorbance of the protein moieties) and 380 nm ($A_{380\ nm}$, the absorbance of all-trans retinal) (a typical spectrum is shown in Fig. 3a). The $A_{280\ nm}/A_{380\ nm}$ ratio should be 3.69, indicating a 1:1 rhodopsin–transducin complex. If the complex is contaminated with excess G_T or excess rhodopsin, the ratio will be either higher or lower than 3.69 (*see* **Note 14**).
3. To routinely assess the integrity of the complex using analytical SEC, inject 5 μL of 10 mg/mL complex diluted with SEC

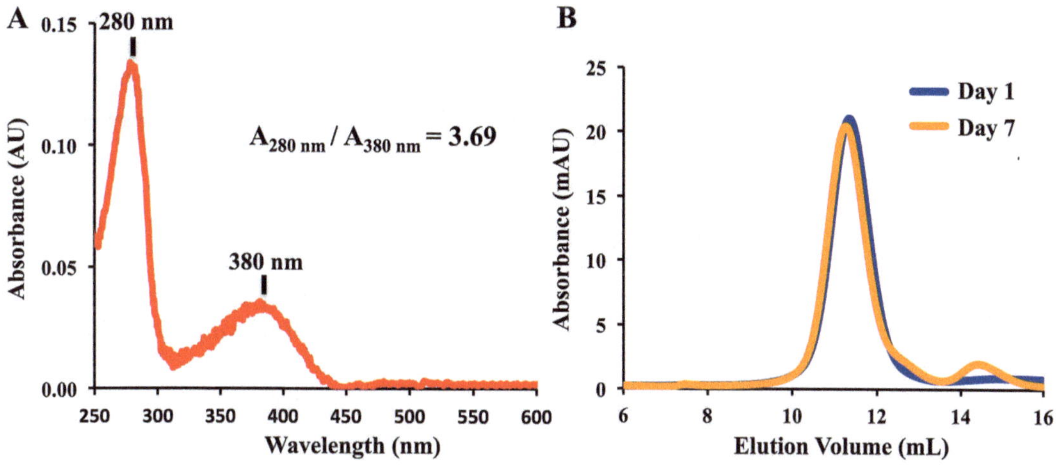

Fig. 3 UV-Vis spectrum and SEC profiles of the purified rhodopsin–transducin complex (**a**) UV-Vis spectrum of purified complex. (**b**) Analytical SEC profiles of complex after 1 day (blue) and 7 days (orange) of storage

buffer to 100 μL onto a Superdex 200 10/300 GL column preequilibrated with SEC buffer at a flow rate of 0.4 mL/min. The complex will elute as a sharp symmetrical peak if it is intact (typical analytical SEC profiles shown in Fig. 3b). The purified complex is generally stable for over a week when kept at 4 °C wrapped in foil.

4 Notes

1. Both dark-state rhodopsin (in the form of urea-washed rod outer segment membrane) and transducin $\beta_1\gamma_1$ subunits can be purified from bovine retinae (W L Lawson Company (Omaha, NE)) as described previously [24–25] and are stored in HMN buffer with 10% glycerol at −80 °C. Typically about 300 mg of rhodopsin and 15 mg of $\beta_1\gamma_1$ can be purified from 300 bovine retinae, at concentrations of 280 μM and 40 μM respectively.
2. The N-terminally His$_6$-tagged α_T/α_{i1} chimera (α_T*), in which residues from 215 to 295 in α_T, except for residues 244 and 247, are replaced with corresponding residues from α_{i1}, can be expressed and purified from *E. coli* BL21(DE3) as described previously [21] and stored in HMN buffer with 10% glycerol at 44 μM concentration at −80 °C. The 10% molar excess of α_T* ensures that all $\beta_1\gamma_1$ is utilized since $\beta_1\gamma_1$ is harder to purify and more valuble than α_T*. Native α_T is N-myristoylated at its amino terminus, which facilitates its association with membranes. The recombinant α_T* purified after expression in *E. coli* is unmodified, but through its interaction with farnesylated $\beta_1\gamma_1$ will associate with the retinal membrane to form the rhodopsin–transducin complex.

3. The large molar excess of rhodopsin is necessary because rhodopsin is very densely packed on retinal membrane and as a result, a large portion of it is not easily accessible to transducin. This large excess of receptor is not necessary if using membranes from insect cells or mammalian cells.
4. Extra care must be taken to ensure that rhodopsin is not exposed to light prior to the light activation step in Subheading 3.1, **step 4**.
5. To provide a controlled light exposure, all other lights in the cold room should be turned off. All following steps should be conducted in the dark with red lights.
6. Prolonged incubation in 1% LMNG will cause complex dissociation.
7. The FPLC should be kept in a cold box set at 4 °C with the glass doors covered with foil to prevent light exposure inside. A red light kept inside the cold box will facilitate visualization of the instrument.
8. The addition of 0.2 mL buffer B results in a final imidazole concentration of 10 mM which prevents the binding of excess rhodopsin to the column.
9. This step further removes contamination from excess rhodopsin and also lowers the detergent concentration from 0.2% to 0.02%.
10. The complex will start to elute from the column at ~10% HisTrap buffer B and peak at 40% HisTrap buffer B.
11. Prolonged incubation with imidazole can destabilize the complex. Therefore, it is necessary to remove imidazole if the complex is to be stored overnight. It is recommended to concentrate the fractions starting from those with high imidazole concentrations, and as a result the final imidazole concentration before dilution with HisTrap buffer A will be much lower. During concentration and cycles of concentration and dilution, it is recommended to invert the concentrators several times to prevent protein aggregation during centrifugation.
12. The complex peak will be contained in four 0.4 mL fractions. During concentration and cycles of concentration and dilution, it is recommended to invert the concentrators several times to prevent protein aggregation during centrifugation.
13. Concentration (in mg/mL) = $A_{280\text{ nm}} \times M_w/\varepsilon_{280\text{ nm}}$.
14. The extinction coefficients used are as follows: rhodopsin ($\varepsilon_{280\text{ nm}}$ = 61,800 M^{-1} cm^{-1}, $\varepsilon_{380\text{ nm}}$ = 42,000 M^{-1} cm^{-1}), α_T* ($\varepsilon_{280\text{ nm}}$ = 35,870 M^{-1} cm^{-1}), and $\beta_1\gamma_1$ ($\varepsilon_{280\text{ nm}}$ = 57,400 M^{-1} cm^{-1}).

References

1. Bjarnadóttir TK et al (2006) Comprehensive repertoire and phylogenetic analysis of the G protein-coupled receptors in human and mouse. Genomics 88:263–273
2. Santos R et al (2017) A comprehensive map of molecular drug targets. Nat Rev Drug Discov 16:19–34
3. Fredriksson R, Lagerstrom MC, Lundin LG, Schiöth HB (2003) The G-protein-coupled receptors in the human genome form five main families. Phylogenetic analysis, paralogon groups, and fingerprints. Mol Pharmacol 63:1256–1272
4. Simon MI, Strathmann MP, Gautam N (1991) Diversity of G proteins in signal transduction. Science 252:802–808
5. Downes GB, Gautam N (1999) The G protein subunit gene families. Genomics 62:544–552
6. Li F, De Godoy M, Rattan S (2004) Role of adenylate and guanylate cyclases in β1-, β2-, and β3-adrenoceptor-mediated relaxation of internal anal sphincter smooth muscle. J Pharmacol Exp Ther 308:1111–1120
7. Rosenbaum DM, Cherezov V, Hanson MA, Rasmussen SG, Thian FS, Kobilka TS, Choi HJ, Yao XJ, Weis WI, Stevens RC, Kobilka BK (2007) GPCR engineering yields high-resolution structural insights into β2-adrenergic receptor function. Science 318:1266–1273
8. Caffrey M, Cherezov V (2009) Crystallizing membrane proteins using lipidic mesophases. Nat Protoc 4:706–731
9. Xu S, Fischetti RF (2007) Design and Performance of a Compact Collimator on GM/CA-CAT At the Advanced Photon Source. Proc SPIE 6665:66650X1–66650X8
10. Bai XC, McMullan G, Scheres SH (2015) How cryo-EM is revolutionizing structural biology. Trends Biochem Sci 40:49–57
11. Rasmussen SG, DeVree BT, Zou Y, Kruse AC, Chung KY, Kobilka TS, Thian FS, Chae PS, Pardon E, Calinski D, Mathiesen JM, Shah ST, Lyons JA, Caffrey M, Gellman SH, Steyaert J, Skiniotis G, Weis WI, Sunahara RK, Kobilka BK (2011) Crystal structure of the β2 adrenergic receptor-Gs protein complex. Nature 477:549–555
12. Liang Y-L, Khoshouei M, Radjainia M, Zhang Y et al (2017) Phase-plate cryo-EM structure of a class B GPCR–G-protein complex. Nature 546(7656):118–123
13. Zhang Y et al (2017) Cryo-EM structure of the activated GLP-1 receptor in complex with a G protein. Nature 546:248–253
14. Stryer L (1991) Visual excitation and recovery. J Biol Chem 266:10711–10714
15. Palczewski K, Kumasaka T, Hori T, Behnke CA, Motoshima H, Fox BA, Le Trong I, Teller DC, Okada T, Stenkamp RE, Yamamoto M, Miyano M (2000) Crystal structure of rhodopsin: a G protein-coupled receptor. Science 289:739–745
16. Li J, Edwards PC, Burghammer M, Villa C, Schertler GFX (2004) Structure of bovine rhodopsin in a trigonal crystal form. J Mol Biol 343:1409–1436
17. Park JH, Scheerer P, Hofmann KP, Choe HW, Ernst OP (2008) Crystal structure of the ligand-free G-protein-coupled receptor opsin. Nature 454:183–187
18. Choe HW, Kim YJ, Park JH, Morizumi T, Pai EF, Krauss N, Hofmann KP, Scheerer P, Ernst OP (2011) Crystal structure of metarhodopsin II. Nature 471:651–655
19. Scheerer P, Park JH, Hildebrand PW, Kim YJ, Krauss N, Choe HW, Hofmann KP, Ernst OP (2008) Crystal structure of opsin in its G-protein-interacting conformation. Nature 455:497–502
20. Gao Y et al (2017) Isolation and structure-function characterization of a signaling-active rhodopsin-G protein complex. J Biol Chem 292:14280–14289
21. Skiba NP, Bae H, Hamm HE (1996) Mapping of effector binding sites of transducin alpha-subunit using G alpha t/G alpha i1 chimeras. J Biol Chem 271:413–424
22. Ballesteros JA, Shi L, Javitch JA (2001) Structural mimicry in G protein-coupled receptors: implications of the high-resolution structure of rhodopsin for structure–function analysis of rhodopsin-like receptors. Mol Pharmacol 60:1–19
23. Zhang FL, Casey PJ (1996) Protein prenylation: molecular mechanisms and functional consequences. Annu Rev Biochem 65:241–269
24. Min KC, Gravina SA, Sakmar TP (2000) Reconstitution of the vertebrate visual cascade using recombinant transducin purified from Sf9 cells. Protein Expr Purif 20:514–526
25. Ramachandran S, Cerione RA (2011) A dominant-negative Gα mutant that traps a stable rhodopsin-Gα-GTP-βγ complex. J Biol Chem 286:12702–12711

References

Chapter 24

Reconstitution of the Rhodopsin–Transducin Complex into Lipid Nanodiscs

Yang Gao, Jon W. Erickson, Richard A. Cerione, and Sekar Ramachandran

Abstract

Transmembrane proteins, such as G protein-coupled receptors (GPCR), require solubilization in detergents prior to purification. The recent development of novel detergents has allowed for the stabilization of GPCRs, which typically have a high degree of structural flexibility and are otherwise subject to denaturation. However, the detergent micelle environment is still very different from the native lipid membrane and the activity of GPCRs can be profoundly affected by interactions with annular lipid molecules. Moreover, GPCRs are often palmitoylated at their intracellular side, and a lipid bilayer environment would allow for proper orientation of these lipid modifications. Therefore, a reconstituted lipid bilayer environment would best mimic the physiological receptor microenvironment for biophysical studies of GPCRs and nanodiscs provide a methodology to address this aim. Nanodiscs are lipid bilayer discs stabilized by amphipathic membrane scaffolding proteins (MSP) where detergent-solubilized transmembrane proteins can be incorporated into them through a self-assembly process. Here we present a method for reconstituting the purified detergent-solubilized rhodopsin–transducin complex, the GPCR–G protein complex in visual phototransduction, into nanodiscs. The resulting complex incorporated into lipid nanodiscs can be used in biophysical studies including small-angle X-ray scattering and electron microscopy. This method is applicable to integral membrane proteins that mediate protein lipidation, including the zDHHC-family of *S*-acyltransferases and membrane-bound *O*-acyltransferases.

Key words G protein-coupled receptor (GPCR), G protein, Rhodopsin, Transducin, GPCR–G protein complex, Nanodiscs, Lipid bilayer, Membrane scaffolding protein

1 Introduction

Transmembrane proteins, such as G protein-coupled receptors (GPCRs), are not soluble in aqueous solutions due to the presence of large hydrophobic surfaces, which are necessary for their insertion and stability in lipid bilayers. Therefore, in order to maintain proper folding, detergents are required during the extraction of these proteins from the membrane. And in the case of GPCRs, as they possess a high degree of structural flexibility [1], most conventional detergents, for example octyl glucoside (OG) and dodecyl maltoside (DDM), are not sufficient for maintaining their

Maurine E. Linder (ed.), *Protein Lipidation: Methods and Protocols*, Methods in Molecular Biology, vol. 2009,
https://doi.org/10.1007/978-1-4939-9532-5_24, © Springer Science+Business Media, LLC, part of Springer Nature 2019

stability and signaling activities. Recently, a series of novel detergents featuring a more rigid architecture, such as maltose neopentyl glycol (MNG) [2] and glyco-diosgenin (GDN) [3], have been developed and these appear to place subtle constraints on protein conformational flexibility. The development of these detergents represents a breakthrough in protein structural studies as they have proven to be very successful at sustaining the structural integrity and activity of GPCRs, resulting in the recent rapid growth in the number of high-resolution GPCR and GPCR–G protein complex structures. However, as noted earlier, despite the high stability of GPCRs in MNG and GDN, the hydrophobic environment provided by the detergent micelles is very different from the native membrane environment, as these detergents tend to have very large micelles that are much thicker than lipid membranes and may distort the conformation of hydrophilic residues close to the hydrophobic "belt." In addition, GPCRs have been shown to interact with lipids in the membrane in a way that dictates their interactions with and activity toward signaling partners. For example, the photochemical properties [4] of the GPCR rhodopsin in the visual phototransduction pathway and its coupling efficiency to the heterotrimeric G protein transducin [5] are sensitive to its lipid environment composition where tightly bound phospholipids molecules have been seen in a high-resolution rhodopsin crystal structure [6]. Moreover, GPCRs are often palmitoylated on one or more cysteines at the intracellular (or luminal) side [7], where the lipid bilayer environment allows for optimal orientation of these lipid modifications. Therefore, a reconstituted lipid bilayer environment may provide an optimal milieu in which the essential characteristics of the active signaling complex are preserved, making nanodiscs an attractive methodology for structural studies.

Nanodiscs are composed of a circular lipid bilayer center, into which a transmembrane protein can be incorporated, and two molecules of membrane scaffold protein (MSP), a modified form of human high-density lipoprotein apoA-1 [8]. MSP contains a series of amphipathic α helices that can wrap around the lipids and thus stabilize the bilayer disc in aqueous solutions. The size of the nanodiscs is very monodisperse and can be easily adjusted by varying the number of amphipathic helices in MSP [9–10]. In the case of rhodopsin, the GPCR in the visual phototransduction pathway, nanodiscs have been used to study the stoichiometry of its interaction with the G protein transducin [11–12], rhodopsin kinase, and arrestin [13]. More recently, double electron–electron resonance (DEER) has been used to characterize rhodopsin incorporated into nanodiscs, revealing that the activated receptor assumes multiple conformations in contrast to the single conformation observed in DDM micelles [14]. Moreover, with the recent advances in cryo-electron microscopy technology [15], nanodiscs have proven to be a successful strategy for obtaining high-resolution structures of transmembrane proteins, maintaining them in an environment that allows for interactions between annular phospholipids and proteins [16]. Here we describe in detail a method for reconstituting the purified detergent-solubilized rhodopsin–transducin complex [17], the GPCR-G protein complex in visual phototransduction, (as described in the previous chapter) into nanodiscs. A truncated version of MSP (MSP1D1ΔH5) [18] was chosen in order to restrict the size of the resulting nanodiscs such that they would accommodate only one receptor complex per nanodisc, thus ensuring a homogeneous preparation. The process starts with mixing purified complex with MSP and lipids solubilized in detergents in a specified ratio with subsequent removal of detergent by incubation with Bio-Beads. During the detergent removal, the nanodiscs self-assemble and incorporate the receptor complex. The resulting complex-embedded nanodiscs can then be further purified with size exclusion chromatography (SEC). As the receptor rhodopsin has a covalently bound agonist all-trans retinal, which has a distinct UV absorption at 380 nm, the 1:1 ratio between nanodisc and complex can be verified by UV-Vis spectroscopy. The purified complex-embedded nanodiscs can be used in further biophysical characterizations, such as small-angle X-ray scattering (SAXS), negative-stain electron microscopy, and possibly be used for cryo-electron microscopy and high-resolution structures of the rhodopsin–transducin complex. In addition, as the composition of the lipids used in nanodisc formation can be varied, the resulting nanodiscs can also be used for studying the influence of different lipids on the activity of this GPCR–G protein complex.

2 Materials

2.1 Incorporation of a Rhodopsin–Transducin Complex into Nanodiscs

1. 1-Palmitoyl-2-oleoyl-*sn*-glycero-3-phosphocholine (POPC) (Avanti Polar Lipids).
2. HMN buffer: 20 mM HEPES pH 7.5, 2 mM $MgCl_2$, 100 mM NaCl, 100 μM TCEP.
3. HMNG buffer: HMN buffer, 10% glycerol.
4. 10% (w/v) sodium cholate in HMN buffer.
5. Methanol.
6. Purified rhodopsin–transducin complex in HMN buffer plus 0.003% lauryl maltose neopentyl glycol as described in the previous chapter.
7. Purified membrane scaffolding protein MSP1D1ΔH5 stored in HMN buffer with 10% glycerol at 300 μM concentration at −80 °C (*see* **Note 1**).
8. Bio-Beads SM-2 resin (Bio-Rad).
9. A sonic dismembrator.

10. A balance accurate to 0.1 mg.
11. A tabletop microcentrifuge and an end-to-end rocker kept in a cold room at 4 °C.
12. 0.7and 1.7 mL microcentrifuge tubes.
13. 1 mL disposable syringes with Luer-Lok tips.
14. 0.22 μm PVDF syringe filters.
15. Aluminum foil.
16. Floor lamps covered with 3M 616 lithographer's tape.

2.2 Purification and Characterization of Rhodopsin–Transducin Complex-Embedded Nanodiscs

1. An FPLC equipped with a UV absorbance monitor set at 280 nm and a fraction collector and kept in a cold box at 4 °C. A desk lamp covered with 3M 616 lithographer's tape.
2. A Superdex 200 10/300 GL column (GE Life Sciences).
3. HMN buffer: 20 mM HEPES pH 7.5, 2 mM $MgCl_2$, 100 mM NaCl, 100 μM TCEP.
4. Amicon Ultra-0.5 mL centrifugal filters with 100 kD molecular weight cutoff (EMD Millipore).
5. A tabletop microcentrifuge kept in a cold room at 4 °C.
6. Aluminum foil.
7. Floor lamps covered with 3M 616 lithographer's tape.
8. UV-Vis spectrophotometer.

3 Methods

3.1 Incorporation of Rhodopsin–Transducin Complex into Nanodiscs

1. Weigh out 5 mg POPC in a 0.7 mL microcentrifuge tube. Add 75 μL HMN buffer to the lipid and vortex for 1 min to achieve a uniform suspension. Sonicate the tube at maximum intensity in water bath for 1 min. Add 56.6 μL 10% (w/v) sodium cholate to the suspension and vortex the mixture for 1 min. Sonicate the tube at maximum intensity in a water bath for 1 min. This results in 50 mM POPC solubilized in HMN buffer containing 100 mM sodium cholate (*see* **Note 2**).
2. Weigh out 200 mg Bio-Beads in a 1.7 mL microcentrifuge tube. Incubate with 1.5 mL methanol for 1 min. Centrifuge the tube in a tabletop microcentrifuge for 0.5 min at maximum speed. Gently aspirate and discard the supernatant. Repeat this process with 1.5 mL HMN buffer three times and store the equilibrated Bio-Beads on ice.
3. The ratio between MSP, lipids and the rhodopsin–transducin complex is critical for obtaining homogeneous complex-embedded nanodiscs. The optimal molar ratio is 1:3:50 complex:MSP1D1ΔH5:POPC (*see* **Note 3**).

4. Mix 50 μL of 300 μM MSP1D1ΔH5 with 5 μL 50 mM POPC from **step 1**, resulting in 3:50 ratio of MSP1D1ΔH5:POPC. Incubate on ice for 20 min.
5. *All the following steps should be carried out in the dark with red light.* Add 5.3 μL 10% sodium cholate (232.6 mM) and 63.1 μL 10 mg/mL (79.3 μM) of the purified rhodopsin--transducin complex to the mixture and incubate on ice for another 20 min (*see* **Note 4**).
6. Weigh out 123 mg wet Bio-Beads in a 0.7 mL microcentrifuge tube. Add the mixture from **step 5** to the tube, wrap the tube light-tight with foil and incubate it on an end-to-end rocker in a cold room set at 4 °C overnight (*see* **Note 5**).
7. Centrifuge the tube from **step 6** in a tabletop microcentrifuge kept in a dark cold room with red light for 0.5 min at maximum speed. Gently aspirate the supernatant and filter it with a 0.22 μm PVDF syringe filter.

3.2 SEC Purification of Rhodopsin–Transducin Complex Embedded Nanodiscs

1. Equilibrate a Superdex 200 10/300 GL column with 60 mL HMN buffer at a flow rate of 0.4 mL/min on an FPLC equipped with a UV absorbance monitor set at 280 nm and a fraction collector (*see* **Note 6**).
2. Inject the complex-embedded nanodiscs from the previous section onto the column run at 0.4 mL/min with HMN buffer. Collect 0.4 mL fractions (a typical SEC profile is shown in Fig. 1).
3. Pool fractions containing the complex-embedded nanodiscs from **step 2** and concentrate the fractions with one 0.5 mL 100 kD molecular weight cutoff Amicon concentrator to about 100 μL by centrifuging at 14,000 × *g* for 3 min each time in a tabletop microcentrifuge kept in a dark cold room with red light. This is the final purified complex-embedded nanodiscs and the concentration is typically about 3 mg/mL. Store the nanodiscs wrapped in foil at 4 °C for over a week (*see* **Note 7**).

3.3 Characterization of the Complex-Embedded Nanodiscs Using UV-Vis Spectroscopy

1. Measure the concentration of the complex-embedded nanodiscs with a UV-Vis spectrophotometer at 280 nm ($A_{280\ \mathrm{nm}}$) using the extinction coefficient $\varepsilon_{280\ \mathrm{nm}} = 197{,}930\ \mathrm{M}^{-1}\ \mathrm{cm}^{-1}$, and molecular weight Mw = 170,240 Da (*see* **Note 8**).
2. Assess the number of rhodopsin–transducin complexes in each nanodisc by UV-Vis spectroscopy by measuring the absorbance at 280 nm ($A_{280\ \mathrm{nm}}$, the absorbance of the protein component) and 380 nm ($A_{380\ \mathrm{nm}}$, the absorbance of rhodopsin bound all-trans retinal). A typical spectrum is shown in Fig. 2. The $A_{280\ \mathrm{nm}}/A_{380\ \mathrm{nm}}$ ratio should be 4.70, indicating a 1:1 molar ratio of complex–nanodisc (*see* **Note 9**).

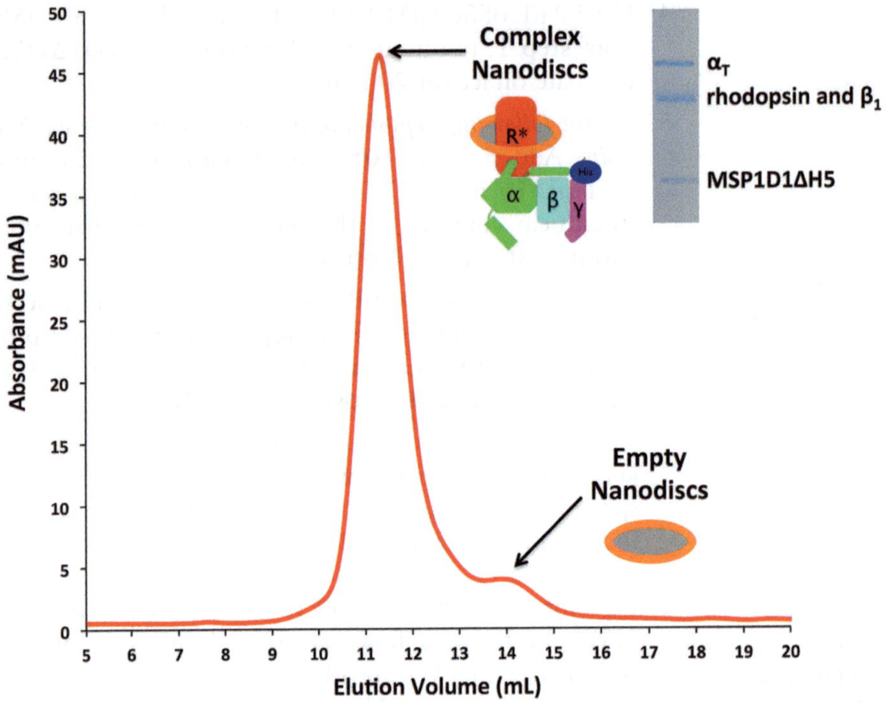

Fig. 1 SEC profile of complex-embedded nanodiscs and SDS-PAGE gel of concentrated peak fractions

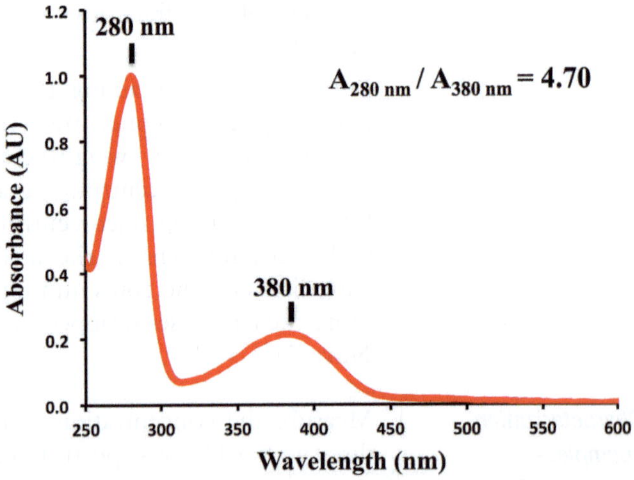

Fig. 2 UV-Vis spectrum of the purified complex-embedded nanodiscs

4 Notes

1. The membrane scaffolding protein MSP1D1ΔH5 can be expressed and purified from *E. coli* BL21(DE3) as described previously [19] using a pET28a vector harboring an *N*-terminal His_6-tag DNA sequence followed by the MSP1D1 gene

with DNA sequence corresponding to residues 121–142 deleted. The protein sequence of MSP1D1ΔH5 is as follows:

MGHHHHHHDYDIPTTENLYFQGSTFSKLREQLGPVTQ EFWDNLEKETEGLRQEMSKDLEEVKAKVQPYLDDFQKK WQEEMELYRQKVEPLGEEMRDRARAHVDALRTHLAPY SDELRQRLAARLEALKENGGARLAEYHAKATEHLSTLSE KAKPALEDLRQG LLPVLESFKVSFLSALEEYTKKLNTQ.

2. The lipids (stored at −20 °C) are hygroscopic and should be allowed to come to room temperature before opening the vial. The detergent-solubilized lipids can be stored at −80 °C for up to 6 months.
3. The excess MSP and lipids ensures only one complex is incorporated into each nanodisc. If MSP1D1 is used, the ratio should be 1:3:145 complex–MSP1D1–POPC.
4. The addition of 10% sodium cholate maintains the cholate concentration at 14 mM, above its critical micelle concentration of 9.5 mM. As the rhodopsin–transducin complex is light sensitive, all steps onward should be conducted in the dark with red light.
5. The ratio of Bio-Beads to nanodisc mixture is 1 mg wet Bio-Beads to 1 μL solution.
6. The FPLC should be kept in a cold box set at 4 °C with the glass doors covered with foil to prevent light exposure inside. A red light can be kept inside the cold box to facilitate visualization.
7. The complex-embedded nanodiscs peak will be contained in four 0.4 mL fractions. During concentration, it is recommended to invert the concentrators several times to prevent protein aggregation during centrifugation.
8. Concentration (in μM) = $A_{280\ \mathrm{nm}} \times 10^6/\varepsilon_{280\ \mathrm{nm}}$. Concentration (in mg/mL) = $A_{280\ \mathrm{nm}} \times \mathrm{Mw}/\varepsilon_{280\ \mathrm{nm}}$.
9. The extinction coefficients used are as follows:

   rhodopsin ($\varepsilon_{280\ \mathrm{nm}}$ = 61,800 $\mathrm{M^{-1}\ cm^{-1}}$, $\varepsilon_{380\ \mathrm{nm}}$ = 42,000 $\mathrm{M^{-1}\ cm^{-1}}$), transducin ($\varepsilon_{280\ \mathrm{nm}}$ = 93,270 $\mathrm{M^{-1}\ cm^{-1}}$), and MSP1D1ΔH5($\varepsilon_{280\ \mathrm{nm}}$ = 21,430 $\mathrm{M^{-1}\ cm^{-1}}$).

References

1. Preininger AM, Meiler J, Hamm H (2013) Conformational flexibility and structural dynamics in GPCR-mediated G protein activation: a perspective. J Mol Biol 425:2288–2298
2. Chae PS, Rasmussen SGF, Rana R et al (2010) Maltose-neopentyl glycol (MNG) amphiphiles for solubilization, stabilization and crystallization of membrane proteins. Nat Methods 7:1003–1008
3. Chae PS, Rasmussen SGF, Rana R et al (2012) A new class of amphiphiles bearing rigid hydrophobic groups for solubilization and stabilization of membrane proteins. Chemistry 18:9485–9490

4. Brown MF (1994) Modulation of rhodopsin function by properties of the membrane bilayer. Chem Phys Lipids 73:159–180
5. Kaya AI, Thaker TM, Preininger AM et al (2011) Coupling efficiency of rhodopsin and transducin in bicelles. Biochemistry 50:3193–3203
6. Li J, Edwards PC, Burghammer M et al (2004) Structure of bovine rhodopsin in a trigonal crystal form. J Mol Biol 343:1409–1438
7. Goddard AD, Watts A (2012) Regulation of G protein-coupled receptors by palmitoylation and cholesterol. BMC Biol 10:27
8. Bayburt TH, Grinkova YV, Sligar SG (2002) Self-assembly of discoidal phospholipid bilayer nanoparticles with membrane scaffold proteins. Nano Lett 2:853–856
9. Ritchie TK, Grinkova YV, Bayburt TH et al (2009) Reconstitution of membrane proteins in phospholipid bilayer nanodiscs. Methods Enzymol 464:211–231
10. Bayburt TH, Sligar SG (2010) Membrane protein assembly into nanodiscs. FEBS Lett 584:1721–1727
11. Bayburt TH, Leitz AJ, Xie G et al (2007) Transducin activation by nanoscale lipid bilayers containing one and two rhodopsins. J Biol Chem 282:14875–14881
12. Whorton MR, Jastrzebska B, Park PSH et al (2008) Efficient coupling of transducin to monomeric rhodopsin in a phospholipid bilayer. J Biol Chem 283:4387–4394
13. Bayburt TH, Vishnivetskiy SA, McLean MA et al (2011) Monomeric rhodopsin is sufficient for normal rhodopsin kinase (GRK1) phosphorylation and arrestin-1 binding. J Biol Chem 286:1420–1428
14. Van Eps N, Caro LN, Morizumi T et al (2017) Conformational equilibria of light-activated rhodopsin in nanodiscs. Proc Natl Acad Sci U S A 114:E3268–E3275
15. Bai XC, McMullan G, Scheres SH (2015) How cryo-EM is revolutionizing structural biology. Trends Biochem Sci 40:49–57
16. Gao Y, Cao E, Julius D et al (2016) TRPV1 structures in nanodiscs reveal mechanisms of ligand and lipid action. Nature 534:347–351
17. Gao Y, Westfield G, Erickson JW et al (2017) Isolation and structure-function characterization of a signaling-active rhodopsin-G protein complex. J Biol Chem 292:14280–14289
18. Hagn F, Etzkorn M, Raschle T et al (2013) Optimized phospholipid bilayer nanodiscs facilitate high-resolution structure determination of membrane proteins. J Am Chem Soc 135:1919–1925
19. Hagn F, Etzkorn M, Raschle T, Wagner G (2013) Facilitate high-resolution structure determination of membrane proteins. J Am Chem Soc 135(5):1919–1925

Index

Maurine E. Linder (ed.), *Protein Lipidation: Methods and Protocols*, Methods in Molecular Biology, vol. 2009, https://doi.org/10.1007/978-1-4939-9532-5, © Springer Science+Business Media, LLC, part of Springer Nature 2019

MIX
Papier aus verantwortungsvollen Quellen
Paper from responsible sources
FSC® C105338

If you have any concerns about our products,
you can contact us on
ProductSafety@springernature.com

In case Publisher is established outside the EU,
the EU authorized representative is:
Springer Nature Customer Service Center GmbH
Europaplatz 3, 69115 Heidelberg, Germany

Printed by Libri Plureos GmbH
in Hamburg, Germany